1986

N

Appendix

ESSENTIALS OF
TRIGONOMETRY

Fourth Edition

ESSENTIALS OF TRIGONOMETRY

Irving Drooyan
Walter Hadel
Charles C. Carico
Emeriti, Los Angeles Pierce College

Macmillan Publishing Company
New York

Collier Macmillan Publishers
London

Macmillan Publishing Company
866 Third Avenue, New York, New York 10022

Collier Macmillan Canada, Inc.

Library of Congress Cataloging in Publication Data

Drooyan, Irving.
    Essentials of trigonometry.

    Includes index.
    1. Trigonometry, Plane.   I. Hadel, Walter.
II. Carico, Charles C.   III. Title.
QA533.D74   1986   516.2′4   85–5143
ISBN 0–02–330570–3

Printing: 1 2 3 4 5 6 7 8   Year: 6 7 8 9 0 1 2 3 4 5

ISBN 0-02-330570-3

PREFACE

This fourth edition of *Essentials of Trigonometry* reflects fifteen years of classroom experience with the earlier editions. Like its predecessors, this edition is designed for use in a one-semester or one-quarter course in trigonometry. The topics included in this text are listed in the Contents.

Prerequisites

Although it is assumed that students undertaking the study of trigonometry have had prerequisite work in algebra and geometry, Appendix A includes a review of topics in these subjects that are needed as a background for the course. Students are referred to a section of this review when the material can be helpful in understanding a particular trigonometric concept.

Changes in This Edition

- Trigonometric function values that were obtained in the previous editions by using tables are obtained by using a calculator.

- Our emphasis on obtaining exact function values of angles with measures of 0°, 30°, and 45°, and integral multiples thereof, is reenforced by always specifying in the instructions when *exact values* are expected.

- The treatment of vectors has been expanded to include some algebra of vectors in addition to geometric vectors and now constitutes a separate chapter.

- A new section has been added on graphing equations in polar coordinates.
- Work in simplifying expressions involving trigonometric function values is used to motivate proving trigonometric identities.
- Exercise Sets have been revised and expanded.
- A discussion of significant digits, rounding-off procedures, and scientific notation is included in Appendix B.
- A discussion of how tables can be used instead of a calculator is now included in Appendix C. These tables also give students a visual representation of how values of the trigonometric functions increase/ decrease over their ranges and provide insight into the cofunction relationships between pairs of these functions.

Pedagogical Features

As in the previous editions, the subject matter of the text is continually reviewed through the use of Chapter Summaries and Review Exercises. Items in both the Summaries and Reviews are cross-referenced to the appropriate sections of the text, as are the symbols listed inside the front cover. A detailed summary of the important topics in the text that appears inside the back cover can serve as a convenient reference for students completing assignments or studying for examinations.

All Exercise Sets include basic exercises designated A; many also include more challenging exercises designated B, which provide the instructor with flexibility in making assignments, depending on the time available and the objectives of the course.

Answers, including graphs, for odd-numbered exercises in each Exercise Set and for all of the Review Exercises follow the Appendixes.

A Student Solution Key containing *solutions* to the even-numbered exercises in each section is also available for this edition.

Acknowledgments

We gratefully acknowledge the encouraging and helpful comments from the reviewers of this fourth edition: Kathy Gosa, Washburn University of Topeka; Roger Harms, Iowa Lakes Community College; J. T. Oldknow, retired professor, Santa Ana College; Yoram Sagher, University of Illinois—Chicago; David Shotwell, Sul Ross State University, and Faye Tamakawa, Honolulu Community College.

Irving Drooyan
Walter Hadel
Charles C. Carico

CONTENTS

APPENDIXES

TO THE STUDENT

Comment on Calculators

If you are not already familiar with the calculator that you are using, you should read the appropriate sections of the instruction manual that pertain to the functions or operations that are being considered in the text at a particular time.

As you use your calculator, you will eventually be able to perform many complicated combined operations involving trigonometric function values by using a single chain of key strokes. Until you become proficient in performing combined operations, however, you can simply use your calculator in a sequence of steps:

1. Obtain approximations for trigonometric function values instead of using tables.
2. Write these values in expressions that involve one or more computations.
3. Complete the computations using the simple function keys for the basic operations.

ESSENTIALS OF
TRIGONOMETRY

1 TRIGONOMETRIC FUNCTIONS

Historically, the subject of trigonometry developed primarily as a means of finding the lengths of sides and the measures of angles of triangles for use in such practical areas as surveying and navigation. In more recent times, trigonometry has become important in describing phenomena that are periodic in nature. In this book we are concerned with presenting the necessary concepts and developing the necessary skills to solve problems in both areas.

In this chapter we first introduce some notions about angles and their measures. Then we shall consider six important functions whose domains are sets of angles.

1.1 Angles and Their Radian Measures

In geometry, an angle is considered to be two rays with a common endpoint, such as $\overrightarrow{OA}$ and $\overrightarrow{OB}$ in Figure 1.1a (see Appendix A.7).* Let us consider $\angle AOB$ and visualize it as being formed by rotating side $\overrightarrow{OA}$, with O as the pivotal point, to side $\overrightarrow{OB}$; side $\overrightarrow{OA}$ is called the **initial side,** and $\overrightarrow{OB}$ is called the **terminal side.** Angles viewed in this way, for which rotation is considered, are called **directed angles.**

Definition 1.1 *An angle is in **standard position** if one of its sides lies in the positive x-axis and its vertex is the origin (see Figure 1.1b).*

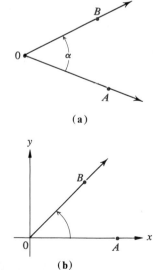

(a)

(b)

Figure 1.1

* Appendix A (page 271) provides a review of prerequisite topics of algebra and geometry.

1

The terminal side of an angle in standard position may lie in any quadrant or on any axis. The angle is said to be in that quadrant in which the terminal side lies. If the terminal side lies on an axis, the angle is called a **quadrantal angle.** For example, in Figure 1.2, α (alpha) is in Quadrant I, β (beta) is in Quadrant II, γ (gamma) is in Quadrant III, δ (delta) is in Quadrant IV, and θ (theta) is a quadrantal angle.

Measure of an Angle

The measure of an angle can be specified in different ways. The two kinds of angle measure most commonly used are **degree measure,** with which you are undoubtedly familiar (see Appendix A.7), and **radian measure,** which is widely used in advanced work in mathematics and the physical sciences.

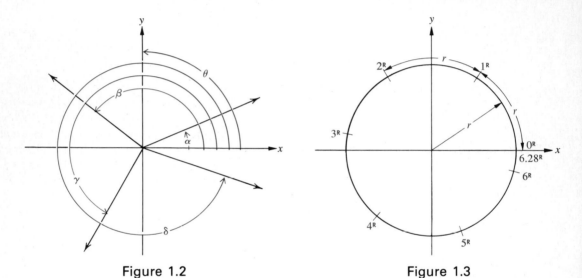

Figure 1.2   Figure 1.3

In radian measure, the circle is divided into unit arcs, each having a length equal to the radius of the circle, as shown in Figure 1.3. The circumference of a circle with radius r is $2\pi r$;* hence, the circle will include 2π, or *approximately* 6.28, of these arcs. The number of such arcs intercepted by a central angle is the measure of the angle in radian units. For this measure we use the notation α^R.

Now if the measure in radians of the angle associated with the entire circle is $2\pi^R$, then the measure of the angle associated with a semicircle is π^R. Measures of other angles that are rational number multiples of π^R such as $(\pi/2)^R$, $(\pi/4)^R$, $(\pi/6)^R$, $(\pi/3)^R$, etc., can be readily obtained.

---

* π is an irrational number that *cannot* be expressed exactly as a decimal. An approximation that can be obtained on most scientific calculators is 3.1415927.

Positive numbers are assigned for the measures of angles in which the rotation from the initial to the terminal side is in a counterclockwise direction, and negative numbers are assigned for the measure if the rotation is in a clockwise direction. In Figure 1.4a, the measuring circle is scaled in degree units and radian units for positive measures that are often used. In Figure 1.4b, the measuring circle is scaled for similar negative units.

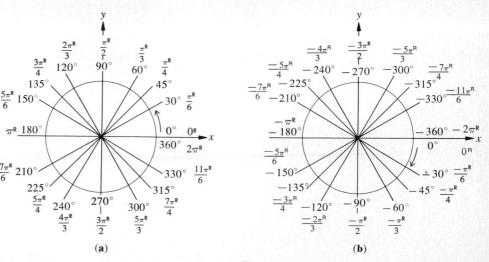

Figure 1.4

For example, in Figure 1.5a,

$$\alpha = 150° \qquad \text{and} \qquad \alpha = \frac{5\pi^R}{6},$$

and in Figure 1.5b,

$$\alpha = -60° \qquad \text{and} \qquad \alpha = -\frac{\pi^R}{3}.$$

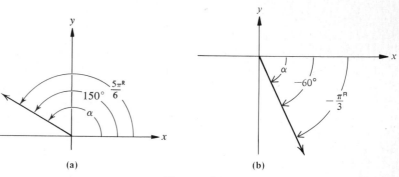

Figure 1.5

As is customary, we shall use the terms "least positive angle" to mean that angle of a set of angles which has the least positive measure and "negative angle" to mean an angle with a negative measure.

Degree–Radian Conversion Formulas

The measure of an angle in radians can be related to the measure of the angle in degrees. Since the ratio of the measure of a given central angle to the measure of the central angle intercepting the entire circle is the same for any unit of measurement, we see that for degree measure and radian measure in particular

$$\frac{\alpha^\circ}{360} = \frac{\alpha^R}{2\pi}.$$

This equation is sometimes written in either of the following forms.

$$\textbf{I.} \qquad \alpha^\circ = \frac{180}{\pi}\,\alpha^R$$

$$\textbf{II.} \qquad \alpha^R = \frac{\pi}{180}\,\alpha^\circ$$

Examples Find the degree measure (to the nearest tenth of a degree)* of each angle α with the given radian measure.

a. $\alpha = \dfrac{3\pi^R}{5}$

b. $\alpha = 0.81^R$

Solutions

a. $\alpha^\circ = \left(\dfrac{180}{\pi} \cdot \dfrac{3\pi}{5}\right)^\circ = 108.0^\circ$

b. $\alpha^\circ = \left(\dfrac{180}{\pi} \cdot 0.81\right)^\circ \approx 46.4^\circ$

In any chain of statements involving both the symbols $=$ and $\approx$ as in Example b, it should be understood that either relationship is valid only for the expressions on either side of the particular symbol.

Examples Find the radian measure (to the nearest hundredth of a radian) of each angle α with the given degree measure.

a. $\alpha = 60^\circ$

b. $\alpha = 330^\circ$

---

* See Appendix B for a brief review of rounding-off procedures.

Solutions

a. $\alpha^R = \left(\dfrac{\pi}{180} \cdot 60 \right)^R$

$\qquad = \dfrac{\pi^R}{3} \approx 1.05^R$

b. $\alpha^R = \left(\dfrac{\pi}{180} \cdot 330 \right)^R$

$\qquad = \dfrac{11\pi^R}{6} \approx 5.76^R$

Many exercises in this text, like the examples above, require some approximation for the value of π. Our results in the answers to the examples and exercises in this text were obtained using a calculator value for π that contained more than two decimal places. If you use an approximation for π with less accuracy, such as 3.1 or 3.14, your results may differ slightly.

From conversion formulas I and II on page 4, we observe that the measure in degrees of an angle α with a measure of 1^R is given by

$$\alpha^\circ = \left(\frac{180}{\pi} \cdot 1 \right)^\circ \approx 57.30^\circ$$

and that the measure in radians of an angle α with a measure of 1° is given by

$$\alpha^R = \left(\frac{\pi}{180} \cdot 1 \right)^R \approx 0.01745^R.$$

Coterminal Angles

If an angle is viewed as being formed by rotation of the initial side to the terminal side, then it is possible to have an angle whose initial side is rotated more than 360°. For example, in Figure 1.6, $\alpha_3 > 360^\circ$. As indicated in the figure, angles may have the same initial side and the same terminal side but different measures. These measures can be positive or negative.

Definition 1.2 **Coterminal angles** *are angles that have the same initial side and the same terminal side.*

In Figure 1.6, α_1, α_2, and α_3 are coterminal.

Examples Find the least positive angle that is coterminal with each of the following.

a. 6.59^R

b. 520°

Solutions

a. Sketch the angle, showing its initial and terminal sides. Since the angle generated in one revolution is $2\pi^R \approx 6.28^R$ to the nearest hundredth, the

(*continued*)

Figure 1.6

least positive angle having the same initial and same terminal sides is approximately $(6.59 - 6.28)^R = 0.31^R$.

b. Sketch the angle. In this case, the least positive angle having the same initial and terminal sides is $(520 - 360)^\circ = 160^\circ$.

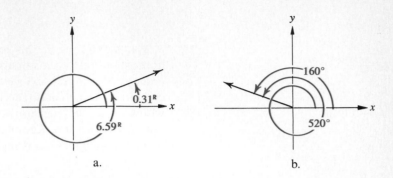

a. b.

Example Write a general expression for all the angles that are coterminal with 45°.

Solution Since one revolution from a terminal side of any angle brings us back to that terminal side, then all angles coterminal with an angle of 45°, or $(\pi/4)^R$, are given by

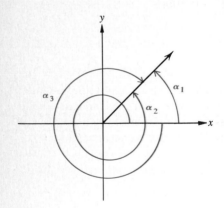

$$(45 + k \cdot 360)^\circ, \quad k \in J \qquad \text{or} \qquad \left(\frac{\pi}{4} + k \cdot 2\pi\right)^R, \quad k \in J.*$$

Note that

for $k = 0$, $\quad \alpha_1 = (45 + 0)^\circ = 45^\circ$ or $\left(\frac{\pi}{4} + 0\right)^R = \frac{\pi^R}{4}$;

for $k = 1$, $\quad \alpha_2 = (45 + 360)^\circ = 405^\circ$ or $\left(\frac{\pi}{4} + 2\pi\right)^R = \frac{9\pi^R}{4}$;

for $k = -1$, $\quad \alpha_3 = (45 - 360)^\circ = -315^\circ$ or $\left(\frac{\pi}{4} - 2\pi\right)^R = -\frac{7\pi^R}{4}$.

EXERCISE SET 1.1

A

Complete the following degree–radian conversion wheels for special angles without referring to Figure 1.4. (These wheels are a handy reference.)

---

* The letter J is used to name the set of integers; the symbol $\in$ is read "is an element of" (see Appendix A.2).

1.

2.

Find the degree measure to the nearest tenth of a degree of an angle α with the given radian measure. Specify the quadrant in which the terminal side of α lies.

Examples   a. $\dfrac{6\pi^{\text{R}}}{5}$

b. -1.24^{R}

Solutions   a. $= \left(\dfrac{180}{\pi} \cdot \dfrac{6\pi}{5}\right)^{\circ}$

b. $\alpha^{\circ} = \left[\dfrac{180}{\pi}(-1.24)\right]^{\circ}$

$= 216.0^{\circ}$;   Quad. III

$\approx -71.0^{\circ}$;   Quad. IV

3. $\dfrac{\pi^{\text{R}}}{3}$ **4.** $\dfrac{2\pi^{\text{R}}}{3}$ **5.** $\dfrac{14\pi^{\text{R}}}{3}$ **6.** $\dfrac{15\pi^{\text{R}}}{4}$

7. $\dfrac{-8\pi^{\text{R}}}{5}$ **8.** $\dfrac{-11\pi^{\text{R}}}{8}$ **9.** $\dfrac{-18\pi^{\text{R}}}{8}$ **10.** $\dfrac{-12\pi^{\text{R}}}{5}$

11. 0.18^{R} **12.** 0.94^{R} **13.** 5.16^{R} **14.** 4.55^{R}

15. -0.27^{R} **16.** -0.86^{R} **17.** -4.04^{R} **18.** -3.61^{R}

Find the radian measure to the nearest hundredth of a radian of an angle α with the given degree measure. Specify the quadrant in which the terminal side of α lies.

Examples   a. $118°$

b. $-213°$

(continued)

Solutions a. $\alpha^R = \left(\dfrac{\pi}{180} \cdot 118\right)^R$ b. $\alpha^R = \left[\dfrac{\pi}{180} \cdot (-213)\right]^R$

$\approx 2.06^R$; Quad. II $\approx -3.72^R$; Quad. II

19. $30°$ **20.** $45°$ **21.** $75°$ **22.** $125°$

23. $143°$ **24.** $137°$ **25.** $402°$ **26.** $560°$

27. $-37°$ **28.** $-51°$ **29.** $-80°$ **30.** $-130°$

31. $-240°$ **32.** $-120°$ **33.** $-510°$ **34.** $-400°$

Find the least positive angle to the nearest hundredth that is coterminal with each of the following. Sketch each angle.

Examples   a. 7.34^R b. $-620°$

Solutions   a. $[7.34 + (-1) \cdot 2\pi]^R$ b. $(-620 + 2 \cdot 360)°$

$\approx (7.34 - 6.28)^R$ $= (-620 + 720)°$

$= 1.06^R$ $= 100°$

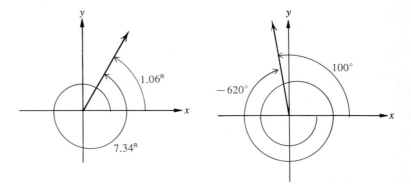

35. 6.72^R **36.** 8.91^R **37.** $379°$ **38.** $423°$

39. -11.11^R **40.** -13.48^R **41.** $-792°$ **42.** $-863°$

Write a general expression for all angles coterminal with each of the following.

Examples   a. $16°$ b. $\dfrac{\pi}{2}^R$

Solutions   a. $(16 + k \cdot 360)°,\quad k \in J$ b. $\left(\dfrac{\pi}{2} + k \cdot 2\pi\right)^{R},\quad k \in J$

43. $21°$ **44.** $128°$ **45.** $242°$ **46.** $310°$

47. $\dfrac{\pi^{R}}{6}$ **48.** $\dfrac{\pi^{R}}{3}$ **49.** $\dfrac{5\pi^{R}}{3}$ **50.** $\dfrac{7\pi^{R}}{6}$

1.2 Trigonometric Ratios

Consider the three real numbers x, y, and $\sqrt{x^2 + y^2}$ that can be associated with any point on the terminal side of an angle in standard position, as shown in Figure 1.7. The components of the ordered pair (x, y) are the coordinates of the point, and from the distance formula (see Appendix A.4) or the Pythagorean theorem, $\sqrt{x^2 + y^2}$ is the length of the line segment from the origin to the point. For each angle α, six possible ratios can be formed using the numbers x, y, and $\sqrt{x^2 + y^2}$. These ratios are called **trigonometric ratios** and are named as follows.

Definition 1.3 *For all angles α in standard position, if x and y are the coordinates of any point on the terminal side of α [where $(x, y) \neq (0, 0)$] and $r = \sqrt{x^2 + y^2}$,*

$$\textbf{sine } \alpha = \frac{y}{r}, \qquad\qquad \textbf{cosecant } \alpha = \frac{r}{y} \;\; (y \neq 0),$$

$$\textbf{cosine } \alpha = \frac{x}{r}, \qquad\qquad \textbf{secant } \alpha = \frac{r}{x} \;\; (x \neq 0),$$

$$\textbf{tangent } \alpha = \frac{y}{x} \;\; (x \neq 0), \qquad \textbf{cotangent } \alpha = \frac{x}{y} \;\; (y \neq 0).$$

Note in Definition 1.3 that r is the positive square root of $x^2 + y^2$. Thus, r is never negative or zero. Furthermore, it can be shown by using similar triangles, and this will be left as an exercise, that if (x_1, y_1) and (x_2, y_2) correspond to any two points (excluding the origin) on the terminal side of an angle in standard position, as shown in Figure 1.8, then

$$\frac{y_1}{r_1} = \frac{y_2}{r_2}, \quad \frac{x_1}{r_1} = \frac{x_2}{r_2}, \quad \text{and} \quad \frac{y_1}{x_1} = \frac{y_2}{x_2}.$$

Similar equalities for the reciprocals of these ratios also hold. Thus, the six trigonometric ratios are determined only by the position of the terminal side of an angle and not by the point selected on the terminal side. The names of the trigonometric ratios in Definition 1.3 are abbreviated,

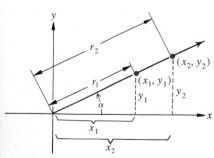

Figure 1.7

Figure 1.8

respectively, as

$$\sin \alpha, \qquad \csc \alpha,$$

$$\cos \alpha, \qquad \sec \alpha,$$

$$\tan \alpha, \qquad \cot \alpha.$$

Example Given that the terminal side of an angle α contains the point $(2, -3)$, find each of the six trigonometric ratios of α.

Solution By Definition 1.3,

$$r = \sqrt{x^2 + y^2}$$
$$= \sqrt{(2)^2 + (-3)^2} = \sqrt{13}.$$

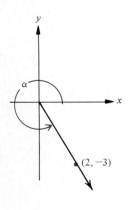

Hence,

$$\sin \alpha = \frac{y}{r} = \frac{-3}{\sqrt{13}}, \qquad \csc \alpha = \frac{r}{y} = \frac{\sqrt{13}}{-3} = \frac{-\sqrt{13}}{3},$$

$$\cos \alpha = \frac{x}{r} = \frac{2}{\sqrt{13}}, \qquad \sec \alpha = \frac{r}{x} = \frac{\sqrt{13}}{2},$$

$$\tan \alpha = \frac{y}{x} = \frac{-3}{2}, \qquad \cot x = \frac{x}{y} = \frac{2}{-3} = \frac{-2}{3}.$$

Note that several ratios in the above example are shown as a fraction in which the denominator is a radical expression. Of course we could rationalize the denominators as follows.

$$\frac{-3 \cdot \sqrt{13}}{\sqrt{13} \cdot \sqrt{13}} = \frac{-3\sqrt{13}}{13} \qquad \text{and} \qquad \frac{2 \cdot \sqrt{13}}{\sqrt{13} \cdot \sqrt{13}} = \frac{2\sqrt{13}}{13}.$$

However, for our work it is just as convenient to use fractions in which the radicals are retained in the denominator.

Basic Identities

Notice that from Definition 1.3 $\csc \alpha$, $\sec \alpha$, and $\cot \alpha$ are the reciprocals of $\sin \alpha$, $\cos \alpha$, and $\tan \alpha$, respectively. That is,

$$\csc \alpha = \frac{1}{\sin \alpha}, \quad \sec \alpha = \frac{1}{\cos \alpha}, \quad \text{and} \quad \cot \alpha = \frac{1}{\tan \alpha}.$$

These three equations are **identities;** *each is true for all the replacements of α for which each member is defined.* Identities are basic to the study of trigonometry.

Another particularly useful identity is closely related to the Pythagorean Theorem (see Appendix A.8). For any point $P(x, y)$ on the terminal side of an angle α in standard position

$$x^2 + y^2 = r^2.$$

Multiplying each member by $1/r^2$ $(r \neq 0)$ yields

$$\left(\frac{x}{r}\right)^2 + \left(\frac{y}{r}\right)^2 = 1$$

from which we obtain

$$\cos^2 \alpha + \sin^2 \alpha = 1.$$

You will be introduced to other important identities in the work that follows.

Trigonometric Functions

Notice that each equation in Definition 1.3 defines a function because any angle α is paired with only one number named by $\sin \alpha$, $\cos \alpha$, $\tan \alpha$, $\csc \alpha$, $\sec \alpha$, or $\cot \alpha$ (see Appendix A.3 for a review of the concept of a function). These six functions are called **trigonometric functions.** An element in the range of any one of these functions is called a **function value.** If a function value is denoted by **trig α,**

$$\text{trig } \alpha \in \{\sin \alpha, \cos \alpha, \tan \alpha, \csc \alpha, \sec \alpha, \cot \alpha\}.$$

The functions can be specified by $\{(\alpha, \text{trig } \alpha)\}$. The **domain** of each function is a set of angles determined by Definition 1.3. Note that, since $r \neq 0$, $\sin \alpha$ and $\cos \alpha$ are defined for all angles. However, $\tan \alpha$ and $\sec \alpha$ are undefined if $x = 0$. From this it follows that the domains of the tangent and the secant functions are the set of all angles α such that

$$\alpha \neq (90 + k \cdot 180)°, \quad k \in J.$$

Furthermore, because $\cot \alpha$ and $\csc \alpha$ are undefined if $y = 0$, the domains of the cotangent and the cosecant functions are the set of all angles α such that

$$\alpha \neq (0 + k \cdot 180)° = k \cdot 180°, \quad k \in J.$$

Hence, if α is in standard position and the terminal side is on the y-axis (see Figure 1.9a), then $\tan \alpha$ and $\sec \alpha$ are undefined; if the terminal side is on the x-axis (see Figure 1.9b), then $\cot \alpha$ and $\csc \alpha$ are undefined.

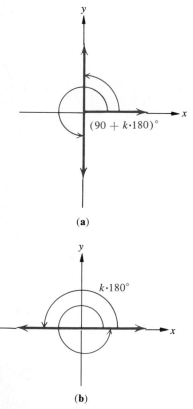

(a)

(b)

Figure 1.9

EXERCISE SET 1.2

A

Find each of the six trigonometric function values of α if the terminal side of α (in standard position) contains the given point. Sketch each angle.

Example $(-5, 12)$

Solution   By Definition 1.3,

$$r = \sqrt{x^2 + y^2} = \sqrt{(-5)^2 + (12)^2} = 13.$$

Hence,

$$\sin \alpha = \frac{y}{r} = \frac{12}{13}, \qquad\qquad \csc \alpha = \frac{r}{y} = \frac{13}{12},$$

$$\cos \alpha = \frac{x}{r} = \frac{-5}{13}, \qquad\qquad \sec \alpha = \frac{r}{x} = \frac{13}{-5} = \frac{-13}{5},$$

$$\tan \alpha = \frac{y}{x} = \frac{12}{-5} = \frac{-12}{5}, \qquad \cot \alpha = \frac{x}{y} = \frac{-5}{12}.$$

1. $(4, 3)$	**2.** $(12, 5)$	**3.** $(8, -15)$	**4.** $(2, -1)$
5. $(-1, 3)$	**6.** $(5, 2)$	**7.** $(-2, 2)$	**8.** $(-\sqrt{3}, 1)$
9. $(-2, -2\sqrt{3})$	**10.** $(0, 2)$	**11.** $(0, -4)$	**12.** $(-2, 0)$

Given one function value of an acute angle α, determine the other five trigonometric function values of α.

Example $\sin \alpha = \dfrac{3}{4}$

Solution   A sketch of a right triangle is shown in the figure. The acute angle α and $\overline{QP}$ and $\overline{OP}$ are sketched so that the lengths of these sides are approximately in the ratio of 3 to 4. From the Pythagorean theorem,

$$(OQ)^2 + 3^2 = 4^2,$$

$$(OQ)^2 = 16 - 9 = 7,$$

$$OQ = \pm\sqrt{7}.$$

Since α is in Quadrant I, we use the positive number, $\sqrt{7}$. The other five ratios can now be specified directly from Definition 1.3. Thus,

$$\sin \alpha = \frac{3}{4}, \qquad \csc \alpha = \frac{4}{3},$$

$$\cos \alpha = \frac{\sqrt{7}}{4}, \qquad \sec \alpha = \frac{4}{\sqrt{7}},$$

$$\tan \alpha = \frac{3}{\sqrt{7}}, \qquad \cot \alpha = \frac{\sqrt{7}}{3}.$$

13. $\tan \alpha = \dfrac{4}{3}$ **14.** $\cot \alpha = \dfrac{5}{12}$ **15.** $\sin \alpha = \dfrac{2}{5}$ **16.** $\cos \alpha = \dfrac{1}{4}$

17. $\sec \alpha = \dfrac{7}{3}$ **18.** $\csc \alpha = 2$ **19.** $\cos \alpha = \dfrac{1}{2}$ **20.** $\sin \alpha = \dfrac{2}{3}$

21. $\csc \alpha = 4$ **22.** $\sec \alpha = \dfrac{5}{4}$ **23.** $\cot \alpha = 1$ **24.** $\tan \alpha = 2$

Give the domain of each function.

25. $\{(\alpha, \sin \alpha)\}$ **26.** $\{(\alpha, \cos \alpha)\}$ **27.** $\{(\alpha, \tan \alpha)\}$

28. $\{(\alpha, \csc \alpha)\}$ **29.** $\{(\alpha, \sec \alpha)\}$ **30.** $\{(\alpha, \cot \alpha)\}$

B

31. Prove that if (x_1, y_1) and (x_2, y_2) correspond to any two points (excluding the origin) on the terminal side of an angle α in the first quadrant, then for $r_1 = \sqrt{x_1{}^2 + y_1{}^2}$ and $r_2 = \sqrt{x_2{}^2 + y_2{}^2}$,

$$\frac{x_1}{y_1} = \frac{x_2}{y_2}, \quad \frac{x_1}{r_1} = \frac{x_2}{r_2}, \quad \text{and} \quad \frac{y_1}{r_1} = \frac{y_2}{r_2}.$$

32. Show that the statement in Problem 31 is valid for an angle α in Quadrant II.

33. Use the fact that $\cos^2 \alpha + \sin^2 \alpha = 1$ to show that

$$1 + \tan^2 \alpha = \sec^2 \alpha.$$

34. Use the fact that $\cos^2 \alpha + \sin^2 \alpha = 1$ to show that

$$1 + \cot^2 \alpha = \csc^2 \alpha.$$

1.3 Function Values for Acute Angles and Quadrantal Angles

Function values for each of the six trigonometric functions for 30°, 45°, and 60° angles can be obtained from geometric considerations of right triangles.

Function Values for 30°, 45°, and 60°

First let us consider an equilateral triangle whose sides are each two units long. Recall from geometry (see Appendix A.8) that the measure of each angle is 60° and that an altitude of the triangle forms two congruent right triangles as shown in Figure 1.10a. From the Pythagorean theorem we can obtain the length of the altitude

$$h = \sqrt{2^2 - 1^2} = \sqrt{3}.$$

Hence, lengths of the sides of this triangle, and of course any 30°–60°–90° triangle, are proportional to the numbers 1, $\sqrt{3}$, and 2 as shown in Figure 1.10b.

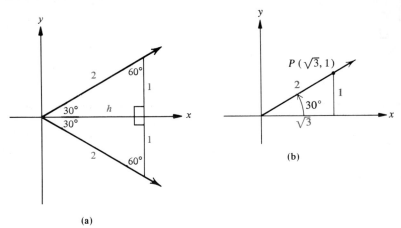

(a)

(b)

Figure 1.10

From Definition 1.3,

$$\sin 30° = \frac{1}{2}, \qquad \csc 30° = \frac{2}{1} = 2,$$

$$\cos 30° = \frac{\sqrt{3}}{2}, \qquad \sec 30° = \frac{2}{\sqrt{3}},$$

$$\tan 30° = \frac{1}{\sqrt{3}}, \qquad \cot 30° = \frac{\sqrt{3}}{1} = \sqrt{3}.$$

Figure 1.11

Reorienting the right triangle in Figure 1.10b, where the angle whose measure is 60° is now in standard position, results in a drawing as in Figure 1.11. From Definition 1.3,

$$\sin 60° = \frac{\sqrt{3}}{2}, \qquad \csc 60° = \frac{2}{\sqrt{3}},$$

$$\cos 60° = \frac{1}{2}, \qquad \sec 60° = \frac{2}{1} = 2,$$

$$\tan 60° = \frac{\sqrt{3}}{1} = \sqrt{3}, \qquad \cot 60° = \frac{1}{\sqrt{3}}.$$

Now let us consider a 45° right triangle whose sides opposite the 45° degree angles are one unit long. In this case, from the Pythagorean theorem, the hypotenuse is given by $\sqrt{1^2 + 1^2} = \sqrt{2}$. Hence, the lengths of the sides of this triangle, and of course any 45°–45°–90° triangle, are proportional to the numbers 1, 1, and $\sqrt{2}$, as shown in Figure 1.12. From Definition 1.3,

$$\sin 45° = \frac{1}{\sqrt{2}}, \qquad \csc 45° = \frac{\sqrt{2}}{1} = \sqrt{2},$$

$$\cos 45° = \frac{1}{\sqrt{2}}, \qquad \sec 45° = \frac{\sqrt{2}}{1} = \sqrt{2},$$

$$\tan 45° = 1, \qquad \cot 45° = 1.$$

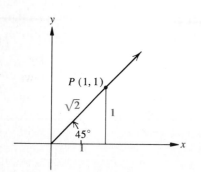

Figure 1.12

Function Values for Quadrantal Angles

As we have noted, the terminal side of a quadrantal angle coincides with a coordinate axis, and one of the coordinates of any point on the terminal side is zero. Hence, some of the trigonometric ratios for quadrantal angles are not defined. For convenience in finding the trigonometric ratios that are defined, we select the point on the terminal side with the nonzero coordinate 1 or −1, as shown in Figure 1.13. For 0°, 90°, 180°, and 270°,

$$r = \sqrt{x^2 + y^2} = \sqrt{1^2} = 1.$$

Figure 1.13

The values of the trigonometric ratios that are defined for the angles shown can be determined directly from Definition 1.3. For example,

$$\cos 90° = \frac{x}{r} \quad \text{and} \quad \sin 270° = \frac{y}{r}$$

$$= \frac{0}{1} \qquad\qquad\qquad = \frac{-1}{1}$$

$$= 0 \qquad\qquad\qquad = -1.$$

The other function values for the quadrantal angles are obtained in a similar way. These values and the function values (trigonometric ratios) for the special angles 30°, 45°, and 60° that have been discussed are summarized in Table 1.1.

Although the function values for angles that are often used are available in Table 1.1, we strongly encourage you to construct a triangle (as we did to obtain these values) each time you want a particular trigonometric ratio for one of the special angles.

TABLE 1.1

$\alpha°$	α^R	$\sin \alpha$	$\cos \alpha$	$\tan \alpha$	$\csc \alpha$	$\sec \alpha$	$\cot \alpha$
0°	0^R	0	1	0	undef.	1	undef.
30°	$\frac{\pi}{6}^R$	$\frac{1}{2}$	$\frac{\sqrt{3}}{2}$	$\frac{1}{\sqrt{3}}$	2	$\frac{2}{\sqrt{3}}$	$\sqrt{3}$
45°	$\frac{\pi}{4}^R$	$\frac{1}{\sqrt{2}}$	$\frac{1}{\sqrt{2}}$	1	$\sqrt{2}$	$\sqrt{2}$	1
60°	$\frac{\pi}{3}^R$	$\frac{\sqrt{3}}{2}$	$\frac{1}{2}$	$\sqrt{3}$	$\frac{2}{\sqrt{3}}$	2	$\frac{1}{\sqrt{3}}$
90°	$\frac{\pi}{2}^R$	1	0	undef.	1	undef.	0
180°	π^R	0	−1	0	undef.	−1	undef.
270°	$\frac{3\pi}{2}^R$	−1	0	undef.	−1	undef.	0

Example Find the value of the product $(\csc 30°)(\cos 30°)(\tan 30°)^2$.

Solution First construct a 30°–60°–90° triangle. The function values can then be read directly from the figure.

$$(\csc 30°)(\cos 30°)(\tan 30°)^2 = (2)\left(\frac{\sqrt{3}}{2}\right)\left(\frac{1}{\sqrt{3}}\right)^2 = \frac{\sqrt{3}}{3}.$$

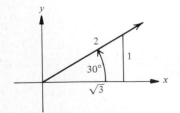

Calculators with trigonometric function value capability make it possible to obtain the table entries quite readily in decimal form. However, the definition of the trigonometric ratios will be more meaningful to you if you obtain these values in fractional form directly from the triangles in Figures 1.10, 1.11, and 1.12, using radicals for exact values rather than obtaining decimal approximations. *In our work in this text we shall express the trigonometric ratios for these special angles in such form.*

A symbol such as $(\tan 30°)^2$, as in the example above, is usually written in more concise form as $\tan^2 30°$. Similarly,

$$\sin^2 \alpha \text{ means } (\sin \alpha)^2,$$

$$\cos^2 \alpha \text{ means } (\cos \alpha)^2,$$

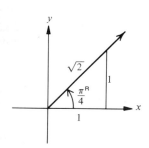

Example   Find the value of $\sin^2 \dfrac{\pi^{\text{R}}}{4} + \tan^2 \dfrac{\pi^{\text{R}}}{3}$.

Solution   First construct appropriate triangles. The function values can then be read directly from the figures.

$$\sin^2 \frac{\pi^{\text{R}}}{4} + \tan^2 \frac{\pi^{\text{R}}}{3} = \left(\frac{1}{\sqrt{2}}\right)^2 + \left(\frac{\sqrt{3}}{1}\right)^2$$

$$= \frac{1}{2} + 3 = \frac{7}{2}.$$

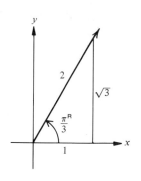

Function Values for Other Acute Angles

The function values for the special angles were obtained using properties of geometry. Approximations for function values of other acute angles can be obtained by formulas that are introduced in the study of the calculus. (Two such formulas are introduced in the exercises on page 22.) However, we can use tables or a scientific calculator to obtain approximations for trigonometric function values.

Using a Calculator

Finding the function values for trigonometric functions by using a calculator is a simple process. However, because the steps to be taken vary somewhat for different types of calculators, we will not consider specific procedures. The instruction booklet for your particular calculator can provide this information.

Calculators vary in the number of digits that are shown in the display; however, we shall round off all readings to four decimal places.

Example Set Degree/Radian switch on RAD:

$$\sin 0.714^R \approx 0.654862.$$

Rounding off the result to four decimal places gives

$$\sin 0.714^R \approx 0.6549.$$

Example Set Degree/Radian switch on DEG:

$$\sin 21° \; 42' = \sin 21.7° \approx 0.369747.$$

Rounding off the result to four decimal places gives

$$\sin 21° \; 42' \approx 0.3697.$$

Note that in the above example it was necessary first to express the angle in decimal form before using the calculator. Since there are 60′ in each degree,

$$42' = \left(\frac{42}{60}\right)° = 0.7°.$$

Hence, $21° \; 42' = 21.7°$.

Values for $\sin \alpha$, $\cos \alpha$, and $\tan \alpha$ can be obtained directly on scientific calculators. Values for $\sec \alpha$, $\csc \alpha$, and $\cot \alpha$ can be obtained by using the reciprocal relationships

$$\sec \alpha = \frac{1}{\cos \alpha}, \qquad \csc \alpha = \frac{1}{\sin \alpha}, \qquad \text{and} \qquad \cot \alpha = \frac{1}{\tan \alpha}$$

introduced in Section 1.2.

Example Set Degree/Radian switch on DEG:

$$\sec 14° \; 18' = \sec 14.3°$$

$$= \frac{1}{\cos 14.3°} \approx 1.03197.$$

Rounding off to the nearest four places gives

$$\sec 14° \; 18' \approx 1.0320.$$

Although we can obtain approximations for function values of all angles using calculators, we shall continue to use exact function values for angles that are integral multiples of 30° and 45°.

Using Tables

Before the advent of scientific calculators, approximations for function values were obtained from extensive tables prepared for this purpose. Although such tables have lost their primary purpose, you may wish to examine the tables in Appendix C which explicitly list function values, to get an overview of the behavior of the trigonometric function values. In particular, they readily suggest the ranges of each of the trigonometric functions and the limiting behavior of the tangent, cotangent, secant, and cosecant function values that are undefined for certain angles.

Cofunctions

A careful observation of these tables also indicates that for any α for which the function is defined

$$\sin \alpha = \cos(90° - \alpha)$$

$$\tan \alpha = \cot(90° - \alpha)$$

$$\sec \alpha = \csc(90° - \alpha).$$

The pairs of functions involved in these relationships are called **cofunctions,** which their names readily suggest.

EXERCISE SET 1.3

A

Find the exact value of each expression, if it exists, by making an appropriate sketch (see Figures 1.10–1.13) and reading the ratio directly from the figure.

Examples   a. $\cos 60°$ b. $\tan \dfrac{\pi^{R}}{6}$

a.

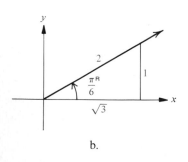

b.

Solutions

a. A 30°–60° right triangle is sketched with the 60° angle in standard position and with sides of length, 1, $\sqrt{3}$, and 2. From Definition 1.3,

$$\cos 60° = \frac{x}{r} = \frac{1}{2}.$$

b. A 30°–60° right triangle is sketched with the $\pi^{R}/6$ angle in standard position and with sides of length, 1, $\sqrt{3}$, and 2. From Definition 1.3,

$$\tan \frac{\pi^{R}}{6} = \frac{y}{x} = \frac{1}{\sqrt{3}}.$$

1. $\sin 60°$	**2.** $\tan 30°$	**3.** $\cos 45°$	**4.** $\csc 60°$
5. $\cos 30°$	**6.** $\cot 45°$	**7.** $\cot 60°$	**8.** $\sin 45°$
9. $\csc \dfrac{\pi}{4}^R$	**10.** $\cot \dfrac{\pi}{6}^R$	**11.** $\sec \dfrac{\pi}{6}^R$	**12.** $\sec \dfrac{\pi}{3}^R$
13. $\sin 90°$	**14.** $\sec 180°$	**15.** $\tan 0°$	**16.** $\cos 270°$
17. $\csc \pi^R$	**18.** $\tan \dfrac{\pi}{2}^R$	**19.** $\sin \dfrac{3\pi}{2}^R$	**20.** $\cos \pi^R$
21. $\sin 0°$	**22.** $\csc 0°$	**23.** $\sec 90°$	**24.** $\cos 0°$

Find the exact value of each expression if the expression represents a real number.

Example $\quad \sin \dfrac{\pi}{4}^R + \left(\cos^2 \dfrac{\pi}{6}^R \right)\left(\tan \dfrac{\pi}{3}^R \right)$

Solution By sketching the appropriate triangles, we find

$$\sin \frac{\pi}{4}^R = \frac{1}{\sqrt{2}}, \quad \cos^2 \frac{\pi}{6}^R = \left(\frac{\sqrt{3}}{2}\right)^2 = \frac{3}{4}, \quad \text{and} \quad \tan \frac{\pi}{3}^R = \sqrt{3}.$$

  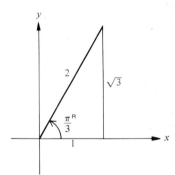

Hence,

$$\sin \frac{\pi}{4}^R + \left(\cos^2 \frac{\pi}{6}^R \right)\left(\tan \frac{\pi}{3}^R \right) = \frac{1}{\sqrt{2}} + \left(\frac{3}{4}\right)\left(\frac{\sqrt{3}}{1}\right)$$

$$= \frac{1(4)}{\sqrt{2}(4)} + \frac{3\sqrt{3}(\sqrt{2})}{4(\sqrt{2})}$$

$$= \frac{4 + 3\sqrt{6}}{4\sqrt{2}}.$$

25. $\cos^2 30°$

26. $\tan^2 30°$

27. $\sin^2 \dfrac{\pi^R}{4}$

28. $\csc^2 \dfrac{3\pi^R}{2}$

29. $(\cos 180°)(\tan^2 60°)$

30. $(\sin 90°)(\cos^2 45°)$

31. $\left(\sin^2 \dfrac{\pi^R}{6}\right)\left(\cot^2 \dfrac{\pi^R}{3}\right)$

32. $\left(\tan^2 \dfrac{\pi^R}{6}\right)(\sec^2 0^R)$

33. $\sec 60° - \cos^2 45°$

34. $\csc 30° - \tan^2 30°$

35. $\tan^2 \dfrac{\pi^R}{4} - \cot^2 \dfrac{\pi^R}{4}$

36. $\tan^2 \dfrac{\pi^R}{3} - \cot^2 \dfrac{\pi^R}{6}$

37. $(\sin 45°)(\cos 30°)(\tan^2 60°)$

38. $(\sec^2 \pi^R)\left(\sin \dfrac{3\pi^R}{2}\right)(\tan 0^R)$

39. $(\cos 0^R)\left(\sin \dfrac{\pi^R}{2}\right) + \tan \dfrac{\pi^R}{4}$

40. $\cot 45° + (\tan 60°)(\cot 60°)$

41. $\cos^2 60° - (\sec 270°)(\tan 45°)$

42. $2(\csc^2 180°)(\sin^2 45°) + \tan 45°$

43. $\dfrac{1 + \tan^2 30°}{\sec^2 30°}$

44. $\dfrac{1 - \cos^2 45°}{\sin^2 45°}$

45. $\dfrac{\left(\sin^2 \dfrac{\pi^R}{6}\right)(\cos \pi^R)}{\sin \dfrac{\pi^R}{6}}$

46. $\dfrac{\tan \dfrac{\pi^R}{3}}{\left(\sin \dfrac{\pi^R}{3}\right)\left(\tan^2 \dfrac{\pi^R}{3}\right)}$

47. $\dfrac{(\sin 90°)(\cos 45°)}{\cos 90°}$

48. $\dfrac{\tan^2 30° + \sec^2 45°}{\cos 270°}$

Find an approximation to four decimal places for each function value.

Examples   a. $\cos 63.7°$ b. $\cot 17° \; 6'$

Solutions   a. $\cos 63.7° \approx 0.4431$ b. $\cot 17° \; 6' = \cot 17.1°$

$$= \frac{1}{\tan 17.1°} \approx 3.2506$$

49. $\sin 16.4°$ **50.** $\cos 21.5°$ **51.** $\tan 69.2°$ **52.** $\sin 82.6°$

53. $\cos 18° \; 48'$ **54.** $\sin 27° \; 6'$ **55.** $\tan 6° \; 12'$ **56.** $\cos 63° \; 30'$

57. $\cot 14.3°$ **58.** $\sec 79.6°$ **59.** $\csc 12° \; 24'$ **60.** $\cot 62° \; 12'$

Examples   a. $\tan 0.43^R$ b. $\sec 1.07^R$

Solutions   a. $\tan 0.43^R \approx 0.4586$ b. $\sec 1.07^R = \dfrac{1}{\cos 1.07^R} \approx 2.0828$

61. $\sin 0.91^R$ **62.** $\cos 1.23^R$ **63.** $\tan 0.42^R$ **64.** $\sin 1.05^R$

65. $\sec 0.63^R$ **66.** $\csc 1.42^R$ **67.** $\cot 0.31^R$ **68.** $\sec 1.15^R$

69. For which angles are the tangent function values undefined? For which are they equal to 0?

70. For which angles are the cotangent function values undefined? For which are they equal to 0?

71. For which angles are the secant function values undefined? Are there any angles for which the secant function value is 0?

72. For which angles are the cosecant function values undefined? Are there any angles for which the cosecant function value is 0?

B

73. Given that $\sin \alpha = y$, find expressions in terms of y for the other five trigonometric ratios.   *Hint:* Consider $\sin \alpha = y/1$.

74. Given that $\tan \alpha = y$, find expressions in terms of y for the other five trigonometric ratios.

In calculus it is shown that for all $x = \alpha^R$ and $n \in N$,

$$\cos x = 1 - \frac{x^2}{2!} + \frac{x^4}{4!} + \cdots + (-1)^{n-1}\frac{x^{2n-2}}{(2n-2)!} + \cdots, \tag{1}$$

$$\sin x = x - \frac{x^3}{3!} + \frac{x^5}{5!} + \cdots + (-1)^{n-1}\frac{x^{2n-1}}{(2n-1)!} + \cdots, \tag{2}$$

where

$$2! = 1 \cdot 2, \qquad 3! = 1 \cdot 2 \cdot 3, \qquad 4! = 1 \cdot 2 \cdot 3 \cdot 4, \qquad \text{etc.}$$

*The expression in the right member of each equation is called an **alternating infinite power series in x**. If any finite number of terms in either of these power series is used to obtain an approximation for the function value, the error that exists is less than the absolute value of the next term in the series.*

Example   Find $\cos 0.3^R$ correct to four decimal places.

Solution   Replacing x in Equation (1) by 0.3 yields

$$\cos 0.3^R = 1 - \frac{(0.3)^2}{2!} + \frac{(0.3)^4}{4!} - \frac{(0.3)^6}{6!} + \cdots$$

$$= 1 - \frac{0.09}{2} + \frac{0.0081}{24} - \frac{0.000729}{720} + \cdots$$

$$= 1 - 0.045 + 0.00034 - 0.000001 + \cdots.$$

Since we wish accuracy to four decimal places, and since the error introduced by taking the first three terms is less than the absolute value of the fourth term (0.000001), then correct to at least four decimal places,

$$\cos 0.3^R \approx 1 - 0.045 + 0.00034 = 0.9553.$$

Note that this is the value for $\cos 0.3^R$ shown in Table III in Appendix C.

Use Equation (1) or (2) above to find each function value correct to four decimal places.

75. $\sin 0.01^R$ **76.** $\cos 0.50^R$ **77.** $\sin 1.00^R$

78. $\cos 0.02^R$ **79.** $\sin 0.60^R$ **80.** $\cos 1.10^R$

81. Two functions f and g are called cofunctions if

$$f(s) = g(t) \quad \text{and} \quad s + t = 90°.$$

Use the definitions of the trigonometric functions to show that for $0° \leq s \leq 90°$ the sine and cosine, the tangent and cotangent, and the secant and cosecant are cofunctions.

1.4 Function Values for All Angles

In Section 1.3 we obtained trigonometric function values for the acute angles $30°, 45°$, and $60°$ by using some geometric notions, and we obtained approximations for other acute angles by using a calculator. In this section we shall obtain function values for all angles α for which a function is defined. We shall first introduce several notions that will be helpful.

Signs of Function Values

For an angle α in standard position and x and y the coordinates of a point on the terminal side, $r = \sqrt{x^2 + y^2}$ is always positive. Thus, the *sign* of a trigonometric ratio depends on the values of x and y, and hence on the quadrant in which the terminal side of the angle lies.

Example   If the terminal side of an angle α in standard position lies in the second quadrant, the x-coordinate of a point on the terminal side is negative and the y-coordinate is positive. Hence,

$$\sin \alpha \left(\text{equal to } \frac{y}{r} \right) \text{ is positive,}$$

$$\cos \alpha \left(\text{equal to } \frac{x}{r} \right) \text{ is negative, and}$$

$$\tan \alpha \left(\text{equal to } \frac{y}{x} \right) \text{ is negative.}$$

The signs of the trigonometric function values of nonquadrantal angles in the four quadrants are shown in Table 1.2.

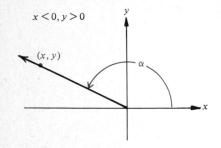

TABLE 1.2

Quadrant	I $x, y > 0$	II $x < 0, y > 0$	III $x, y < 0$	IV $x > 0, y < 0$
$\sin \alpha$ or $\csc \alpha$	+	+	−	−
$\cos \alpha$ or $\sec \alpha$	+	−	−	+
$\tan \alpha$ or $\cot \alpha$	+	−	+	−

Reference Angles

Definition 1.4   *The* **reference angle** *of any nonquadrantal angle α in standard position is the positive acute angle $\tilde{\alpha}$ whose sides are the terminal side of α and a ray of the x-axis.*

The reference angles for angles in each of the four quadrants are shown in Figure 1.14.

Examples   Sketch $\tilde{\alpha}$ for each angle and find $\tilde{\alpha}°$.

a. $\alpha = 135°$ b. $\alpha = 240°$

Figure 1.14

a.

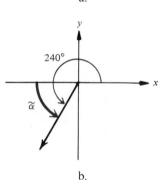

b.

Solutions

a. $\tilde{\alpha} = (180 - 135)°$

   $= 45°$

b. $\tilde{\alpha} = (240 - 180)°$

   $= 60°$

Using Reference Angles

Using geometric considerations, we can establish that *the absolute values of the trigonometric ratios of any angle* α *are equal, respectively, to the values of the corresponding trigonometric ratios for* $\tilde{\alpha}$.

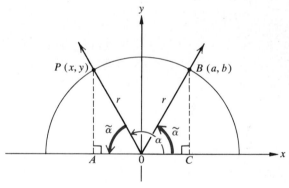

Figure 1.15

For the case in which α is in Quadrant II, as shown in Figure 1.15, we can construct:

1. $\overline{PA}$ perpendicular to the *x*-axis for any point P on the terminal side of α.
2. Semicircle with radius $\overline{OP}$.
3. Radius $\overline{OB}$, making an angle $\tilde{\alpha}$ with the *x*-axis.
4. $\overline{BC}$ perpendicular to the *x*-axis.

Because the hypotenuse and an acute angle of triangle *POA* are congruent, respectively, to the hypotenuse and an acute angle of triangle *BOC*, the triangles are congruent, and hence $PA = BC$ and $OA = OC$.

Thus, $x = -a$ and $y = b$. We then have

$$|\sin \alpha| = \left|\frac{y}{r}\right| = \frac{b}{r} = \sin \tilde{\alpha}, \qquad |\csc \alpha| = \left|\frac{r}{y}\right| = \frac{r}{b} = \csc \tilde{\alpha},$$

$$|\cos \alpha| = \left|\frac{x}{r}\right| = \frac{a}{r} = \cos \tilde{\alpha}, \qquad |\sec \alpha| = \left|\frac{r}{x}\right| = \frac{r}{a} = \sec \tilde{\alpha},$$

$$|\tan \alpha| = \left|\frac{y}{x}\right| = \frac{b}{a} = \tan \tilde{\alpha}, \qquad |\cot \alpha| = \left|\frac{x}{y}\right| = \frac{a}{b} = \cot \tilde{\alpha}.$$

The above relationships hold for an angle in any quadrant, as can be shown by using congruent triangles as above. Hence, for all α for which trig α is defined,

$$|\text{trig } \alpha| = \text{trig } \tilde{\alpha}.$$

Function Values for Integral Multiples of 30°, 45°, and 60°

The notion of a reference angle facilitates finding *exact* function values of angles with measures that are integral multiples of 30°, 45°, and 60°. For such an angle α, we proceed as follows:

1. Find trig $\tilde{\alpha}$.
2. Prefix the algebraic sign, $+$ or $-$, depending on whether the original function value, trig α, is positive or negative; use Table 1.2 as necessary.

Example   Find tan 210°.

Solution   A sketch showing the reference angle is helpful. Observe that

$$\tilde{\alpha} = (210 - 180)° = 30°.$$

Since α is in Quadrant III, tan 210° is positive (see Table 1.2). Therefore,

$$\tan 210° = \tan \tilde{\alpha} = \tan 30° = \frac{1}{\sqrt{3}}.$$

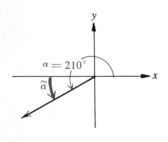

Example   Find $\cos \dfrac{3\pi^R}{4}$.

Solution   A sketch showing the reference angle is helpful. Observe that

$$\tilde{\alpha} = \left(\pi - \frac{3\pi}{4}\right)^R = \frac{\pi^R}{4}.$$

Since α is in Quadrant II, $\cos \dfrac{3\pi^{R}}{4}$ is negative (see Table 1.2). Therefore,

$$\cos \frac{3\pi^{R}}{4} = -\cos \tilde{\alpha}$$

$$= -\cos \frac{\pi}{4} = -\frac{1}{\sqrt{2}}.$$

The trigonometric function values are ratios of numbers associated with a point on the terminal side of an angle in standard position. Therefore, the trigonometric function values for an angle α are equal to the analogous function values for any angle coterminal with α, and function values for *all* integral multiples of $30°, 45°,$ and $60°$ can be obtained by using the methods we used above to obtain function values for $0° \leq \alpha \leq 360°$.

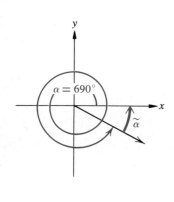

Example   Find the exact value of $\tan 690°$.

Solution   Since $690° - 360° = 330°$, the reference angle

$$\tilde{\alpha} = 360° - 330° = 30°.$$

Because $690°$ is in Quadrant IV, we have

$$\tan 690° = -\tan 30°$$

$$= -\frac{1}{\sqrt{3}}.$$

Example   Find the exact value of $\cos \dfrac{16\pi^{R}}{3}$.

Solution   Since

$$\frac{16\pi^{R}}{3} - 4\pi^{R} = \frac{16\pi^{R}}{3} - \frac{12\pi^{R}}{3} = \frac{4\pi^{R}}{3},$$

the reference angle

$$\tilde{\alpha} = \frac{4\pi^{R}}{3} - \pi^{R} = \frac{\pi^{R}}{3}.$$

Because $(16\pi/3)^{R}$ is in Quadrant III, we have

$$\cos \frac{16\pi^{R}}{3} = -\cos \frac{\pi^{R}}{3}$$

$$= -\frac{1}{2}.$$

Function values for angles with negative measures can also be found using similar procedures.

Example   Find the exact value of $\sin(-45°)$.

Solution   The reference angle is

$$\tilde{\alpha} = 45°$$

Because $-45°$ is in Quadrant IV, we have

$$\sin(-45°) = -\sin 45° = -\frac{1}{\sqrt{2}}$$

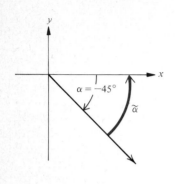

Using a Calculator

Calculators are programmed to give approximations for $\sin \alpha$, $\cos \alpha$, and $\tan \alpha$ directly for all values of α.

Examples
   a. Set Degree/Radian switch on RAD:

$$\sin 5.84^{\text{R}} \approx -0.4288.$$

   b. Set Degree/Radian switch on DEG:

$$\cos 284.3° \approx 0.2470.$$

Again, values for $\sec \alpha$, $\csc \alpha$, and $\cot \alpha$ can be obtained on a calculator by using the reciprocal relationships.

Example   Set Degree/Radian switch on DEG:

$$\sec 163° \ 18' = \sec 163.3°$$

$$= \frac{1}{\cos 163.3°} \approx -1.0440.$$

As noted on page 18, we shall only use a calculator to obtain function values for angles that are not integral multiples of $30°$ and $45°$.

Using Tables

The use of tables to obtain trigonometric function values outside the interval 0^{R} to $(\pi/2)^{\text{R}}$ is discussed in Appendix C. Reference angles in conjunction with tables are used for such values, and 3.14 is used as an approximation for π when the angle is given in radian measure. Hence the approximations obtained are less accurate than the approximations obtained by using a calculator programmed for a closer approximation to π.

EXERCISE SET 1.4

Determine the quadrant in which the terminal side of each angle α lies.

Example $\sin \alpha < 0$ and $\tan \alpha > 0$

Solution   For $\sin \alpha < 0$, the terminal side of α lies in Quadrant III or IV, and for $\tan \alpha > 0$, the terminal side of α lies in Quadrant I or III. Therefore, the terminal side of α must lie in Quadrant III.

1. $\sin \alpha < 0$ and $\cos \alpha > 0$ 2. $\sin \alpha < 0$ and $\tan \alpha < 0$

3. $\sin \alpha > 0$ and $\sec \alpha > 0$ 4. $\sin \alpha > 0$ and $\cot \alpha < 0$

5. $\cos \alpha < 0$ and $\tan \alpha > 0$ 6. $\cos \alpha < 0$ and $\csc \alpha < 0$

7. $\cos \alpha < 0$ and $\cot \alpha < 0$ 8. $\tan \alpha > 0$ and $\csc \alpha > 0$

Given one function value and the sign of another, determine the other five function values of α.

Example $\sin \alpha = \dfrac{3}{5}$,   if $\cos \alpha < 0$

Solution   From Table 1.2, if $\sin \alpha > 0$ and $\cos \alpha < 0$, then α must be in Quadrant II. From the Pythagorean theorem,

$$(OQ)^2 + 3^2 = 5^2,$$

$$(OQ)^2 = 25 - 9 = 16,$$

$$OQ = 4.$$

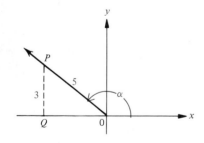

Hence the coordinates of point P are $(-4, 3)$, and the six ratios are

$$\sin \alpha = \frac{3}{5}, \qquad \csc \alpha = \frac{5}{3},$$

$$\cos \alpha = -\frac{4}{5}, \qquad \sec \alpha = -\frac{5}{4},$$

$$\tan \alpha = -\frac{3}{4}, \qquad \cot \alpha = -\frac{4}{3}.$$

9. $\tan \alpha = -\dfrac{5}{12}$, if $\sin \alpha > 0$ **10.** $\cos \alpha = -\dfrac{3}{4}$, if $\sin \alpha > 0$

11. $\sin \alpha = -\dfrac{2}{3}$, if $\tan \alpha > 0$ **12.** $\tan \alpha = \dfrac{8}{15}$, if $\cos \alpha > 0$

13. $\cos \alpha = \dfrac{4}{5}$, if $\cot \alpha < 0$ **14.** $\sin \alpha = -\dfrac{1}{3}$, if $\tan \alpha < 0$

Sketch $\tilde{\alpha}$ and find $\tilde{\alpha}$ in the same units in which each α is given.

Examples a. $150°$ b. $\dfrac{11\pi^R}{6}$

Solutions a. $\tilde{\alpha} = 180° - 150°$ b. $\tilde{\alpha} = 2\pi^R - \dfrac{11\pi^R}{6}$
$\qquad\qquad\quad = 30°$ $\qquad\quad = \dfrac{\pi^R}{6}$

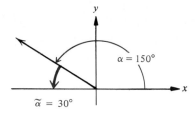

15. $120°$ **16.** $210°$ **17.** $225°$ **18.** $315°$

19. $\dfrac{5\pi^R}{4}$ **20.** $\dfrac{7\pi^R}{4}$ **21.** $\dfrac{5\pi^R}{3}$ **22.** $\dfrac{4\pi^R}{3}$

Find the exact value for each expression.

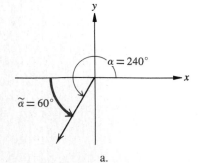

a.

Examples a. $\sin 240°$ b. $\tan \dfrac{7\pi^R}{4}$

Solutions
   a. The reference angle

$$\tilde{\alpha} = (240 - 180)° = 60°.$$

Since 240° is in Quadrant III, sin 240° is negative. Therefore,

$$\sin 240° = -\sin 60° = -\frac{\sqrt{3}}{2}.$$

b. The reference angle

$$\tilde{\alpha} = 2\pi^R - \frac{7\pi^R}{4} = \frac{\pi^R}{4}.$$

Since $(7\pi/4)^R$ is in Quadrant IV, $\tan (7\pi/4)^R$ is negative. Therefore,

$$\tan \frac{7\pi^R}{4} = -\tan \frac{\pi^R}{4} = -1.$$

b.

23. sin 330°	**24.** cot 120°	**25.** tan 225°	**26.** sin 315°
27. cos 120°	**28.** sec 330°	**29.** cot 225°	**30.** csc 150°
31. $\sin \dfrac{5\pi^R}{4}$	**32.** $\cos \dfrac{3\pi^R}{4}$	**33.** $\tan \dfrac{2\pi^R}{3}$	**34.** $\cot \dfrac{5\pi^R}{3}$
35. $\sec \dfrac{7\pi^R}{6}$	**36.** $\csc \dfrac{5\pi^R}{6}$		

Find the exact value of each function value.

Example   tan 480°

Solution   Since $480° - 360° = 120°$, the reference angle

$$\tilde{\alpha} = 180° - 120° = 60°.$$

Because 480° is in Quadrant II, tan 480° is negative. Using Table 1.1 or by constructing an appropriate triangle, we have that

$$\tan 480° = -\tan 60° = -\sqrt{3}.$$

37. tan 420°	**38.** sin 570°	**39.** cos 675°	**40.** sec 510°
41. cot 930°	**42.** tan 870°	**43.** $\sin 4\pi^R$	**44.** $\tan 5\pi^R$
45. $\cos \dfrac{11\pi^R}{4}$	**46.** $\sec \dfrac{8\pi^R}{3}$	**47.** $\tan \dfrac{10\pi^R}{3}$	**48.** $\csc \dfrac{17\pi^R}{6}$

Find an approximation for each function value.

Example   sec 128° 12′

Solution   Set Degree/Radian setting on DEG:

$$\sec 128° \, 12' = \sec 128.2°$$

$$= \frac{1}{\cos 128.2°} \approx -1.6171.$$

49. cos 310°   **50.** sin 230°   **51.** tan 263.4°   **52.** cos 138.7°

53. cot 342.5°   **54.** csc 273.6°   **55.** csc 322° 24′   **56.** sec 269° 42′

Example   cot 5.36ᴿ

Solution   Set Degree/Radian setting on RAD:

$$\cot 5.36^R = \frac{1}{\tan 5.36^R}$$

$$\approx -0.7564.$$

57. tan 2.10ᴿ   **58.** sin 4.22ᴿ   **59.** cos 1.62ᴴ   **60.** tan 3.48ᴿ

61. sec 4.87ᴿ   **62.** csc 1.84ᴿ   **63.** cot 2.93ᴿ   **64.** sec 6.20ᴿ

1.5 Finding Acute Angles

Given a positive element, Trig α, in the range of any of the six trigonometric functions, we can sketch a triangle or use Table 1.1, or a calculator, to find the corresponding elements in the domain $0° \le \alpha \le 90°$ or $0^R \le \alpha \le (\pi/2)^R$.

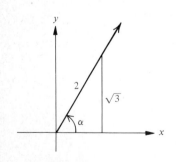

Example   If $\sin \alpha = \sqrt{3}/2$, we can sketch an acute right triangle where the side opposite α is $\sqrt{3}$ and the hypotenuse is 2 (or use Table 1.1) to note that $\alpha = 60°$ or $\alpha = (\pi/3)^R$.

Inverse Notation

Special notations called **inverse notations** are often used to express the element α in the domain of a trigonometric function explicitly in terms of the corresponding element in its range. For example,

$$\sin \alpha = \frac{\sqrt{3}}{2}$$

can be expressed equivalently as

$$\alpha = \mathrm{Sin}^{-1} \frac{\sqrt{3}}{2} \qquad \text{or} \qquad \alpha = \mathrm{Arcsin} \frac{\sqrt{3}}{2},$$

which are both read as "α equals Arcsin $\sqrt{3}/2$." The phrase "inverse sine of $\sqrt{3}/2$" is also used in this context. In each case we can simply view each of the expressions

$$\mathrm{Sin}^{-1} \frac{\sqrt{3}}{2} \qquad \text{and} \qquad \mathrm{Arcsin} \frac{\sqrt{3}}{2}$$

as the *acute angle* whose sine is $\sqrt{3}/2$. Hence, from the example above,

$$\alpha = \mathrm{Sin}^{-1} \frac{\sqrt{3}}{2} = 60°$$

or

$$\alpha = \mathrm{Sin}^{-1} \frac{\sqrt{3}}{2} = \frac{\pi^{\mathrm{R}}}{3}.$$

Note that a capital letter is used for the first letter in each symbol, Sin^{-1} and Arcsin. This is a usual practice.

In the above example, the element in the range was the positive number $\sqrt{3}/2$. At this time we will only consider inverse notation Trig^{-1} y or Arctrig y for $y > 0$ as specified in the following definition. We will consider inverse notation in more detail in Chapter 3 where values of y can also be negative.

Definition 1.5 *For $y > 0$, the inverse notation* **Trig^{-1} y,** *or* **Arctrig y,** *names an acute angle α such that* trig $\alpha = y$.

Example Express tan $\alpha = 1.4882$ using inverse notation.

Solution $\alpha = \mathrm{Tan}^{-1} 1.4882 \qquad \text{or} \qquad \alpha = \mathrm{Arctan}\ 1.4882$

It may happen that a problem involves an equation of the form $\alpha = \mathrm{Trig}^{-1}\ y$. In such a case, it is often helpful to rewrite the equation as trig $\alpha = y$, since this form sometimes provides more insight regarding the problem.

Examples

a. Express $\alpha = \mathrm{Cos}^{-1}\ 0.9461$ without using inverse notation.

b. Find the value of the acute angle.

Solutions

a. $\cos \alpha = 0.9461$.

b. Using a calculator (set on DEG), we find that

$$\alpha \approx 18.9°.$$

Using a calculator (set on RAD), we find that

$$\alpha \approx 0.33^R.$$

EXERCISE SET 1.5

Express each of the following using the inverse notation $Trig^{-1}\ y$.

Examples   a. $\cos \alpha = \dfrac{1}{\sqrt{2}}$ b. $\csc \alpha = 2.4123$

Solutions   a. $\alpha = \mathrm{Cos}^{-1} \dfrac{1}{\sqrt{2}}$ b. $\alpha = \mathrm{Csc}^{-1}\ 2.4123$

1. $\tan \alpha = \dfrac{1}{\sqrt{3}}$ **2.** $\sin \alpha = \dfrac{1}{\sqrt{2}}$ **3.** $\sec \alpha = \sqrt{2}$

4. $\csc \alpha = 1$ **5.** $\cos \alpha = \dfrac{\sqrt{3}}{2}$ **6.** $\cot \alpha = \sqrt{3}$

7. $\sin \alpha = 0.2169$ **8.** $\tan \alpha = 3.1242$ **9.** $\cot \alpha = 2.0941$

10. $\cos \alpha = 0.6929$ **11.** $\sec \alpha = 2.2013$ **12.** $\csc \alpha = 1.0491$

Express each of the following without using inverse notation.

Examples   a. $\alpha = \mathrm{Sin}^{-1} \dfrac{1}{2}$ b. $\alpha = \mathrm{Arctan}\ 1.4321$

Solutions   a. $\sin \alpha = \dfrac{1}{2}$ b. $\tan \alpha = 1.4321$

13. $\alpha = \mathrm{Cot}^{-1}\sqrt{3}$ **14.** $\alpha = \mathrm{Cos}^{-1}\dfrac{1}{2}$

15. $\alpha = \mathrm{Arcsec}\ 1.4214$ **16.** $\alpha = \mathrm{Arccsc}\ 2.0031$

17. $\alpha = \mathrm{Arcsin}\ 0.4112$ **18.** $\alpha = \mathrm{Arctan}\ 1.0341$

Find the exact value for the measure of angle α, in degrees, where $0° < \alpha < 90°$.

Example $\tan \alpha = \sqrt{3}$

Solution   Sketch an acute right triangle where the side opposite α is $\sqrt{3}$ and the side adjacent to α is 1 (or use Table 1.1) to note that

$$\alpha = \mathrm{Arctan}\ \sqrt{3} = 60°.$$

19. $\cos \alpha = \dfrac{1}{2}$ **20.** $\sin \alpha = \dfrac{1}{\sqrt{2}}$ **21.** $\tan \alpha = \dfrac{1}{\sqrt{3}}$

22. $\sec \alpha = \sqrt{2}$ **23.** $\csc \alpha = \dfrac{2}{\sqrt{3}}$ **24.** $\cot \alpha = \dfrac{1}{\sqrt{3}}$

Find an approximation to the nearest tenth for the measure of angle α, in degrees, where $0° \le \alpha \le 90°$.

Example $\sec \alpha = 1.1937$

Solution   Calculators are not programmed for the secant function directly. However, a value for $\cos \alpha$ can be obtained as the reciprocal of the given value $\sec \alpha$. Thus, set Degree/Radian setting on DEG:

$$\cos \alpha = \frac{1}{\sec \alpha} = \frac{1}{1.1937};$$

$$\alpha = \mathrm{Cos}^{-1}\frac{1}{1.1937} \approx 33.1°.$$

25. $\sin \alpha = 0.3616$ **26.** $\cos \alpha = 0.9164$

27. $\tan \alpha = 0.4877$ **28.** $\sin \alpha = 0.8695$

29. $\sec \alpha = 1.7655$ **30.** $\cot \alpha = 1.6775$

31. $\csc \alpha = 1.7348$ **32.** $\sec \alpha = 1.5590$

Find an approximation to the nearest hundredth for the measure of angle α, in radians, where $0^R \leq \alpha \leq 1.57^R$.

Example $\cot \alpha = 1.560$

Solution First express $\tan \alpha$ as the reciprocal of the given value $\cot \alpha$. Set Degree/Radian setting on RAD. Thus,

$$\tan \alpha = \frac{1}{\cot \alpha} = \frac{1}{1.560};$$

$$\alpha = \text{Tan}^{-1} \frac{1}{1.560} \approx 0.57^R.$$

33. $\sin \alpha = 0.2667$	**34.** $\cos \alpha = 0.9211$	**35.** $\tan \alpha = 0.5334$
36. $\tan \alpha = 1.592$	**37.** $\cot \alpha = 0.2669$	**38.** $\sec \alpha = 1.022$
39. $\csc \alpha = 1.355$	**40.** $\cot \alpha = 0.0910$	

In Exercises 41–44 use the fact that the path of a light ray changes direction as it passes from one medium to another in accordance with the following law:

$$n_i \sin i = n_r \sin r,$$

where i and r are angles with the vertical of the incident (initial) and refracted ray, and n_i and n_r are the respective constants for specific media (see figure). In particular, $n_i = 1.00$ for air and $n_r = 1.33$ for water. Express answers to the nearest tenth.

41. Find the angle of refraction if the angle of incidence is 28.1°.

42. Find the angle of incidence if the angle of refraction is 14.7°.

43. Find the angle of incidence if the angle of refraction is 42° 24′.

44. Find the angle of refraction if the angle of incidence is 36° 18′.

Chapter Summary

1.1 An **angle** is the union of two rays that have a common endpoint. The rays are called the **sides of the angle.**

An angle that has its initial side along the positive *x*-axis and its vertex at the origin is said to be in **standard position.** Such an angle is said to be in that quadrant in which its terminal side lies.

Two kinds of angle measure most commonly used are called **degree measure** and **radian measure.**

The measure of an angle in radians is related to the measure of the angle in degrees by the equations

$$\alpha° = \frac{180}{\pi} \alpha^R \quad \text{and} \quad \alpha^R = \frac{\pi}{180} \alpha°.$$

In particular, $1^R \approx 57.30°$ and $1° \approx 0.01745^R$.

Angles having the same initial side and the same terminal side are called **coterminal angles.**

1.2 For all angles α in standard position, if x and y are the coordinates of any point on the terminal side of α $[(x, y) \neq (0, 0)]$ and $r = \sqrt{x^2 + y^2}$,

$$\text{sine } \alpha = \frac{y}{r}, \qquad\qquad \text{cosecant } \alpha = \frac{r}{y} \quad (y \neq 0),$$

$$\text{cosine } \alpha = \frac{x}{r}, \qquad\qquad \text{secant } \alpha = \frac{r}{x} \quad (x \neq 0),$$

$$\text{tangent } \alpha = \frac{y}{x} \quad (x \neq 0), \qquad \text{cotangent } \alpha = \frac{x}{y} \quad (y \neq 0).$$

The following identities follow from the definitions of the trigonometric ratios.

$$\csc \alpha = \frac{1}{\sin \alpha}, \quad \sec \alpha = \frac{1}{\cos \alpha}, \quad \cot \alpha = \frac{1}{\tan \alpha},$$

and

$$\cos^2 \alpha + \sin^2 \alpha = 1.$$

1.3 Function values for each of the six trigonometric functions for certain angles can be obtained from geometric considerations of right triangles. A calculator can be used to find decimal approximations for function values of all angles. Alternatively, tables in Appendix C can be used to obtain such function values.

1.4 The **reference angle** of any nonquadrantal angle α in standard position is the positive acute angle $\tilde{\alpha}$ whose sides are the terminal side of α and a ray of the x-axis. The absolute values of the trigonometric ratios of any angle α are equal, respectively, to the analogous trigonometric ratios for $\tilde{\alpha}$. The use of a reference angle facilitates finding function values of nonquadrantal angles with measures greater than $90°$ or less than $0°$.

The sign (plus or minus) of a trigonometric function value depends on the quadrant in which the terminal side of the angle lies.

The trigonometric function values for an angle α are equal to the analogous function values for *any* angle coterminal with α, including angles with negative measures.

1.5   A calculator or the tables can also be used to find a positive acute angle α, where trig α is given. The **inverse notation** Trig^{-1} can be used to name an acute angle α such that trig $\alpha = y$.

Review Exercises

1.1   *Find the degree measure of α to the nearest tenth of a degree.*

1. $\alpha = \dfrac{2\pi^{\text{R}}}{5}$ **2.** $\alpha = 2.41^{\text{R}}$

Find the radian measure of α to the nearest hundredth of a radian.

3. $\alpha = 75°$ **4.** $\alpha = 257°$

Find the least positive angle coterminal with each angle α.

5. $\alpha = 938°$ **6.** $\alpha = 7.23^{\text{R}}$

Write a general expression for all angles coterminal with α.

7. $\alpha = 131°$ **8.** $\alpha = \dfrac{7\pi^{\text{R}}}{4}$

1.2   *Find the element in the range of each of the six trigonometric functions of α if the terminal side of α (in standard position) contains the given point.*

9. $(-4, 3)$ **10.** $(-1, 2)$ **11.** $(\sqrt{3}, -1)$ **12.** $(-8, -15)$

Given one function value of an acute angle α, determine the other five trigonometric function values of α.

13. $\cos \alpha = \dfrac{3}{5}$ **14.** $\tan \alpha = \dfrac{4}{7}$ **15.** $\sin \alpha = \dfrac{1}{3}$ **16.** $\sec \alpha = \dfrac{5}{2}$

1.3   *Find the exact value of each expression, if it exists, by making an appropriate sketch and reading the ratio directly from the figure.*

17. csc 60°   **18.** sec 45°   **19.** tan 270°   **20.** sin 180°

Compute the exact value of each expression if the expression represents a real number.

21. $\sec^2 30°$

22. $(\cos 180°)(\tan^2 45°)$

23. $\left(\sec \dfrac{\pi^{\text{R}}}{4}\right)\left(\sin \dfrac{\pi^{\text{R}}}{6}\right)\left(\cot^2 \dfrac{\pi^{\text{R}}}{3}\right)$

24. $\dfrac{2\left(\tan^2 \dfrac{\pi^{\text{R}}}{3}\right) + \cos^2 \dfrac{\pi^{\text{R}}}{4}}{\sin^2 \dfrac{\pi^{\text{R}}}{2}}$

Find an approximation for each function value.

25. tan 79.5°   **26.** sec 44° 18′   **27.** csc 0.34^{R} **28.** cos 0.73^{R}

1.4   *Determine the quadrant in which the terminal side of each angle α lies if α is in standard position.*

29. $\sin \alpha < 0$ and $\sec \alpha < 0$ **30.** $\cos \alpha < 0$ and $\cot \alpha > 0$

Given one function value and the sign of another, determine the other five function values of α.

31. $\sin \alpha = \dfrac{2}{5}$,   if $\tan \alpha < 0$.   **32.** $\tan \alpha = \dfrac{3}{4}$,   if $\sin \alpha < 0$.

Sketch $\tilde{\alpha}$ and find the measure of $\tilde{\alpha}$ in the same units in which each α is given.

33. 210°   **34.** 330°   **35.** $\dfrac{7\pi^{\text{R}}}{4}$ **36.** $\dfrac{3\pi^{\text{R}}}{4}$

Find the exact value for each of the following.

37. sin 150°

38. tan 330°

39. $\cos \dfrac{11\pi^{\text{R}}}{4}$

40. $\sec \left(-\dfrac{3\pi}{4}\right)^{\text{R}}$

Find an approximation for each function value.

41. sin 163°   **42.** cot 331° 48′   **43.** tan 3.27^{R} **44.** sec 2^{R}

1.5   **45.** Express $\tan \alpha = 1.4321$ equivalently using inverse notation.

46. Express $\alpha = \text{Cos}^{-1}\, 0.4232$ equivalently without inverse notation.

Find an approximation for the measure of each angle α in degree measure, where $0° \le \alpha \le 90°$.

47. $\tan \alpha = 1.4770$ **48.** $\sec \alpha = 1.6502$

Find an approximation for the measure of each angle α in radian measure, where $0^R \le \alpha \le 1.57^R$.

49. $\cos \alpha = 0.7712$ **50.** $\csc \alpha = 4.582$

2 APPLICATIONS

When we use given information about the lengths of one, two, or three sides of a triangle and the measure of one or two angles to find the remaining, or unknown, lengths and measures, we are said to be *solving a triangle*. In this chapter you will first solve different kinds of triangles. Then, you will be introduced to the notion of a geometric vector, an important prerequisite for the study of advanced topics in mathematics. You may want to review some pertinent topics of geometry in Appendices A.7–A.9 before you begin this chapter.

2.1 Right Triangles

Right triangles can be solved by using the trigonometric ratios of the the acute angles of the triangle. Figure 2.1 shows that, in the first quadrant, any point (x, y) on the terminal side of an angle α in standard position determines a right triangle with sides of length x and y and with hypotenuse of length $\sqrt{x^2 + y^2}$. Therefore, in the special case where α is an acute angle in a right triangle,

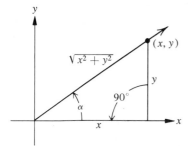

Figure 2.1

$$\sin \alpha = \frac{\text{length of side opposite } \alpha}{\text{length of hypotenuse}}, \qquad \csc \alpha = \frac{\text{length of hypotenuse}}{\text{length of side opposite } \alpha},$$

$$\cos \alpha = \frac{\text{length of side adjacent to } \alpha}{\text{length of hypotenuse}}, \qquad \sec \alpha = \frac{\text{length of hypotenuse}}{\text{length of side adjacent to } \alpha},$$

$$\tan \alpha = \frac{\text{length of side opposite } \alpha}{\text{length of side adjacent to } \alpha}, \qquad \cot \alpha = \frac{\text{length of side adjacent to } \alpha}{\text{length of side opposite } \alpha},$$

(a)

(b)

Figure 2.2

It is not necessary that α be in standard position. The above relations involving the ratios of the lengths of the sides of a right triangle are equally applicable to a right triangle in any position. For example, in Figure 2.2a, the six trigonometric ratios of the angle α are

$$\sin \alpha = \frac{a}{c}, \qquad \csc \alpha = \frac{c}{a},$$

$$\cos \alpha = \frac{b}{c}, \qquad \sec \alpha = \frac{c}{b},$$

$$\tan \alpha = \frac{a}{b}, \qquad \cot \alpha = \frac{b}{a}.$$

In Figure 2.2b, the trigonometric ratios of the angle β are

$$\sin \beta = \frac{b}{c}, \qquad \csc \beta = \frac{c}{b},$$

$$\cos \beta = \frac{a}{c}, \qquad \sec \beta = \frac{c}{a},$$

$$\tan \beta = \frac{b}{a}, \qquad \cot \beta = \frac{a}{b}.$$

Generally, the sides of a triangle are labeled with their lengths, a, b, and c, and the angles opposite a, b, and c are labeled α, β, and γ, or A, B, and C, respectively, where γ or C names the right angle in a right triangle. The measures of angles in triangles are usually given in degree units.

Solution of Triangles

Right triangles can be solved by using the trigonometric ratios on page 41 and the fact that *the sum of the measures of the angles of any triangle equals* 180°.

Example   One angle of a right triangle measures 34° and the side opposite this angle has a length of 12 inches. Solve the triangle.

Solution   Sketch the right triangle, showing the given information. First, from the fact that $\alpha + \beta + \gamma = 180°$ and $\gamma = 90°$,

$$\beta = 90° - \alpha$$

$$= 90° - 34° = 56°.$$

By definition,

$$\tan 34° = \frac{a}{b} = \frac{12}{b} \quad \text{or} \quad b = \frac{12}{\tan 34°}$$

and

$$\sin 34° = \frac{a}{c} = \frac{12}{c} \quad \text{or} \quad c = \frac{12}{\sin 34°}.$$

Thus,

$$b \approx \frac{12}{0.6745} \approx 17.8 \quad \text{and} \quad c \approx \frac{12}{0.5592} \approx 21.5.$$

Hence, the measure of β is 56°, $b \approx 17.8$ in., and $c \approx 21.5$ in.

In the above example we have shown values for tan 34° and sin 34° to four decimal places as intermediate steps in the computation. The final results which have been rounded off to three significant digits* will be the same if a single key-stroke sequence is used to obtain values for b and c. We will continue to show function values in intermediate steps even though as soon as you develop calculator proficiency, you will be able to make a single key-stroke sequence to obtain most results.

Example   The hypotenuse of a right triangle is 14.7 centimeters long and the length of a side is 6.8 centimeters. Solve the triangle.

Solution   Sketch the right triangle showing the given information. From the two given dimensions we could first obtain the measure of α using the ratios

$$\sin \alpha = \frac{6.8}{14.7} \quad \text{or} \quad \csc \alpha = \frac{14.7}{6.8}.$$

Alternatively we could first obtain the measure of β using the ratios

$$\cos \beta = \frac{6.8}{14.7} \quad \text{or} \quad \sec \beta = \frac{14.7}{6.8}.$$

We will use

$$\sin \alpha = \frac{6.8}{14.7} \approx 0.4626,$$

from which

$$\alpha \approx \text{Sin}^{-1} \, 0.4626 \approx 27.6°.$$

(continued)

---

* See Appendix B for a discussion of rounding-off procedures and significant digits.

Then we have that

$$\beta \approx 90° - 27.6° = 62.4°.$$

We can find an approximation for b by using the Pythagorean theorem:

$$b^2 = \sqrt{(14.7)^2 - (6.8)^2};$$
$$b = \sqrt{216.09 - 46.24} \approx 13.0.$$

Alternatively, we can find b by using the results we obtained for the measure of α or β. If we use $\alpha \approx 27.6°$, we can set up an equation using a variety of trigonometric ratios. One ratio is

$$\cos \alpha = \frac{b}{14.7},$$

from which

$$b = 14.7 \cos \alpha$$
$$\approx 14.7 \cos 27.6°$$
$$\approx 14.7(0.8862)$$
$$\approx 13.0.$$

Hence, the length b is approximately 13.0 cm.

Accuracy of Results

In the foregoing example the lengths b and c were expressed by an approximation using three significant digits. Unless otherwise stated, we shall continue to specify lengths of line segments to this degree of accuracy when solutions are given in decimal notation, and we shall specify measures of angles to the *nearest tenth of a degree or nearest six minutes*. Furthermore, we shall assume that all data have similar accuracy. Thus, $a = 12$ in the above example implies that $a = 12.0$ and $\alpha = 34°$ implies that $\alpha = 34.0° = 34° \, 00'$.

Special Names for Angles

Certain angles are given special names in the fields of surveying and navigation. For example, the angle formed by a horizontal ray and an observer's "line of sight" to any object above the horizontal is called an **angle of elevation** (Figure 2.3a). The angle formed by a horizontal ray and an observer's "line of sight" to any object below the horizontal is called an **angle of depression** (Figure 2.3b).

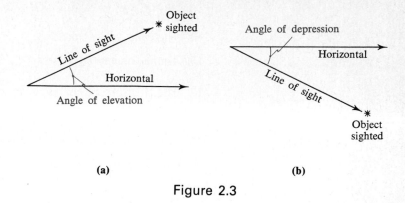

(a)   **(b)**

Figure 2.3

Example   On level ground at point A the measure of the angle of elevation of the top of a tower is $57°\ 24'$. The distance from A to the base of the tower is 50 meters. How high is the tower?

Solution   By definition,

$$\tan 57°\ 24' = \frac{h}{50},$$

where h represents the height of the tower. Equivalently,

$$h = 50 \tan 57°\ 24' = 50 \tan 57.4°$$

$$\approx 50(1.5637) \approx 78.2.$$

Thus, the height of the tower is approximately 78.2 meters.

EXERCISE SET 2.1

A

Solve each right triangle. In each case, $\gamma = 90°$.

1. $a = 2$ and $b = 3$ 2. $c = 41$ and $b = 9$

3. $c = 17$ and $a = 15$ 4. $c = 14$ and $\alpha = 17°$

5. $c = 23.5$ and $\beta = 51°$ 6. $a = 6.1$ and $\alpha = 73°$

7. $a = 11$ and $\beta = 22.2°$ 8. $b = 5$ and $\alpha = 64.7°$

9. $b = 15$ and $\alpha = 50°\ 18'$ 10. $c = 20$ and $\beta = 27°\ 30'$

11. $b = 30$ and $\beta = 41°\ 36'$ 12. $a = 32$ and $\beta = 75°\ 6'$

Find the area of the indicated triangle.

13. The triangle in Exercise 3.   **14.** The triangle in Exercise 4.

15. The triangle in Exercise 7.   **16.** The triangle in Exercise 8.

17. The measure of an angle of elevation from a surveyor's position (ignoring the surveyor's height) to the top of a hill is $37°\ 24'$. The top of the hill is 275 meters above a level line through the surveyor's position. How far is the surveyor from a point directly below the top of the hill?

18. From the top of a vertical cliff, 80 meters above the surface of the ocean, the measure of the angle of depression to a marker on the surface of the ocean is $18°\ 12'$. How far is the marker from the foot of the cliff?

19. A 40-foot ladder is used to reach the top of a 24-foot wall. If the ladder extends 10 feet past the top of the wall, find the measure of the angle that the ladder forms with the horizontal ground.

20. The lengths of the sides of an isosceles triangle are 17, 17, and 30 centimeters. Find the measures of the angles in the triangle. *Hint:* Use the altitude of the triangle.

21. A ladder 30 decimeters long is leaning against the side of a building. If the ladder makes an angle of $25°\ 18'$ with the side of the building, how far up from the ground does the ladder touch the building?

22. A tower 120 feet high casts a shadow 60 feet long. Find the angle of elevation of the sun.

23. A railroad track makes an angle of $4°\ 48'$ with the horizontal. How far must a train go up the track to gain 20 meters in altitude?

24. A rectangle is 20 centimeters long and 14 centimeters wide. Find the angles that a diagonal makes with the sides.

25. A parachutist is descending vertically. How far does the parachutist fall as the angle of elevation from A changes from $50°$ to $30°$?

Ex. 25   Ex. 26

26. A surveyor measures AC as 35.3 meters and α as $57.2°$. How wide is the river?

Ex. 27   Ex. 28

27. An observer at P notes that the angle of elevation to the top of a mountain is 16.5°. If the mountain is 3500 feet high, how far is P from the top of the mountain?

28. In the figure, the angle of depression from the top of one tower to the top of the other is 20°. If the height of the taller tower is 150 feet, how high is the shorter?

B

29. A regular pentagon is inscribed in a circle of radius 5 centimeters. Find the length of a side of the pentagon.

Ex. 29

30. The angle of elevation of the top of the tower is 62.3° from A and 37.1° from B. If AB is 60 feet, find the height of the tower.

Ex. 30

Ex. 31

31. If $\alpha = 30°$, $\beta = 60°$, and $BC = 2$, find AD.

32. Show that (see drawing for Exercise 31)

$$\cos \beta = \frac{BC \cot x}{AD}.$$

Ex. 33

33. Prove that when a simple pendulum of length l is inclined at an angle α with the vertical, the bob is at a height h above its lowest position, where $h = l(1 - \cos \alpha)$.

34. Show that the length h of the perpendicular from C to the side c of any triangle ABC can be found by

$$h = \frac{c}{\cot \alpha + \cot \beta}.$$

Hint: $AD = h \cot \alpha$.

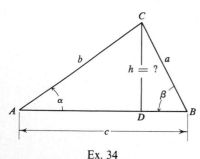

Ex. 34

2.2 The Law of Sines

In Section 2.1 we solved right triangles using trigonometric ratios expressed in terms of the lengths of the sides of the triangle. In this section and in the next, special formulas will be derived to enable us to solve any triangle, right, acute, or obtuse.

Consider $\triangle ABC$ in Figure 2.4 (β may be either acute or obtuse), and introduce the altitude CD. Thus, $\triangle ADC$ and $\triangle BDC$ are right triangles. In either figure, in $\triangle ADC$,

$$\sin \alpha = \frac{h}{b}, \qquad \text{from which} \qquad h = b \sin \alpha.$$

In $\triangle BDC$ in Figure 2.4a,

$$\sin \beta = \frac{h}{a}, \qquad \text{from which} \qquad h = a \sin \beta.$$

In $\triangle BDC$ in Figure 2.4b, $\angle CBD$ is equal to the reference angle for β; therefore,

$$\sin \beta = \sin \angle CBD = \frac{h}{a},$$

$$h = a \sin \beta.$$

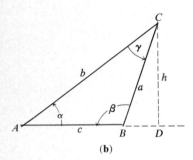

(b)

Figure 2.4

Hence, we have in each case

$$h = b \sin \alpha = a \sin \beta.$$

Dividing each member of $b \sin \alpha = a \sin \beta$ by $a \cdot b$, we have

$$\frac{\sin \alpha}{a} = \frac{\sin \beta}{b}. \tag{1}$$

If in Figure 2.4 we use the altitude from vertex A and a similar argument, we will have

$$\frac{\sin \beta}{b} = \frac{\sin \gamma}{c}. \tag{2}$$

Comparing Equations (1) and (2), we find that

$$\frac{\sin \alpha}{a} = \frac{\sin \beta}{b} = \frac{\sin \gamma}{c}.$$

In many applications this relationship, called the **Law of Sines,** is conveniently used in the form

$$\frac{a}{\sin \alpha} = \frac{b}{\sin \beta} = \frac{c}{\sin \gamma}.$$

Two Angles and a Side

The Law of Sines can be used to solve triangles when several kinds of data are given. We consider first the case in which the measures of two angles and the length of one side of a triangle are given.

Example   Solve the triangle for which $\alpha = 30°$, $\beta = 45°$, and $a = 20$.

Solution   A sketch of the triangle showing the given information is helpful. First, from the fact that $\alpha + \beta + \gamma = 180°$, it follows that

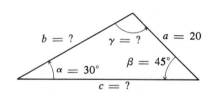

$$\gamma = 180° - \alpha - \beta$$

$$= 180° - 30° - 45° = 105°.$$

From the Law of Sines,

$$\frac{20}{\sin 30°} = \frac{b}{\sin 45°},$$

or, equivalently,

$$b = \frac{20 \sin 45°}{\sin 30°} \approx \frac{20(0.7071)}{0.5000} \approx 28.3.$$

Then, again from the Law of Sines,

$$\frac{20}{\sin 30°} = \frac{c}{\sin 105°},$$

or, equivalently,

$$c = \frac{20 \sin 105°}{\sin 30°} = \frac{20 \sin 75°}{\sin 30°} \approx \frac{20(0.9659)}{0.5000}$$

$$= 40(0.9659) \approx 38.6.$$

Hence, $b \approx 28.3$, $c \approx 38.6$,   and $\gamma = 105°$.

Two Sides and an Opposite Angle

A second case in which the Law of Sines is used is one in which the lengths of two sides of a triangle, say a and b, and the measure of an angle opposite one of them, say α, are given. This case is called the *ambiguous case* because ambiguity may arise, depending on a in relation to b and the measure of α. Figure 2.5, in which α is an *acute angle* and h is the length of a perpendicular segment, illustrates the following possibilities that may occur.

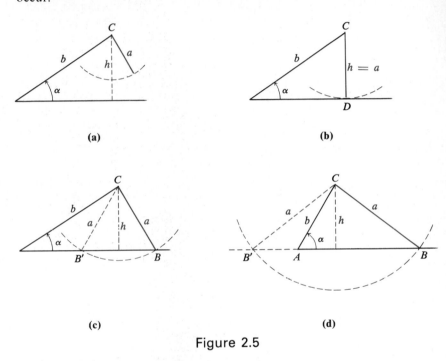

Figure 2.5

1. If $a < h$, the side with length a cannot intersect the side opposite C and *no triangle is possible,* as shown in Figure 2.5a. In this case, since $h = b \sin \alpha$, we have $a < b \sin \alpha$.

2. If $a = h = b \sin \alpha$, the side with length a intersects the side opposite C at D. In this case *exactly one right triangle is possible,* as shown in Figure 2.5b.

3. If $a > h$ (or $b \sin \alpha$) and $a < b$, or $b \sin \alpha < a < b$, *two triangles are possible,* as shown in Figure 2.5c.

4. If $a \geq b$, the triangle $AB'C$, shown in Figure 2.5d, does not contain the given angle α and therefore does not lead to a solution. Hence, in this case, *only one triangle is possible.* In particular, if $a = b$, the triangle is isosceles.

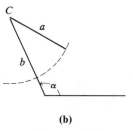

(a)

(b)

Figure 2.6

Figure 2.6, in which α is an *obtuse angle,* illustrates the following two possibilities.

1. If $a > b$, then one triangle is possible, as shown in Figure 2.6a.

2. If $a \leq b$, then no triangle is possible as shown in Figure 2.6b.

While we have concerned ourselves only with the variables a, b, and α in the above discussion, similar results are valid when other variables are involved in the given information. A sketch of the triangle using the given information will help to indicate the possibilities that exist. In particular, sketches of triangles like those shown in Figures 2.5 and 2.6, in which the given angle is drawn on the left, are usually the most helpful in deciding which possibility exists for given data.

Example   Solve the triangle(s) for which $a = 7$, $b = 5$, and $\beta = 30°$.

Solution   Sketch the triangle(s) using the given information. Since the lengths of two sides and the measure of an angle β, opposite the side with measure b, are given, a check must be made to see how many triangles are possible. Since

$$h = a \sin \beta = 7 \sin 30°$$

$$= 7(0.5000) = 3.5,$$

it follows that $h < b < a$ and that two triangles are possible, as indicated in the figure.

From the Law of Sines,

$$\frac{\sin \alpha}{7} = \frac{\sin 30°}{5}$$

$$\sin \alpha = \frac{7 \sin 30°}{5} = \frac{7(0.5000)}{5} = 0.7000.$$

Hence,

$$\alpha = \operatorname{Sin}^{-1} 0.7000 \approx 44.4°.$$

Since α is the base angle of an isosceles triangle, and α' is the supplement

(*continued*)

of the other base angle,

$$\alpha' \approx 180° - 44.4° = 135.6°.$$

For $\alpha \approx 44.4°$, the solution of one of the two possible triangles is completed as follows. Since the sum of the measures of the angles of a triangle equals $180°$,

$$\gamma = 180° - \alpha - \beta$$
$$\approx 180° - 44.4° - 30° = 105.6°.$$

From the Law of Sines and the fact that $\sin 105.6° = \sin 74.4°$,

$$\frac{5}{\sin 30°} = \frac{c}{\sin 105.6°};$$

$$c = \frac{5 \sin 74.4°}{\sin 30°} \approx \frac{5(0.9632)}{0.5000} \approx 9.63.$$

For $\alpha' \approx 135.6°$, the solution of the second possible triangle can be completed. Since

$$\gamma' \approx 180° - 135.6° - 30° = 14.4°,$$

then by the Law of Sines

$$\frac{5}{\sin 30°} = \frac{c'}{\sin 14.4°};$$

$$c' \approx \frac{5(0.2487)}{0.5000} \approx 2.49.$$

Thus, the two solutions are

$$\alpha \approx 44.4°, \quad \gamma \approx 105.6°, \quad \text{and} \quad c \approx 9.63;$$
$$\alpha' \approx 135.6°, \quad \gamma' \approx 14.4°, \quad \text{and} \quad c' \approx 2.49.$$

EXERCISE SET 2.2

A

Solve each triangle.

1. $a = 5$, $\alpha = 20°$, and $\beta = 75°$
2. $a = 32$, $\alpha = 26.3°$, and $\gamma = 81.8°$
3. $a = 58.4$, $\beta = 37.2°$, and $\gamma = 100°$
4. $b = 1.9$, $\alpha = 111.7°$, and $\beta = 5.1°$

5. $b = 0.42$, $\alpha = 35° 36'$, and $\gamma = 91° 30'$

6. $b = 0.88$, $\beta = 63° 54'$, and $\gamma = 34° 12'$

7. $c = 13.6$, $\alpha = 30° 24'$, and $\beta = 72° 6'$

8. $c = 1.4$, $\alpha = 135° 12'$, and $\gamma = 34° 48'$

Determine the number of triangles that satisfy each set of conditions and solve each triangle.

9. $a = 7$, $b = 5$, and $\alpha = 30°$ **10.** $a = 10$, $b = 9$, and $\beta = 60°$

11. $a = 4.2$, $c = 6.1$, and $\alpha = 32.2°$ **12.** $a = 38$, $c = 45$, and $\gamma = 35.6°$

13. $b = 16$, $c = 32$, and $\beta = 30°$ **14.** $b = 0.3$, $c = 0.4$, and $\gamma = 62°$

15. $b = 3.9$, $a = 5$, and $\beta = 42° 42'$ **16.** $b = 37$, $a = 51$, and $\alpha = 135° 30'$

Find the area of each triangle.

17. $a = 3.4$, $\beta = 13°$, and $\alpha = 50°$ **18.** $c = 3.8$, $\beta = 15° 6'$, and $\gamma = 50°$

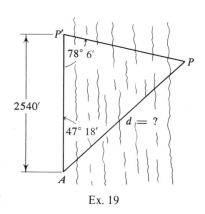

Ex. 19

19. To find the distance from an observer's point A on one side of a river (see figure) to a point P on the other side, a line AP' measuring 2540 feet was laid out on one side and the angles $P'AP$ and $AP'P$ were measured and found to be $47° 18'$ and $78° 6'$, respectively. Find the distance from A to P to the nearest foot.

20. Two men, 400 feet apart, observe a balloon between them that is in a vertical plane with them. The respective angles of elevation of the balloon are observed by the men to be $75° 24'$ and $49° 30'$. Find the height of the balloon above the ground.

Ex. 23

21. One diagonal of a parallelogram is 20 centimeters long and at one end forms angles of 20° and 40° with the sides of the parallelogram. Find the lengths of the sides.

22. A ship at sea is sighted from two points, A and B, on the shore. A and B are 5 kilometers apart, and the line of sight from A to the ship forms an angle of $30° 18'$ with $\overline{AB}$, while the line of sight from B to the ship forms an angle of $41° 12'$ with $\overline{AB}$. How far is the ship from point A?

23. A tower is built vertically on a hillside that slants upward 10° from the horizontal. From a point 200 meters uphill from the foot of the tower, the angle of elevation of the top of the tower is $12° 30'$ (see figure). How high is the tower?

Ex. 24

24. From a window 75 meters above the ground, the measure of the angle of elevation to the top of a nearby building is $29° 42'$ (see figure). From a point on the ground directly below the window, the measure of the

Ex. 25

Ex. 26

angle of elevation of the top of the same building is 52° 24′. Find the height of the building.

25. The angle of elevation of the building is 48° from *A*, and 61° from *B*. If *AB* is 20 feet, find the height of the building.

26. The pole is tilted toward the sun at an angle of 8° from the vertical. When the angle of elevation of the sun is 54°, the shadow of the pole is 12 feet long. How long is the pole?

27. A hill makes an angle of 10° with the horizontal. From a point 50 meters downhill from a vertical tower, the angle of elevation of the tower is 56°. Find the height of the tower.

28. From two points on shore, the angles α and β to a light at *C* are 16.4° and 58.6°, respectively. If *AB* is 1500 feet, find the distance, *d*, of the light from the shore.

Ex. 27

Ex. 28

B

29. Using the figure, show that

$$c = \frac{a \sin \beta \tan \gamma}{\sin(\alpha + \beta)}$$

30. Using the figure and the results for Exercise 29, if $\alpha = 40°$, $\beta = 60°$, $\gamma = 30°$, and $a = 100$ feet, find *c*.

Ex. 29

Show that, in any triangle ABC, each of the following statements is true.   Hint: *In Exercises 31 and 32 start with the Law of Sines.*

31. $\dfrac{a + b}{b} = \dfrac{\sin \alpha + \sin \beta}{\sin \beta}$

32. $\dfrac{a - b}{b} = \dfrac{\sin \alpha - \sin \beta}{\sin \beta}$

33. $\dfrac{a - b}{a + b} = \dfrac{\sin \alpha - \sin \beta}{\sin \alpha + \sin \beta}$

34. $\dfrac{a + b + c}{b} = \dfrac{\sin \alpha + \sin \beta + \sin \gamma}{\sin \beta}$

2.3 The Law of Cosines

The Law of Sines considered in Section 2.2 cannot be used directly to solve triangles in which the lengths of two sides and the measure of the angle formed by their respective rays are specified or in which only the lengths of the three sides are specified. Equations (1), (2), and (3) below can be used directly in such cases.

If α, β, and γ are the angles of any triangle, and a, b, and c are the lengths of the sides opposite α, β, and γ, respectively, then

$$c^2 = a^2 + b^2 - 2ab \cos \gamma, \tag{1}$$

$$b^2 = a^2 + c^2 - 2ac \cos \beta, \tag{2}$$

$$a^2 = b^2 + c^2 - 2bc \cos \alpha. \tag{3}$$

These statements are called the **Law of Cosines** and follow from a consideration of the distance formula (see Appendix A.4). Each of the three formulae expresses the same geometric fact as the other two. Hence, we shall only prove Equation (1).

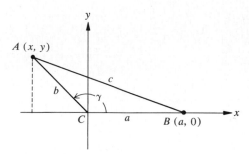

Figure 2.7

Orienting triangle ABC with γ at the origin, as shown in Figure 2.7, we find that

$$\cos \gamma = \frac{x}{b} \qquad \text{and} \qquad \sin \gamma = \frac{y}{b},$$

from which the coordinates of A are

$$b \cos \gamma \qquad \text{and} \qquad b \sin \gamma$$

Of course, statements (1), (2), and (3) above would also be true if γ

were an acute angle. From the distance formula,

$$c^2 = (y - 0)^2 + (x - a)^2$$
$$= (b \sin \gamma - 0)^2 + (b \cos \gamma - a)^2$$
$$= b^2 \sin^2 \gamma + b^2 \cos^2 \gamma - 2ab \cos \gamma + a^2$$
$$= b^2(\sin^2 \gamma + \cos^2 \gamma) - 2ab \cos \gamma + a^2.$$

Since $\sin^2 \gamma + \cos^2 \gamma = 1$ (see page 11), we have

$$c^2 = a^2 + b^2 - 2ab \cos \gamma.$$

If the triangle is oriented with A or B at the origin, Equations (2) and (3) in the Law of Cosines can be obtained in a similar way. The particular form of the equation that should be used will, of course, depend on the given information.

Notice that in Equation (1) if $\gamma = 90°$, then $\cos \gamma = \cos 90° = 0$ and the Law of Cosines reduces to

$$c^2 = a^2 + b^2 - 2ab(0),$$
$$c^2 = a^2 + b^2,$$

the familiar formula applicable to right triangles (the Pythagorean theorem).

Two Sides and the Included Angle

Now consider how the Law of Cosines can be used to solve a triangle when the measure of an angle and the lengths of the adjacent sides are known.

Example   Given the measures $\beta = 75°$, $a = 7$, and $c = 5$ of an oblique triangle, find the length b.

Solution   Sketch a triangle showing the given information. The length b can be found by using the Law of Cosines expressed in the form $b^2 = a^2 + c^2 - 2ac \cos \beta$. Substituting the known values gives

$$b^2 = 7^2 + 5^2 - 2(7)(5) \cos 75°$$
$$\approx 49 + 25 - 70(0.2588) \approx 55.9;$$
$$b \approx \sqrt{55.9} \approx 7.48.$$

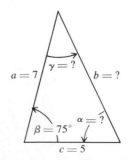

Three Sides

The Law of Sines could be used to complete the solution of the triangle in the above example since the length b, opposite the given angle β, is known. However, we shall use the Law of Cosines in the following example in order to show how it can be used to determine the measure of an angle if the lengths of three sides of a triangle are known.

Example   To find the measure of α in the foregoing example, the equation

$$a^2 = b^2 + c^2 - 2bc \cos \alpha$$

is first written in the form

$$\cos \alpha = \frac{b^2 + c^2 - a^2}{2bc}.$$

From the previous calculations, $b^2 \approx 55.9$ and $b \approx 7.48$. Thus,

$$\cos \alpha \approx \frac{55.9 + 25 - 49}{2(7.48)(5)} \approx 0.4265,$$

from which

$$\alpha = \text{Cos}^{-1} \, 0.4265 \approx 64.8°.$$

Because $\alpha + \beta + \gamma = 180°$,

$$\gamma \approx 180° - 64.8° - 75° = 40.2°.$$

Thus, the solution to the triangle is

$$b \approx 7.48, \quad \alpha \approx 64.8°, \quad \text{and} \quad \gamma \approx 40.2°.$$

If, after first finding the length of the third side, $b = 7.48$ in the above example, the Law of Sines is used to find the measure of a second angle, the angle selected should be the smaller of the two remaining angles (the one opposite the shorter of the two given sides) and hence *an acute angle*. Thus, if the Law of Sines had been used to find the second angle in the above example, we would have found $\sin \gamma$ rather than $\sin \alpha$ because, of the two given sides, $c < a$ and γ is opposite c.

Given three sides of a triangle, any angle can be obtained by using one of the Equations (1), (2), or (3) on page 55. If an equation is used that gives the measure of either of the smaller angles, which will necessarily be acute angles, then the cosine of these angles will be positive. If the equation is used that gives the measure of the largest angle, the angle may be acute or obtuse, and the cosine of the angle may be positive or negative.

Example Solve the triangle for which the lengths of the three sides are 3.4, 2.7, and 1.3.

Solution Draw a sketch showing the given information. We shall first find the measure of the smallest angle. From the Law of Cosines,

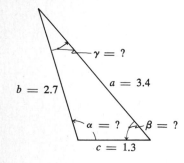

$$\cos \gamma = \frac{a^2 + b^2 - c^2}{2ab} = \frac{(3.4)^2 + (2.7)^2 - (1.3)^2}{2(3.4)(2.7)}$$

$$\approx 0.9346.$$

Hence,

$$\gamma = \text{Cos}^{-1}\, 0.9346 \approx 20.8°.$$

From the Law of Sines,

$$\frac{\sin \beta}{b} = \frac{\sin \gamma}{c},$$

$$\frac{\sin \beta}{2.7} = \frac{\sin 20.8°}{1.3},$$

from which

$$\sin \beta = \frac{2.7 \sin 20.8°}{1.3} \approx \frac{2.7(0.3551)}{1.3} \approx 0.7375.$$

Thus,

$$\beta = \text{Sin}^{-1}\, 0.7375 \approx 47.5°.$$

From the fact that $\alpha = 180° - \beta - \gamma$,

$$\alpha \approx 180° - 47.5° - 20.8°$$

$$\approx 111.7°.$$

Additional applications that require solutions of triangles occur in work with geometric vectors which are introduced in Sections 5.1 and 5.2. These sections can be considered at this time.

EXERCISE SET 2.3

A

Solve each triangle.

   1. $b = 4$, $c = 3.5$, and $\alpha = 71°$

   2. $c = 0.3$, $a = 0.1$, and $\beta = 70°$

   3. $a = 3.2$, $b = 2.2$, and $\gamma = 75.3°$

   4. $b = 6.0$, $a = 5.1$, and $\gamma = 83.5°$

   5. $a = 0.7$, $c = 0.8$, and $\beta = 141.5°$

   6. $b = 3.4$, $a = 2.1$, and $\gamma = 122.2°$

   7. $b = 1.6$, $c = 3.2$, and $\alpha = 100.4°$

   8. $c = 2.1$, $b = 4.3$, and $\alpha = 130.6°$

Solve the triangles for which the lengths of the three sides are given.

   9. $a = 9$, $b = 7$, and $c = 5$ **10.** $a = 2.7$, $b = 5.1$, and $c = 4.4$

   11. $a = 1.2$, $b = 9$, and $c = 10$ **12.** $a = 4.5$, $b = 7.5$, and $c = 5.8$

   13. $a = 6$, $b = 8$, and $c = 12$ **14.** $a = 6$, $b = 12$, and $c = 13$

   15. $a = 4$, $b = 5$, and $c = 6$ **16.** $a = 5$, $b = 7$, and $c = 8$

   17. Find the measure of the smallest angle of the triangle whose sides have lengths 4.3, 5.1, and 6.3.

   18. Find the measure of the smallest angle of the triangle whose sides have lengths 3.0, 4.2, and 3.8.

   19. Find the measure of the largest angle of the triangle whose sides have lengths 2.9, 3.3, and 4.1.

   20. Find the measure of the largest angle of the triangle whose sides have lengths 6.0, 8.2, and 9.4.

   21. Find the area of the triangle in Exercise 17.

   22. Find the area of the triangle in Exercise 18.

   23. A ship is supposed to travel directly from port A to port B, a distance of 10 kilometers. After traveling a distance of 5 kilometers, the captain discovers he has been traveling 15° off course. At this point, how far is the ship from port B?

Ex. 27

Ex. 28

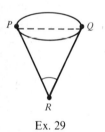

P • - - - - - • Q

R

Ex. 29

24. Two ships leave the same port on courses that form an angle of 37° 12'. When one ship has traveled 12 kilometers, the other has traveled 18 kilometers. How far apart are the ships at that time?

25. The lengths of two adjacent sides of a parallelogram are 6 centimeters and 8 centimeters. The included angle is 67°. Find the length of the longer diagonal.

26. A tower 40 meters high is on top of a hill. A cable 50 meters long runs from the top of the tower to a point 20 meters down the hill. Find the angle the hillside forms with a horizontal line.

27. A crane at the edge of a dock is supported by a cable 11 meters long, attached to the dock 7 meters from the base of the crane (see figure). If the crane arm is 5 meters long, find the acute angle that the crane arm forms with the horizontal.

28. Three circles with radii 1, 3, and 4 centimeters are tangent to each other (see figure). To the nearest 6', find the three angles formed by the lines joining their centers.

29. Points P and Q are on opposite ends of a lake. If point R is 1200 feet from P and 1600 feet from Q, and the angle at R is 50°, how long is the lake?

30. A vertical tower 30 feet high is on a hill that forms an angle of 15° with the horizontal. How long is a guy wire that runs from the top of the tower to a point 20 feet downhill from the base of the tower?

Ex. 30

Ex. 31

31. Two airplanes leave a field at the same time. One flies 30° east of north at 250 kilometers per hour; the other 45° east of south at 300 kilometers per hour. How far apart are they at the end of 2 hours?

32. A pilot is flying from Salt Lake City to Denver, a distance of about 500 miles. When she is 300 miles from Salt Lake City, she discovers she has been flying 16° off course. How far is she from Denver at that time?

B

The formulas in Exercises 33–36 where $s = \frac{1}{2}(a + b + c)$, *are sometimes used in the solution of triangles. Show that each statement is true for any* α *such that* $0° < \alpha < 180°$.   Hint: *Use the fact that* $\cos \alpha = \dfrac{b^2 + c^2 - a^2}{2bc}$.

33. $1 + \cos \alpha = \dfrac{(b + c + a)(b + c - a)}{2bc}$

34. $1 - \cos \alpha = \dfrac{(a - b + c)(a + b - c)}{2bc}$

35. $\dfrac{1 + \cos \alpha}{2} = \dfrac{s(s - a)}{bc}$.   **36.** $\dfrac{1 - \cos \alpha}{2} = \dfrac{(s - b)(s - c)}{bc}$.

8.3"

$\mathcal{A} = ?$

Ex. 40

37. Use the results of Exercises 35 and 36 to prove Hero's (or Heron's) formula, $\mathcal{A} = \sqrt{s(s - a)(s - b)(s - c)}$, which can be used to find the area of any triangle directly from the lengths of its three sides.

38. Use Hero's formula to find the area of the triangle in Exercise 9 above.

39. Use Hero's formula to find the area of the triangle in Exercise 11 above.

40. Two sides of a parallelogram are of lengths 6.8 and 8.3 inches, and one of the diagonals is of length 4.2 inches (see figure). Find the area of the parallelogram.

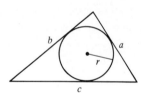

Ex. 41

41. Show that the radius r of an inscribed circle of a triangle (see figure) is given by

$$r = \sqrt{\dfrac{(s - a)(s - b)(s - c)}{s}}$$

where $s = \frac{1}{2}(a + b + c)$.

42. Show that the radius r of a circumscribed circle of a triangle (see figure) is given by

$$r = \dfrac{a}{2 \sin \alpha} = \dfrac{b}{2 \sin \beta} = \dfrac{c}{2 \sin \gamma}.$$

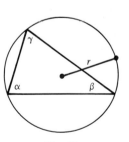

Ex. 42

2.4 Uniform Circular Motion

In this section we will consider a point moving along a circle at a constant (uniform) speed. First, however, we discuss a relationship between the measure of a central angle of a circle, the length of the intercepted arc, and the length of the radius.

Arc Length

Recall from Section 1.1 that α^R indicates how many times the length of the radius r of the circle has been used as a unit length along the arc of a circle. Hence, the length s of an intercepted arc is given by

$$s = r \cdot \alpha^R. \tag{1}$$

The length s is a directed length; it is a positive or negative real number as α^R is a positive or negative real number, respectively. If the measure of an angle is given in degrees, the measure must first be changed to radian measure before Equation (1) can be applied.

Examples   On each circle with the given radius, find the length (to the nearest tenth) of the arc intercepted by the given angle.

a. $r = 4$ cm; $\alpha = 2^R$ b. $r = 6$ cm; $\alpha = -150°$

Solutions

a. From Equation (1) above, $s = 4 \cdot 2 = 8.0$. Therefore the arc is 8 centimeters in length.

b. From Equation II (page 4), the related radian measure of $-150°$ is

$$\alpha = \left[\frac{\pi}{180}(-150) \right]^R = -\frac{5\pi^R}{6}.$$

From Equation (1) above,

$$s = 6\left(-\frac{5\pi}{6} \right) = -5\pi.$$

The arc length to the nearest tenth is -15.7 centimeters.

a.

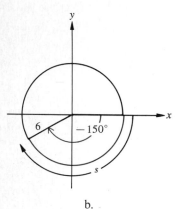

b.

Linear and Angular Velocity

Now consider a point moving along an arc of a circle at a constant (uniform) speed. If the point moves from point P_1 to P_2 in time t, the distance traveled in time t equals the length of the arc $s = r \cdot \alpha^R$, as shown in Figure 2.8. The angle α is called the **angle of rotation.**

Definition 2.1   *The **linear velocity** v of a point moving along an arc of a circle at a constant speed is given by*

$$v = \frac{\text{directed length of the arc}}{\text{time}} = \frac{r \cdot \alpha^R}{t}. \tag{2}$$

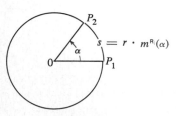

Figure 2.8

Example   The earth rotates $2\pi^R$ in 24 hours. Assuming the radius of the earth to be 4000 miles, the linear velocity of an object on the equator is given by

$$v = \frac{4000 \cdot 2\pi}{24}$$

$$\approx 1047 \text{ miles per hour.}$$

Definition 2.2   *The **angular velocity** ω (the Greek omega) of a point moving along an arc of a circle at a constant speed is given by*

$$\omega = \frac{\textbf{directed measure of the angle of rotation}}{\textbf{time}}.$$

If the measure of the angle is in radians,

$$\omega = \frac{\alpha^R}{t}, \tag{3}$$

where ω is given in radians per unit time. If the measure of the angle is in degrees,

$$\omega = \frac{\alpha^\circ}{t}, \tag{3a}$$

where ω is given in degrees per unit time.

   Notice that the linear velocity and angular velocity are both positive if the motion is in a counterclockwise direction because the length of the arc and the measure of the angle of rotation are both positive. The linear velocity and angular velocity are both negative if the motion is in a clockwise direction.

Example   In the figure, point B revolves in a circular path about A making 10 complete revolutions in 4 seconds. Find the linear and the angular velocities of point B if the radius of the circle is 2 centimeters.

Solution   One revolution corresponds to an angle of rotation of 2π radians. From Definition 2.1, with $r = 2$,

$$v = \frac{2 \cdot 10(2\pi)}{4}$$

$$= 10\pi \approx 31.4.$$

Hence the linear velocity is approximately 31.4 centimeters per second.

(continued)

From Definition 2.2,

$$\omega = \frac{10 \cdot 2\pi}{4} = 5\pi \approx 15.7.$$

Therefore, the angular velocity is approximately 15.7 radians per second.

Substituting ω, the left-hand member of (3), for α^R/t in the right-hand member of (2), we have the linear velocity v of a point moving along the arc of a circle at a constant speed given by

$$v = r\omega, \tag{4}$$

where r is the length of the radius of the circle and ω is the angular velocity of the moving point in *radians per unit time*.

Example   A turbine with a radius of length 0.74 meter turns at the rate of 10,000 revolutions per minute. What is the linear velocity of a point on the rim of the turbine?

Solution   One revolution corresponds to 2π radians. Therefore, 10,000 revolutions per minute corresponds to an angular velocity of $20,000\pi$ radians per minute. Using Equation (4), with $r = 0.74$,

$$v = r\omega = 0.74 \cdot 20,000\pi$$
$$= 14,800\pi.$$

Hence, the linear velocity is approximately 46,500 meters per minute.

EXERCISE SET 2.4

A

On each circle with the given radius, find the length of the arc (to the nearest tenth of a unit) intercepted by each angle with given measure.

Example $r = 9$ centimeters; $240°$

Solution   The related radian measure of $240°$ is

$$m^R(\alpha) = \left(\frac{\pi}{180} \cdot 240 \right)^R$$
$$= \frac{4\pi^R}{3}.$$

Thus,

$$s = r \cdot m^R(\alpha) = 9 \cdot \frac{4\pi}{3}$$

$$= 12\pi \approx 37.7,$$

and the length of the intercepted arc is approximately 37.7 centimeters.

1. $r = 2$ meters; 3.1^R
2. $r = 30$ centimeters; 1.5^R
3. $r = \pi$ decimeters; $150°$
4. $r = 2\pi$ meters; $210°$
5. $r = 3$ feet; $-135°$
6. $r = 20$ inches; $-240°$

Find the measure (to the nearest tenth of a degree) of the central angle that intercepts each arc on a circle with given radius.

Example $s = 15$ centimeters; $r = 5$ centimeters

Solution   Substituting 5 for r and 15 for s in $s = r \cdot m^R(\alpha)$ yields

$$15 = 5 \cdot m^R(\alpha).$$

Thus

$$m^R(\alpha) = \frac{15^R}{5} = 3^R.$$

Thus,

$$m°(\alpha) = \left(\frac{180}{\pi} \cdot 3\right)° \approx 171.9°.$$

7. $s = 31$ decimeters; $r = 14$ decimeters
8. $s = 2.5$ meters; $r = 2.5$ meters
9. $s = 4.9$ centimeters; $r = 1$ centimeter
10. $s = 14.20$ inches; $r = 1.27$ inches

In Exercises 11–18, consider a flywheel with uniform circular motion, where v is the linear velocity of a point on the rim and ω is the angular velocity. Find v, r, or ω to the nearest tenth of a unit.

Example   Find ω in degrees per hour if $v = 0.5$ meter per minute and $r = 0.63$ meter.

(continued)

Solution   0.5 meter per minute is 60(0.5) or 30 meters per hour. Using the fact that $v = r\omega$, we have that

$$30 = 0.63\omega,$$

from which ω equals approximately 47.62 radians per hour. Therefore, in degrees per hour,

$$\omega = 47.62 \left(\frac{180}{\pi}\right) \approx 2728.4.$$

11. Find v if $r = 0.3$ meter and $\omega = 12$ radians per second.

12. Find r if $v = 10$ centimeters per second and $\omega = 10$ radians per second.

13. Find ω in radians per minute if $v = 3$ meters per second and $r = 6$ centimeters.

14. Find ω in degrees per minute if $v = 4$ meters per second and $r = 0.2$ meter.

15. Find v if $r = 9$ decimeters and $\omega = -3$ revolutions per second.

16. Find v if $r = 4$ decimeters and $\omega = -120$ degrees per second.

17. Find ω in degrees per second if $r = 2.5$ feet and $v = -10$ feet per minute.

18. Find r in inches if $\omega = 1/12$ revolution per second and $v = 8.2$ feet per minute.

19. The center line of a freeway curve is laid out as an arc of a circle of radius 300 meters. What is the length of the center line of the curve if it subtends an angle of 20° at the center of the circle?

20. A pendulum 24 decimeters long oscillates 4.2° on each side of its vertical position. Find the length of the arc through which the end of the pendulum swings.

21. Two points on the earth's surface (assuming the earth to be a perfect sphere) are located so that the radii from the center of the earth to each point form an angle of 25°. Assuming the earth's radius to be 4000 miles, find the distance between the points measured along the earth's surface.

22. A wheel 10 inches in diameter is turning at a rate of 200 revolutions per minute. If it is rolling on the ground without slipping, how far *in feet* will it travel in 5 minutes?

23. A belt moving at a rate of 100 centimeters per minute is turning a pulley at an angular velocity of 8 revolutions per minute. Find the pulley diameter.

24. The rear gear of a bicycle has a radius of 10 centimeters and the front gear has a radius of 25 centimeters. Through what angle (in radians) does the front gear turn if the back gear turns through 5 radians?

25. A propeller, 2 meters from tip to tip, is rotating at a rate of 800 revolutions per minute. Find the linear velocity of a tip of the propeller in meters per minute.

26. Find the linear velocity, in centimeters per minute, of the tip of an hour hand of a clock, if the hand has a length of 1.5 decimeters.

27. A racing driver moves around a curved track at a speed of 160 miles per hour. If the radius of the circle, of which the driver's path is an arc, is 1500 feet, what is the driver's angular speed in radians per minute around the curve?

28. Using 8000 miles for the diameter of the earth, find the linear velocity, in miles per hour, of a point on the equator.

29. Assuming that the earth's orbit about the sun is circular, what is the angular velocity of the earth in radians per hour (assuming 365 days per year and 24 hours per day)?

30. Using the information in Exercise 29, what is the linear velocity in miles per hour of the earth in its orbit around the sun if the circular path of the earth has a radius of 9.3×10^7 miles?

31. Assuming that the moon's orbit about the earth is circular with a radius of 2.4×10^5 miles and that it takes 28 days for one orbit, what is the linear velocity of the moon in miles per hour?

32. The planet Mars moves around the sun in 686 earth-days. If the radius of Mars' orbit (assumed to be circular) around the sun is approximately 227,000,000 kilometers, find the linear velocity of Mars in its orbit in kilometers per hour. Assume 1 earth-day equals 24 hours.

B

33. Show that the area $\mathscr{A}$ of the sector of a circle (see figure) with central angle θ is given by $\mathscr{A} = \frac{1}{2}r^2\theta^{\mathrm{R}}$.

34. Use the formula in Exercise 33 to find the area of a sector of a circle whose radius has a length of 1 meter, where the sector has a central angle with measure of 148°.

Ex. 33

93,000,000 mi

Sun Earth

r r

$0°\ 0'\ 1''$

C

Ex. 35

S E_2

θ

E_1

Ex. 36

35. A unit of distance used in astronomy is the **parsec.** It is the distance r shown in the drawing. Assuming point C is the center of a circle through the sun and the earth on which the indicated arc is 93,000,000 miles long, find r, or 1 parsec, in miles. Give the answer to three significant figures.

36. Assuming that the earth travels in a circular orbit about the sun as in the figure, what would be the measure of θ during one week's travel? If the earth is 93,000,000 miles from the sun, how far does it travel in its orbit in one week?

37. If a bicycle wheel has a diameter of 27 inches, find the radian measure of the angle through which the wheel turns as the bicycle travels a distance of 1 mile.

38. If a pulley has a diameter of 20 centimeters, find the radian measure of the angle through which it turns as 4 meters of pulley belt is pulled over it. Assume the belt does not slip on the pulley.

2.5 Circular Functions; Simple Harmonic Motion

In this section we will see how the sine and cosine functions can be used as models for a variety of phenomena that are periodic. First however, we will introduce functions that are closely related to the trigonometric functions.

Circular Functions

The trigonometric functions have been defined so that the *domains are sets of angles.* Let us now consider six new functions in which the *domains are sets of real numbers.* Each of these new functions pairs *real numbers x* with real numbers trig x where the values of x in the domain are *radian measures* of angles and the values of trig x in the range are the same as the corresponding values of trig x^R. These new functions are given the same names as the corresponding trigonometric functions. For example,

$$\sin \frac{\pi}{6} \text{ is paired with the real number } \frac{\pi}{6}$$

and

$$\cos 1.73 \text{ is paired with the real number } 1.73,$$

where

$$\sin \frac{\pi}{6} = \sin \frac{\pi^R}{6} \quad \text{and} \quad \cos 1.73 = \cos 1.73^R.$$

Because the elements in the domains of these new functions $\{(x, \text{trig } x)\}$ are the measures of angles in radians and because the radian measures of angles are determined by the lengths of arcs on a circle (see Section 1.1), these functions are called **circular functions.** Because these functions pair real numbers with real numbers, they are sometimes called **real number functions.**

Since trig $x = $ trig x^R for every element x in their respective domains, we can obtain values trig x in the range of the circular functions by using Table 1.1 (page 16) or a calculator set on the RAD setting, which we have been using to obtain values for the trigonometric functions.

Examples   Find numerical values for each circular function value.

a. $\sin \dfrac{\pi}{3}$ b. $\cos 4$

Solutions
   a. Using Table 1.1, we have that

$$\sin \frac{\pi}{3} = \frac{\sqrt{3}}{2}.$$

   b. Using a calculator set on the RAD setting, we have that

$$\cos 4 \approx -0.6536.$$

The six circular functions (as well as the trigonometric functions) have important properties in common. One such property makes these functions particularly useful.

Periodic Property

Note that the sine function has an infinite number of elements in the domain that are paired with a given element in the range. For example,

$$\sin \frac{\pi}{6} = \frac{1}{2}, \qquad \sin \frac{13\pi}{6} = \frac{1}{2}, \qquad \sin\left(-\frac{11\pi}{6}\right) = \frac{1}{2}.$$

In fact, for all $x \in R$ and $k \in J$,

$$\sin(x + k \cdot 2\pi) = \sin x. \tag{1}$$

The property of being *periodic*, illustrated by the example above, is possessed by many functions, including the circular functions and the trigonometric functions. It is defined as follows.

Definition 2.3 *If f is a function with domain a subset of the set R of real numbers, such that for all x in the domain and for some $a \in R$ $(a \neq 0)$, $x + a$ is in the domain and*

$$f(x + a) = f(x),$$

then f is called a **periodic function** *and a is a* **period.**

In Equation (1), if $k = 1$, we have

$$\sin(x + 2\pi) = \sin x.$$

Hence, a period of the sine function is 2π.

The cosine function is also periodic because

$$\cos(x + k \cdot 2\pi) = \cos x, \qquad k \in J. \tag{2}$$

Since for $k = 1$,

$$\cos(x + 2\pi) = \cos x,$$

it follows that a period of the cosine function is also 2π.

For other values of k in Equations (1) and (2), the sine and cosine functions have different periods. For example, if $k = 2$,

$$\sin(x + 2 \cdot 2\pi) = \sin x \qquad \text{and} \qquad \cos(x + 2 \cdot 2\pi) = \cos x$$

and the period in each case is 4π. The *least positive* value of a in Definition 2.3 is sometimes called the **fundamental period** of the function. We shall simply call this number the period of the function.

Simple Harmonic Motion

The usefulness of the circular functions arises from the fact that they are periodic and that *the real numbers that are elements in the domain can be viewed as measures of things other than angles.* For the latter reason, the variables frequently used to represent unspecified elements in the set R of real numbers $(r, s, t, x, y,$ etc.) are used to represent elements in the domain of the circular functions. For example, in the field of physics, any oscillatory (periodic) motion satisfying an equation of the form

$$d = A \sin Bt \qquad \text{or} \qquad d = A \cos Bt,$$

where d is a measure of displacement, A and B are constants for the particular motion, and t is a measure of time, is called **simple harmonic**

motion. We can now solve some simple problems pertaining to such motion.

At rest

Example   A spring is displaced from its resting position as shown in the figure. The displacement d is given by the equation $d = A \cos \omega t$, where A, the initial displacement, and d are expressed in centimeters, t is expressed in seconds, and $\omega = 6\pi$. If the initial displacement is 12 centimeters, find d when $t = 2$.

Solution   Since $A = 12$, $\omega = 6\pi$, and $t = 2$,

$$d = A \cos \omega t$$
$$= 12 \cos 6\pi(2) = 12 \cos 12\pi.$$

Since $\cos 12\pi = \cos(0 + 6 \cdot 2\pi) = 1$, we have

$$d = 12 \cos 0 = 12(1) = 12.$$

Hence, the displacement when $t = 2$ is 12 centimeters.

Besides mechanical systems such as the example above, certain electromagnetic systems can be analyzed using the same mathematical models.

Example   The voltage e in an electrical circuit can be expressed as $e = E \sin t$, where E is the maximum voltage and t is the time in seconds. If $E = 100$, find e when $t = 2$.

Solution   Since $E = 100$ and $t = 2$,

$$e = E \sin t = 100 \sin 2$$
$$\approx 100(0.9093) \approx 90.9.$$

Hence, e is approximately 90.9 volts when $t = 2$.

In this section we have directed our attention to the periodic properties of the sine and cosine functions. We shall consider these properties again for all the circular functions when we study their graphs in Chapter 3.

EXERCISE SET 2.5

A

First write each of the following expressions in the form $\cos(x + k \cdot 2\pi)$, or $\sin(x + k \cdot 2\pi)$, where $0 \le x < 2\pi$, $k \in J$. Then specify the value of the expression.

Examples

a. $\cos \dfrac{9\pi}{4} = \cos\left(\dfrac{\pi}{4} + \dfrac{8\pi}{4}\right)$ b. $\sin \dfrac{16\pi}{3} = \sin\left(\dfrac{4\pi}{3} + \dfrac{12\pi}{3}\right)$

$\qquad\qquad = \cos\left(\dfrac{\pi}{4} + 2\pi\right);$ $\qquad\qquad = \sin\left(\dfrac{4\pi}{3} + 4\pi\right);$

$\cos\left(\dfrac{\pi}{4} + 2\pi\right) = \cos\dfrac{\pi}{4}$ $\sin\left(\dfrac{4\pi}{3} + 4\pi\right) = \sin\dfrac{4\pi}{3}$

$\qquad\qquad = \dfrac{1}{\sqrt{2}}$ $\qquad\qquad = -\dfrac{\sqrt{3}}{2}$

1. $\cos \dfrac{11\pi}{4}$ **2.** $\sin \dfrac{8\pi}{3}$ **3.** $\sin \dfrac{5\pi}{4}$

4. $\cos \dfrac{7\pi}{3}$ **5.** $\sin \dfrac{13\pi}{4}$ **6.** $\cos \dfrac{17\pi}{6}$

7. $\cos \dfrac{15\pi}{4}$ **8.** $\sin \dfrac{11\pi}{3}$ **9.** $\sin \dfrac{19\pi}{6}$

10. $\cos \dfrac{17\pi}{4}$ **11.** $\cos \dfrac{13\pi}{3}$ **12.** $\sin \dfrac{13\pi}{6}$

Find an approximate value for each expression.

Examples

a. $\sin 4.21$ b. $\cos 2$ c. $\cos 200(0.04)$

Solutions Set calculator on RAD setting.

a. $\sin 4.21 \approx -0.8764$ b. $\cos 2 \approx -0.4161$ c. $\cos 200(0.04)$

$\qquad\qquad\qquad\qquad\qquad\qquad\qquad\qquad = \cos 8 \approx -0.1455$

13. $\sin 1.32$ **14.** $\cos 2.87$ **15.** $\cos 3$

16. $\sin 4$ **17.** $\sin 160(0.02)$ **18.** $\cos 180(0.03)$

Solve Exercises 19–22 for the conditions for the spring in the example on page 71.

19. If the initial displacement is 10, find d when $t = 4$ and when $t = 4\frac{1}{2}$.

20. If the initial displacement is -10, find d when $t = 4$ and when $t = 4\frac{1}{2}$.

21. If the initial displacement is 8, find d when $t = \frac{1}{2}$ and when $t = \frac{1}{4}$.

22. If the initial displacement is -8, find d when $t = \frac{1}{2}$ and when $t = \frac{1}{4}$.

The horizontal displacement d of the bob on an oscillating pendulum is described by the equation $d = K \sin 2\pi t$, where K and d are expressed in centimeters and t is expressed in seconds.

23. Find d when $t = \frac{1}{4}$ and $K = 22$.

24. Find d when $t = \frac{1}{6}$ and $K = 20$.

25. Find K when $d = 20$ and $t = \frac{1}{2}$.

26. Find K when $d = 10$ and $t = \frac{1}{12}$.

Exs. 23–26

The instantaneous voltage e in a circuit at time t is given by the equation $e = 200 \sin 3700t$ where e is expressed in volts and t in seconds. Find the voltage to three significant digits at each of the following times.

Example $t = 0.04$

Solution   Evaluate formula for $t = 0.04$.

$$e = 200 \sin 3700t$$

$$= 200 \sin 3700(0.04) = 200 \sin 148.$$

$$\approx 200(-0.3383) \approx -67.7.$$

Hence, the voltage is approximately -67.7 volts when $t = 0.04$.

27. $t = 0.001$	**28.** $t = 0.002$	**29.** $t = 0.005$
30. $t = 0.008$	**31.** $t = 0.012$	**32.** $t = 0.020$

The instantaneous current i in an alternating system is given by the equation $i = I \sin \omega t$, where i is expressed in amperes and t is expressed in seconds. Find the current to the nearest tenth at each of the following times if $I = 0.05$ and $\omega = 377$.

33. $t = 0.01$	**34.** $t = 0.03$	**35.** $t = 0.2$
36. $t = 0.4$	**37.** $t = 1.0$	**38.** $t = 3.0$

The pressure P in a traveling sound wave is described by the equation $P = 10 \sin 200\pi[t - (x/1000)]$, where x is the distance from source of the sound in centimeters, t is the time expressed in seconds, and P is expressed *in dynes per square centimeter. Find P to three significant digits for each of the following conditions.*

39. $x = 200$ and $t = 0.06$ **40.** $x = 330$ and $t = 0.01$

41. $x = 660$ and $t = 0.02$ **42.** $x = 1650$ and $t = 0.05$

B

43. The displacement d of a point on a vibrating spring under certain conditions is given by $d = (1/50) \sin 50t - t \cos 50t$, where d is in feet and t is in seconds. Find the displacement when $t = 0.3$ second.

44. It is shown in calculus that an object having simple harmonic motion (described on page 20) has a velocity v given by $v = AB \cos Bt$ and an acceleration given by $a = -AB^2 \sin Bt$. The motion of a certain spring is expressed by $d = 3 \sin(\pi t/2)$, where d is a displacement in centimeters and t is in seconds.
   a. Express the velocity and the acceleration of a point on the free end of the spring in terms of t.
   b. Find the velocity and acceleration when t equals 2, 3, and 4 seconds.

Chapter Summary

2.1 The six trigonometric ratios of an acute angle α may be expressed in terms of the sides of a right triangle, where a is the length of the side opposite α, b is the length of the side adjacent to α, and c is the length of the hypotenuse:

$$\sin \alpha = \frac{a}{c}, \qquad \csc \alpha = \frac{c}{a},$$

$$\cos \alpha = \frac{b}{c}, \qquad \sec \alpha = \frac{c}{b},$$

$$\tan \alpha = \frac{a}{b}, \qquad \cot \alpha = \frac{b}{a}.$$

2.2 The **Law of Sines:**
   If α, β, and γ are the angles of any triangle, and if a, b, and c are the lengths of the sides opposite α, β, and γ, respectively, then

$$\frac{\sin \alpha}{a} = \frac{\sin \beta}{b} = \frac{\sin \gamma}{c}.$$

This law is used if the measures of two angles and the length of any side, or if the lengths of two sides and the measure of an angle opposite one of these sides, are given. In the latter case, certain ambiguities may arise. Two, one, or no triangles may exist.

2.3   The **Law of Cosines:**
If α, β, and γ are the angles of any triangle, and a, b, and c are the lengths of the sides opposite α, β, and γ, respectively, then

$$c^2 = a^2 + b^2 - 2ab \cos \gamma,$$
$$b^2 = a^2 + c^2 - 2ac \cos \beta,$$
$$a^2 = b^2 + c^2 - 2bc \cos \alpha.$$

This law is used if the lengths of two sides and the measure of the included angle, or if the lengths of three sides, are given.

2.4   The length s of the arc of a circle of radius of length r intercepted by an angle α whose vertex is at the center of the circle is given by

$$s = r \cdot \alpha^R.$$

The **linear velocity** v of a point moving along an arc of a circle at a constant speed is given by

$$v = \frac{s}{t} = \frac{r \cdot \alpha^R}{t},$$

where t is the time required to move through the distance s. The **angular velocity** ω of the moving point is given by

$$\omega = \frac{\alpha}{t}.$$

The linear velocity v and the angular velocity ω of a point on a circle with radius of length r are related by the equation

$$v = r\omega,$$

where ω is given in radians per unit time.

2.5   Functions in which the elements in the domains are numbers that are measures (in radian units) of angles and whose ranges are the same as the respective ranges of the analogous trigonometric functions are called **circular functions.** A calculator set on RAD or tables in radian measure that are used to obtain elements in the ranges of the trigonometric functions can be used to obtain elements in the ranges of the circular functions

A function f is **periodic** if there exists a number a ($a \neq 0$) such that for each x in the domain, $x + a$ is in the domain and $f(x + a) = f(x)$. The smallest positive value of a for which this is true is the **period** of the function.

Review Exercises

A

2.1 *Solve each right triangle. In each case, $\gamma = 90°$.*

1. $c = 14$, $\beta = 51°$

2. $c = 6.7$, $a = 5.4$

3. $a = 12.9$, $\alpha = 32° \ 36'$

4. $b = 9.72$, $\alpha = 71° \ 18'$

2.2 *Solve each triangle.*

5. $c = 13.6$, $\alpha = 30.3°$, $\beta = 72.2°$

6. $c = 1.4$, $\alpha = 135° \ 6'$, $\gamma = 34° \ 54'$

Determine the number of triangles that satisfy the conditions in each exercise and solve each triangle.

7. $b = 3.9$, $a = 5$, $\beta = 21.7°$

8. $b = 37$, $a = 51$, $\alpha = 135° \ 30'$

2.3 *Solve each triangle.*

9. $a = 1.9$, $b = 2.3$, $\gamma = 58° \ 12'$

10. $b = 2.3$, $c = 5.7$, $\alpha = 23.1°$

11. $a = 3.6$, $b = 4.2$, $c = 6.1$

12. $a = 2.1$, $b = 3.4$, $c = 4.1$

2.4 **13.** Find the length of the arc intercepted on a circle by a central angle α with measure 2.2^R if the radius of the circle is 1.3 meters in length.

14. Find the measure of the central angle that intercepts an arc with length 60.2 centimeters on a circle whose radius is 3.1 meters in length.

15. Find the linear velocity v of a point on a circle moving with constant speed if $r = 2$ decimeters and $\omega = 6$ radians per minute.

16. Find the angular velocity ω of a point on a circle moving with constant speed if $r = 0.5$ meter and $v = 19.7$ centimeters per second.

2.5 *Write each of the following expressions in the form $\cos(x + k \cdot 2\pi)$, or $\sin(x + k \cdot 2\pi)$, where $0 \leq x < 2\pi$, $k \in J$. Specify the value of the expression.*

17. $\cos \dfrac{8\pi}{3}$

18. $\sin \dfrac{17\pi}{4}$

Find an approximate value for each expression.

19. sin 1.26

20. cos 140(0.04)

21. The displacement of a spring is given by $d = 4 \cos 8\pi t$, where d is expressed in centimeters and t is expressed in seconds. Find d when $t = 2\frac{1}{2}$ seconds.

22. The horizontal displacement d of the bob on an oscillating pendulum is given by the equation $d = 18 \sin 2\pi t$, where d is expressed in centimeters and t is expressed in seconds. Find d when $t = 2\frac{1}{2}$ seconds.

23. The instantaneous voltage e in a circuit at time t is given by the equation $e = 1.3 \sin 980\pi t$, where e is expressed in volts and t is expressed in seconds. Find the voltage when $t = 3$.

24. The instantaneous current i in an alternating current system is given by the equation $i = I \sin \omega t$, where i and I are expressed in amperes, ω in radians per second, and t in seconds. Find the current to three significant digits at $t = 0.002$ if $I = 0.04$ and $\omega = 352$.

3 GRAPHS; INVERSE TRIGONOMETRIC FUNCTIONS

The periodic characteristics of the trigonometric and circular functions can be exhibited very effectively by graphing these functions in a Cartesian coordinate system. You may want to review the material on graphing in Appendix A.4 before starting this chapter.

3.1 Sine and Cosine Functions

In graphing the trigonometric and circular functions, it is convenient to view line segments on the x-axis of a Cartesian coordinate system as obtained by unwinding the arc on a unit circle with the same length, as shown in Figure 3.1. A similar association could be made for the lengths of arcs in a clockwise direction and line segments on the negative axis.

Sometimes it is convenient to use integers instead of rational multiples of π as scale numbers on the x-axis. Both of these scale numbers are shown in a Cartesian system in Figure 3.2. The y-axis is scaled in integers.

The graphs of the sine and cosine functions, defined by the two equations $y = \sin x$ and $y = \cos x$, consist of the set of points corresponding to all ordered pairs (x, y) in the respective functions. The replacement set of x is considered to be *either* the set of all angles *or* the set of real numbers.

Although a calculator can be used to find good approximations for the values of $\sin x$ and $\cos x$ function values from Table 1.1 (page 16) with which you are probably now familiar, and the notion of a reference angle, are usually sufficient to obtain good approximations to the graphs of these functions for $0 \le x \le 2\pi$.

Figure 3.1

Figure 3.2

The Sine Graph

We first consider the graph of

$$y = \sin x. \tag{1}$$

In order not to distort the form of the graph of (1) and the following graphs, we shall use the same units of measurement on the x-axis and y-axis. To obtain selected ordered pairs $(x, \sin x)$ in the function shown in Figure 3.3a on page 80, we first note from Table 1.1 (or by construction of an appropriate right triangle, or from memory) that

$$\sin \frac{\pi}{4} = \frac{1}{\sqrt{2}} \approx 0.7;$$

(a)   (b)

Figure 3.3

then using a reference angle, we obtain

$$\sin \frac{3\pi}{4} = \sin \frac{\pi}{4} \approx 0.7,$$

$$\sin \frac{5\pi}{4} = -\sin \frac{\pi}{4} \approx -0.7,$$

$$\sin \frac{7\pi}{4} = -\sin \frac{\pi}{4} \approx -0.7.$$

In calculus it is shown that the trigonometric functions are continuous over the intervals for which they are defined. For us this means that since the sine function is defined for all angles (or real numbers) its graph contains no "breaks." Hence we connect the points in Figure 3.3a with a smooth curve to produce Figure 3.3b. Of course, a calculator can be used to obtain additional points to assist us in graphing the function in the interval 0 to 2π.

Now, because the sine function is periodic with period 2π,

$$\sin(x + k \cdot 2\pi) = \sin x, \quad k \in J.$$

Hence we repeat the graph in Figure 3.3b in both directions and obtain Figure 3.4 which is part of the graph of $y = \sin x$ for all angles x or real

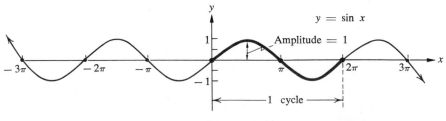

Figure 3.4

numbers x. The graph is called a **sine wave** or **sinusoidal wave.** The graph over one period is called a **cycle** of the wave.

The amplitude of a sine wave is defined as follows.

Definition 3.1   *The* **amplitude** *of a sine wave is the absolute value of half the difference of the maximum and minimum ordinates of the wave. This number is also called the amplitude of the function.*

Thus, the amplitude of the wave in Figure 3.4 is

$$\left| \frac{1}{2} [1 - (-1)] \right| = 1.$$

Several important notions of the sine function are evident from the graph. Observe that the range of the function (see Appendix A.3) is

$$\{ y \,|\, -1 \le y \le 1 \}.$$

Also observe that the zeros of the function (see Appendix A.5), values of x for which $\sin x$ is 0, are $k\pi$, $k \in J$. Note too that these are also the x-coordinates of the points where the curve intersects the x-axis.

The Cosine Graph

We obtain the graph of the cosine function

$$y = \cos x \qquad\qquad (2)$$

in the same way that we obtained the graph of $y = \sin x$. Some ordered pairs in the function for $0 \le x \le 2\pi$ are obtained for special angles. First noting that

$$\cos \frac{\pi}{4} \approx 0.7,$$

we then obtain

$$\cos \frac{3\pi}{4} = -\cos \frac{\pi}{4} \approx -0.7,$$

$$\cos \frac{5\pi}{4} = -\cos \frac{\pi}{4} \approx -0.7,$$

$$\cos \frac{7\pi}{4} = \cos \frac{\pi}{4} \approx 0.7.$$

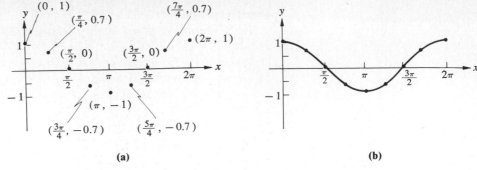

(a)   (b)

Figure 3.5

The graphs of the ordered pairs are shown in Figure 3.5a. In this case function values decrease as x increases from zero to π and they increase as x increases from π to 2π. We connect the points in Figure 3.5a with a smooth curve to produce Figure 3.5b. Again, of course, a calculator can be used to obtain additional points for graphing the curve.

Since the cosine function is also periodic with period 2π,

$$\cos(x + k \cdot 2\pi) = \cos x, \quad k \in J.$$

Hence, we repeat the pattern in Figure 3.5b and obtain Figure 3.6, which is part of the graph of $y = \cos x$ for all angles x or real numbers x. We observe that, like the sine function, the cosine function has the range $\{y \mid -1 \le y \le 1\}$. Also like the sine function, the amplitude of the wave is 1. Furthermore, we note that the zeros of the function, values of x for which $\cos x$ is 0 and which are the x-coordinates of the points where the curve crosses the x-axis, are

$$\frac{\pi}{2} + k\pi, \quad k \in J.$$

Observe that the graphs of the sine and cosine functions have the same "form." For this reason *both* curves are called sine waves or sinusoidal waves.

Figure 3.6

EXERCISE SET 3.1

A

1. Graph the elements of

$$\left\{ -2\pi, \frac{-3\pi}{2}, -\pi, \frac{-\pi}{2}, 0, \frac{\pi}{2}, \pi, \frac{3\pi}{2}, 2\pi \right\}$$

on the number line below that is scaled in integral units. Use $\pi \approx 3.14$.

2. Graph the elements of

$$\left\{ 0, \frac{\pi}{4}, \frac{\pi}{2}, \frac{3\pi}{4}, \pi, \frac{5\pi}{4}, \frac{3\pi}{2}, \frac{7\pi}{4}, 2\pi \right\}$$

on a number line scaled in integral units as shown in Problem 1. Use $\pi \approx 3.14$.

3. Graph the elements of

$$\left\{ 0, \frac{\pi}{4}, \frac{\pi}{2}, \frac{3\pi}{4}, \pi, \frac{5\pi}{4}, \frac{3\pi}{2}, \frac{7\pi}{4}, 2\pi \right\}$$

on the number line below that is scaled in multiples of π.

4. Graph the elements of

$$\left\{ -2\pi, \frac{-7\pi}{4}, \frac{-3\pi}{2}, \frac{-5\pi}{4}, -\pi, \frac{-3\pi}{4}, \frac{-\pi}{2}, \frac{-\pi}{4}, 0 \right\}$$

on a number line scaled in multiples of π as shown in Exercise 3.

5. Graph the elements of

$$\left\{ -2\pi, \frac{-5\pi}{3}, \frac{-4\pi}{3}, -\pi, \frac{-2\pi}{3}, \frac{-\pi}{3}, 0, \frac{\pi}{3}, \frac{2\pi}{3}, \pi, \frac{4\pi}{3}, \frac{5\pi}{3}, 2\pi \right\}$$

on a number line scaled in multiples of π.

6. Graph the elements of

$$\left\{0, \frac{\pi}{6}, \frac{\pi}{3}, \frac{\pi}{2}, \frac{2\pi}{3}, \frac{5\pi}{6}, \pi, \frac{7\pi}{6}, \frac{4\pi}{3}, \frac{3\pi}{2}, \frac{5\pi}{3}, \frac{11\pi}{6}, 2\pi\right\}$$

on a number line scaled in multiples of π.

In Exercises 7–18, (**a**) *graph the function defined by each equation by graphing one cycle and then repeating the pattern and* (**b**) *specify the zeros and the amplitude of the function.*

Example $y = 2 \sin x$

Solution

a. Some ordered pairs in the function defined by the equation are obtained in order to graph one cycle. The systematic tabulation of the data shown is helpful. The ordered pairs (x, y) are then graphed as shown, connected with a smooth curve, and the pattern of the wave duplicated. A calculator can be used to obtain additional points to construct the graph.

x	$\sin x$	$2 \sin x$ or y
0	0	0
$\dfrac{\pi}{2}$	1	2
π	0	0
$\dfrac{3\pi}{2}$	-1	-2
2π	0	0

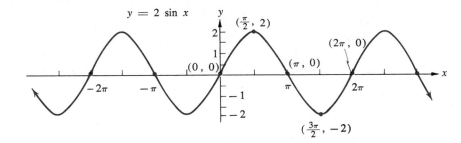

b. Zeros: $k\pi$, $k \in J$;   amplitude: $\left|\dfrac{1}{2}[2 - (-2)]\right| = 2$.

7. $y = 3 \sin x$ **8.** $y = 2 \cos x$ **9.** $y = \dfrac{1}{2} \cos x$

10. $y = \dfrac{1}{3} \sin x$ **11.** $y = -\sin x$ **12.** $y = -\cos x$

Example $y = \cos 2x$

Solution

a. Some ordered pairs in the function defined by the equation are obtained in order to graph one cycle. The ordered pairs (x, y) obtained from

x	$2x$	$\cos 2x$ or y
0	0	1
$\dfrac{\pi}{4}$	$\dfrac{\pi}{2}$	0
$\dfrac{\pi}{2}$	π	-1
$\dfrac{3\pi}{4}$	$\dfrac{3\pi}{2}$	0
π	2π	1

the table are then graphed as shown, connected with a smooth curve, and the pattern of the wave duplicated.

b. Zeros: $x = \dfrac{\pi}{4} + k \cdot \dfrac{\pi}{2}, \ k \in J$;   amplitude: $\left| \dfrac{1}{2}[1-(-1)] \right| = 1.$

13. $y = \sin 2x$ **14.** $y = \sin \dfrac{1}{2}x$ **15.** $y = \cos \dfrac{1}{2}x$

16. $y = \cos \dfrac{1}{3}x$ **17.** $y = \sin\left(x - \dfrac{\pi}{2}\right)$ **18.** $y = \cos\left(x + \dfrac{\pi}{4}\right)$

Graph two cycles of each of the following.

Example $y = \cos \dfrac{\pi}{3}x$

x	$\dfrac{\pi}{3}x$	$\cos \dfrac{\pi}{3}x$ or y
0	0	1
1	$\dfrac{\pi}{3}$	$\dfrac{1}{2}$
2	$\dfrac{2\pi}{3}$	$-\dfrac{1}{2}$
3	π	-1
4	$\dfrac{4\pi}{3}$	$-\dfrac{1}{2}$
5	$\dfrac{5\pi}{3}$	$\dfrac{1}{2}$
6	2π	1

Solution   In such cases where the coefficient of x is a multiple of π, it is helpful to label the x-axis in rational units and to select rational numbers for x, as shown in the table. Some ordered pairs in the function defined by the equation are graphed as shown and connected with a smooth curve. Then the pattern of the wave is duplicated.

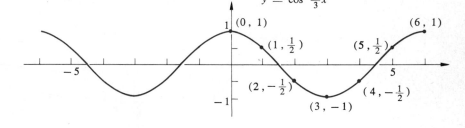

19. $y = \cos \dfrac{\pi}{2}x$ **20.** $y = \sin \dfrac{\pi}{2}x$

21. $y = \sin \pi x$ **22.** $y = \cos \pi x$

B

23. Graph $y = \dfrac{\sin x}{x}$, $-\pi < x < \pi$, and describe the graph as x approaches zero. Is the graph periodic? Why or why not?

24. Graph $y = x \sin x$, $-4\pi \leq x \leq 4\pi$.

25. Graph $y = \dfrac{\cos x}{x}$, $-\pi < x < \pi$, and describe the graph as x approaches zero. Is the graph periodic? Why or why not?

26. Graph $y = x \cos x$, $-4\pi \leq x \leq 4\pi$.

Examples of noncircular periodic functions are given below. The symbol $[x]$ *denotes the number that is the greatest integer not greater than* x *for each* $x \in R$.

Examples

a. $[3] = 3$ b. $[5.91] = [5 + 0.91]$ c. $[-1.21] = [-2 + 0.79]$

$= 5$ $= -2$

Graph the function defined by each equation. Specify the period of the function.

27. $y = x - [x]$, $-4 \leq x < 4$ 28. $y = x + [x]$, $-4 \leq x < 4$

29. $y = 2x - [2x]$, $-3 \leq x < 3$ 30. $y = -2x + [2x]$, $-3 \leq x < 3$

3.2 General Sine Waves

Functions defined by equations of the form

$$y = A \sin B(x + C) \qquad \text{or} \qquad y = A \cos B(x + C),$$

where $A, B, C, x, y \in R$ and $A, B \neq 0$, always have sine waves for their graphs. Depending upon the values for A, B, and C, these sine waves have different amplitudes, different periods, and different horizontal positions with respect to the origin. To facilitate graphing such functions, we now consider these factors in more detail.

Amplitude

First consider how the value of A in the defining equation affects the graph of the sine function or the cosine function. For each value of x, each ordinate of the graph of

$$y = A \sin x$$

is A times the ordinate of the graph of $y = \sin x$. The effect of A, then, is to multiply the amplitude of the graph of $y = \sin x$ by $|A|$. The graph of $y = A \cos x$ is related to the graph of $y = \cos x$ in a similar way.

Example   Graph $y = 2 \cos x$ over the interval $-2\pi \le x \le 4\pi$.

Solution
1. Sketch the graph of $y = \cos x$, over the interval $0 \le x \le 2\pi$, as a reference.
2. The amplitude of the graph of $y = 2 \cos x$ is 2. Therefore, sketch the desired graph on the same coordinate system over the same interval by making each ordinate 2 times the ordinate of the graph of $y = \cos x$.
3. Repeat the cycle over the interval $-2\pi \le x \le 4\pi$.

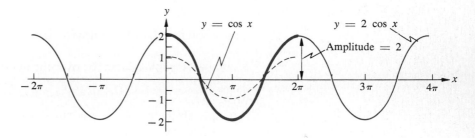

The first cycle is shown sketched with a heavier line for emphasis.

In the above example, the coefficient of $\cos x$ is a positive number. If the coefficient were a negative number, say -2, each ordinate of the graph of $y = -2 \cos x$ would be the negative of the corresponding ordinate of $y = 2 \cos x$. The graph of $y = -2 \cos x$, over the interval $-2\pi \le x \le 4\pi$, is shown in Figure 3.7.

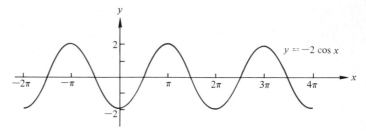

Figure 3.7

Period

Now consider how different values for B affect the graph of

$$y = \sin Bx \qquad \text{or} \qquad y = \cos Bx.$$

A complete cycle of the graph of $y = \sin x$ (or $y = \cos x$) occurs over the period 2π. Hence, a cycle of the graph of $y = \sin Bx$ (or $y = \cos Bx$) is obtained as Bx increases from 0 to 2π. Therefore, the period of either function is the solution of the equation

$$Bx = 2\pi \qquad \text{or} \qquad x = \frac{2\pi}{B}.$$

Hence, the period P of a function defined by an equation of the form $y = \sin Bx$ or $y = \cos Bx$ is

$$P = \frac{2\pi}{|B|}.$$

The absolute value of B is used so that the period is a positive number.

Example   Graph $y = \sin 3x$, over the interval $0 \le x \le 2\pi$.

Solution
1. Sketch the graph of $y = \sin x$, over the interval $0 \le x \le 2\pi$, as a reference.
2. The period

$$P = \frac{2\pi}{|B|} = \frac{2\pi}{3}.$$

Therefore, sketch a complete cycle of the desired graph on the same set of axes over the interval $0 \le x \le 2\pi/3$.
3. Repeat the cycle obtained in step 2 over the interval $0 \le x \le 2\pi$.

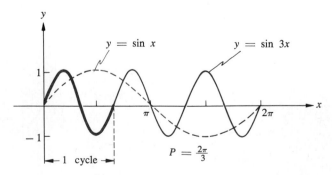

Phase Shift

Finally, consider how different values for C affect the graph of

$$y = \sin(x - C) \qquad \text{or} \qquad y = \cos(x - C).$$

For $x = C$,

$$\sin(x - C) = \sin(C - C) = \sin 0 = 0,$$

and, for any x_1, the value of $\sin(x - C)$ at $x_1 + C$ is the same as the value of $\sin x$ at x_1. That is,

$$\sin[(x_1 + C) - C] = \sin x_1.$$

For this reason, the graph of $y = \sin(x - C)$ is shifted C units to the right of the graph of $y = \sin x$ if $C > 0$ and shifted $|C|$ units to the left of the graph of $y = \sin x$ if $C < 0$.

The graphs of $y = \cos(x - C)$ and $y = \cos x$ are related in a similar way.

Definition 3.2 *The number $|C|$ is the **phase shift** of the graph of periodic functions defined by equations of the form*

$$y = \sin(x - C) \qquad \text{or} \qquad y = \cos(x - C).$$

Example   Graph $y = \sin\left(x - \dfrac{\pi}{3}\right)$ over the interval $-2\pi \le x \le 4\pi$.

Solution

1. Sketch the graph of $y = \sin x$, over the interval $0 \le x \le 2\pi$, as a reference.
2. Since $C = \pi/3$, the desired graph is $\pi/3$ units to the right of the graph of $y = \sin x$.
3. Sketch the graph over the interval $-2\pi \le x \le 4\pi$.

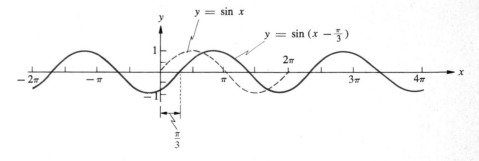

Note that in the above example the graph of $y = \sin(x - \pi/3)$ is $\pi/3$ *units to the right* of the graph of $y = \sin x$ because $\pi/3$, the *value of C*, is *positive*. The graph of $y = \sin(x + \pi/3)$, which is equivalent to $y = \sin[x - (-\pi/3)]$, would be *to the left* of the graph of $y = \sin x$ because $-\pi/3$, the *value of C*, is *negative*.

Graphs of $y = A \sin B(x - C)$ and $y = A \cos B(x - C)$

We can now use all information about the effects of A, B, and C on the graphs of functions defined by the equation $y = A \sin B(x - C)$ or by the equation $y = A \cos B(x - C)$ to facilitate making such graphs.

If an equation were written in the form $y = A \cos(Bx - C)$, it would be necessary first to factor B from $Bx - C$ in order to make a judgment about the period and phase shift of the graph. For example, we would write

$$y = 2 \cos\left(3x + \frac{\pi}{2}\right) = 2 \cos 3\left[x - \left(-\frac{\pi}{6}\right)\right],$$

from which we could observe that the period is $2\pi/3$ and the phase shift is $\pi/6$ unit to the left of the graph of $y = 2 \cos 3x$. The graph is shown in the example in the exercise set on page 93.

The Period P, a Positive Rational Number

If the period P of a graph is a positive rational number, instead of a multiple of π such as we encountered in the examples above, it is helpful to use selected rational values as elements of the domain and to label the x-axis in rational units to facilitate sketching the graph.

Example   Graph $y = \dfrac{1}{2} \cos \pi x$ over the interval $-5 \le x < 5$.

Solution

1. The phase shift $|C|$ is zero. Therefore compare the equation with

$$y = A \cos Bx.$$

It is evident that the graph of $y = \frac{1}{2} \cos \pi x$ has

$$\text{amplitude: } |A| = \frac{1}{2}; \quad \text{period: } P = \frac{2\pi}{|B|} = \frac{2\pi}{\pi} = 2.$$

2. Sketch the graph, marking the x-axis in rational units.

EXERCISE SET 3.2

A

Sketch the graph of each equation over the interval $-2\pi \le x \le 2\pi$.

Example $y = \dfrac{1}{2} \cos 2x$

Solution

1. Compare the equation with $y = A \cos Bx$. It is evident that the graph of $y = \frac{1}{2} \cos 2x$ has

$$\text{amplitude: } |A| = \frac{1}{2}; \quad \text{period: } P = \frac{2\pi}{|B|} = \frac{2\pi}{2} = \pi.$$

2. Sketch the graph.

1. $y = 3 \sin x$ **2.** $y = 3 \cos x$ **3.** $y = -2 \cos x$

4. $y = -5 \sin x$ **5.** $y = \dfrac{1}{2} \sin x$ **6.** $y = \dfrac{1}{3} \cos x$

7. $y = \cos 2x$ **8.** $y = \sin 3x$ **9.** $y = \sin \dfrac{1}{2} x$

10. $y = \cos \dfrac{1}{3} x$ **11.** $y = 2 \sin 2x$ **12.** $y = \dfrac{1}{3} \cos \dfrac{1}{2} x$

Example $y = 3 \sin\left(x + \dfrac{\pi}{2}\right)$

Solution

1. Write the equation as

$$y = 3 \sin\left[x - \left(-\dfrac{\pi}{2}\right)\right].$$

2. Compare this equation with $y = A \sin B(x - C)$. It is evident that the graph has

amplitude: $|A| = 3$;   period: $P = \dfrac{2\pi}{|B|} = 2\pi$;

phase shift: $\pi/2$ units to the left of the graph of $y = 3 \sin x$.

3. Sketch the graph.

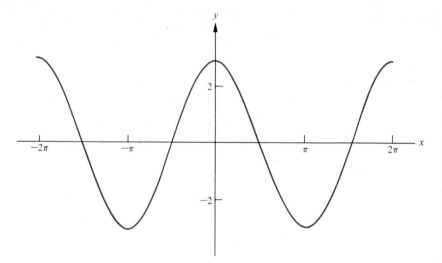

13. $y = \cos(x + \pi)$ **14.** $y = \sin\left(x - \dfrac{\pi}{2}\right)$

15. $y = 2 \sin\left(x - \dfrac{\pi}{4}\right)$ **16.** $y = \dfrac{1}{2} \cos\left(x + \dfrac{\pi}{6}\right)$

17. $y = \dfrac{1}{2} \sin\left(x + \dfrac{\pi}{3}\right)$ **18.** $y = \cos\left(x - \dfrac{\pi}{4}\right)$

Example $y = 2 \cos\left(3x + \dfrac{\pi}{2}\right)$

Solution

1. Factor 3 from the expression in parentheses and then write the equation as

$$y = 2 \cos 3\left[x - \left(-\frac{\pi}{6} \right) \right].$$

2. Compare this equation with $y = A \cos B(x - C)$. It is evident that the graph of $y = 2 \cos 3[x - (-\pi/6)]$ has

amplitude: $|A| = 2$;   period: $P = \dfrac{2\pi}{|B|} = \dfrac{2\pi}{3}$;

phase shift: $\dfrac{\pi}{6}$ unit to the left of the graph of $y = 2 \cos 3x$.

3. Sketch the graph.

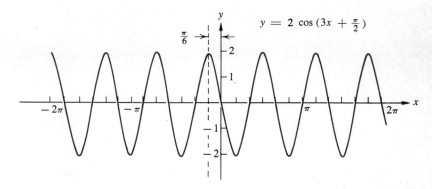

19. $y = 2 \cos\left(2x + \dfrac{\pi}{2} \right)$

20. $y = \dfrac{1}{2} \sin\left(\dfrac{1}{2} x - \dfrac{\pi}{4} \right)$

21. $y = \dfrac{1}{2} \sin\left(\dfrac{1}{2} x + \dfrac{\pi}{2} \right)$

22. $y = 2 \cos(2x - \pi)$

23. $y = 3 \cos\left(\dfrac{1}{3} x - \dfrac{\pi}{3} \right)$

24. $y = 4 \sin\left(3x + \dfrac{3\pi}{4} \right)$

Graph two cycles of each of the following with positive elements in the domain.   Hint: Scale the x-axis in rational units 1, 2, 3, ...

Example $y = \sin \pi x$

(continued)

Solution

1. Compare the equation with $y = A \sin Bx$. It is evident that the graph of $y = \sin \pi x$ has amplitude $|A| = 1$ and period

$$P = \frac{2\pi}{|B|} = \frac{2\pi}{\pi} = 2.$$

2. Sketch the graph on a coordinate system in which the x-axis is scaled in rational units.

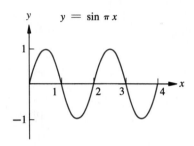

25. $y = \cos \pi x$ **26.** $y = \sin \dfrac{\pi}{2} x$

27. $y = \dfrac{1}{2} \sin \dfrac{\pi}{3} x$ **28.** $y = 2 \cos \dfrac{\pi}{4} x$

B

Graph two cycles of each function.   Hint: *If $y = |f(x)|$, then y is always nonnegative.*

29. $y = |\sin x|$ **30.** $y = |\cos x|$ **31.** $y = 3|\cos 2x|$

32. $y = 2\left|\sin \dfrac{1}{2} x\right|$ **33.** $y = |\sin \pi x|$ **34.** $y = |\cos \pi x|$

3.3 Graphical Addition

Functions defined by equations such as

$$y = 2 \sin x + \cos x \qquad \text{and} \qquad y = x + \cos x$$

can be graphed by methods similar to those we used in the preceding section. We can find some elements in the range for several elements in the domain by first finding values $\cos x$ and $\sin x$ for some values of x in a given interval and then performing the indicated operations to obtain some ordered pairs in the function. Then we can graph the set of ordered pairs and complete the graph by interpolating for other elements in the range, and connecting these points with a smooth curve.

An alternative, and somewhat simpler, method involves the **graphical addition of ordinates.** The following two examples illustrate this method.

Example   Graph $y = 2 \sin x + \cos x$ over the interval $0 \le x \le 2\pi$.

Solution   Sketch the graphs of $y_1 = 2 \sin x$ and $y_2 = \cos x$ on the same set of axes over the given interval. The ordinate of the graph of $y = 2 \sin x + \cos x$ for *any* x is the algebraic sum of the ordinates of the graph of $y_1 = 2 \sin x$ and the graph of $y_2 = \cos x$. For example, for $x = \pi/3$,

$$2 \sin x = 2\left(\frac{\sqrt{3}}{2}\right) \approx 1.7 \qquad \text{and} \qquad \cos x = 0.5,$$

from which

$$y = 2 \sin x + \cos x$$

$$\approx 1.7 + 0.5 = 2.2.$$

The ordinate 2.2 can be approximated graphically by "adding" vertical directed segments, $\overrightarrow{AC}$ and $\overrightarrow{AB}$, from the x-axis to the respective curves at $x = \pi/3$, as shown in the figure. Other ordinates can be found graphically in a similar way. The points in the graph of

$$y = 2 \sin x + \cos x$$

for which $2 \sin x$ or $\cos x$ equals zero are particularly easy to locate. Connect the points with a smooth curve to obtain the graph shown.

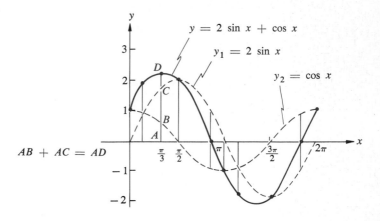

$AB + AC = AD$

Example Graph $y = x + \cos x$ over the interval $-\pi \le x \le 2\pi$.

Solution Sketch the graphs of $y_1 = x$ and $y_2 = \cos x$ on the same set of axes over the given interval. Then "add" directed segments for arbitrary values of x. The points in the graph of $y = x + \cos x$ for which $\cos x = 0$ are easy to locate. Connect the points with a smooth curve to obtain the graph shown.

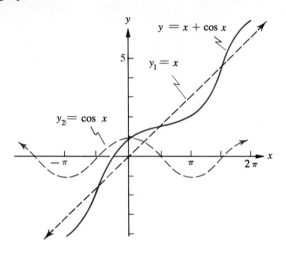

EXERCISE SET 3.3

A

Sketch the graph of each equation over the interval $-\pi \le x \le 2\pi$.

Example $y = \sin x + \cos x$

Solution

1. Graph $y_1 = \sin x$ and $y_2 = \cos x$ on the same set of axes.
2. Add the ordinates of several pairs of points for several values of x and draw a smooth curve.

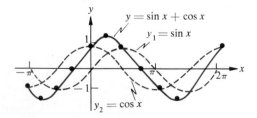

1. $y = 2 \cos x + \sin x$ **2.** $y = \cos 2x + \dfrac{1}{2} \sin x$

3. $y = \dfrac{1}{2}\cos x + \sin \dfrac{1}{2}x$ **4.** $y = \cos \dfrac{1}{2}x + \dfrac{1}{3}\sin x$

5. $y = \sin 2x + \dfrac{1}{2}\cos x$ **6.** $y = \sin \dfrac{1}{2}x + \cos x$

7. $y = 2\sin x + \cos \dfrac{1}{2}x$ **8.** $y = \sin \dfrac{1}{2}x + 2\cos \dfrac{1}{2}x$

9. $y = 2\sin x - \cos x$ **10.** $y = \cos x - \dfrac{1}{2}\sin x$

11. $y = \cos \dfrac{1}{2}x - 2\sin x$ **12.** $y = 2\cos x - \dfrac{1}{3}\sin x$

13. $y = 2 + 2\cos x$ **14.** $y = 1 - 2\sin x$

15. $y = x + \sin x$ **16.** $y = \dfrac{x}{2} - \cos x$

B

In Exercises 17–22, sketch the graph of each equation over the specified interval by using "addition of ordinates."

17. $y = x - 2 + \cos x,\ 0 \le x \le 2\pi$.
    Hint: First graph $y_1 = x - 2$ and $y_2 = \cos x$.

18. $y = 3 - x + \sin x,\ 0 \le x \le 2\pi$.
    Hint: First graph $y_1 = 3 - x$ and $y_2 = \sin x$.

19. $y = 2\cos x + \dfrac{x^2}{12},\ 0 \le x \le 2\pi$

20. $y = 2\sin x + \dfrac{1}{x},\ 0 \le x \le 2\pi$

21. $y = \sin \pi x - x + 4,\ 0 \le x \le 4$

22. $y = \cos \dfrac{\pi}{2}x - \dfrac{x^2}{4},\ 0 \le x \le 4$

3.4 Graphs of Other Trigonometric Functions

We can graph the tangent, cotangent, secant, and cosecant functions on a Cartesian coordinate system in the same way that we graphed the sine and cosine functions in Sections 3.1 and 3.2. Although the graphs of these functions are not sine waves, the graphs do exhibit the periodic nature and other important features of the functions.

Tangent Function

The period of the tangent function is π. Therefore we need consider some special values for x only in the interval 0 to π. We obtain some ordered pairs in the function and graph these ordered pairs as shown in Figure 3.8. Observe that

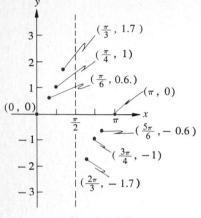

$$\tan \frac{\pi}{6} = \frac{1}{\sqrt{3}} \approx 0.6.$$

and

$$\tan \frac{\pi}{3} = \sqrt{3} \approx 1.7.$$

Figure 3.8

Since $\tan x = \sin x / \cos x$, the function defined by $y = \tan x$ is continuous in the domain.

$$0 \le x < \frac{\pi}{2} \qquad \text{and} \qquad \frac{\pi}{2} < x \le \pi$$

($\cos x \neq 0$ in these intervals), and we connect the points in Figure 3.8 with a smooth curve to produce one cycle (heavy line). Then we repeat this pattern in both directions to obtain Figure 3.9. A calculator can be used to obtain additional points to assist in graphing this function.

Because $\tan x$ is undefined for $x = \pi/2$, there is no point in the graph of the function when $x = \pi/2$. Notice, however, that $|\tan x|$ increases

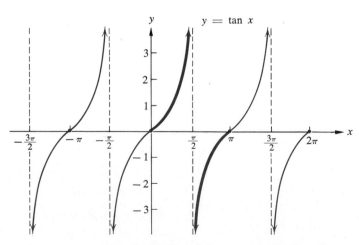

Figure 3.9

indefinitely as x is taken closer and closer to $\pi/2$. The vertical dashed lines in Figures 3.8 and 3.9 are called **asymptotes.** The entire set of asymptotes to the curve is the set of lines which are the graphs of

$$x = \frac{\pi}{2} + k\pi, \quad k \in J.$$

Furthermore, observe that the zeros of the function that are associated with the points where the curve intersects the x-axis are $k\pi$, $k \in J$.

Cotangent Function

The graph of the function defined by $y = \cot x$ can also be obtained by plotting some points. However, for $x \neq k(\pi/2)$, $k \in J$,

$$\cot x = \frac{1}{\tan x}.$$

Thus, each ordinate of the graph of $y = \cot x$, $x \neq k(\pi/2)$, is the reciprocal of the corresponding ordinate of the graph of $y = \tan x$. With this information we can quickly sketch the graph of $y = \cot x$ if we first sketch the graph of $y = \tan x$ on the same set of axes. The graphs appear in Figure 3.10. Note that the zeros of the function defined by $y = \cot x$ are

$$\frac{\pi}{2} + k\pi, \quad k \in J,$$

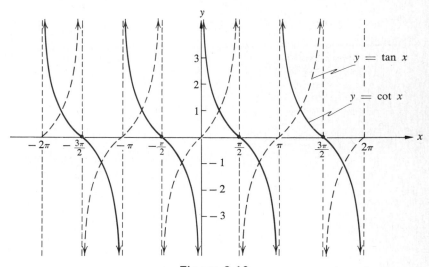

Figure 3.10

and that the asymptotes to the curve are given by

$$x = k\pi, \quad k \in J.$$

Also, the range of the tangent or cotangent function is $\{y \mid y \in R\}$.

Secant and Cosecant Functions

The graphs of the functions defined by $y = \sec x$ and $y = \csc x$ can be obtained by using the facts that

$$\sec x = \frac{1}{\cos x}, \quad x \neq \frac{\pi}{2} + k\pi, \quad k \in J, \tag{1}$$

and

$$\csc x = \frac{1}{\sin x}, \quad x \neq k\pi, \quad k \in J. \tag{2}$$

Hence, each ordinate of the graph of $y = \sec x$ is the reciprocal of the corresponding ordinate of the graph of $y = \cos x$, and each ordinate of the graph of $y = \csc x$ is the reciprocal of the corresponding ordinate of the graph of $y = \sin x$, except for the noted restrictions. Thus, the graphs of

$$y = \sec x \qquad \text{and} \qquad y = \csc x$$

can be obtained directly from the graphs of

$$y = \cos x \qquad \text{and} \qquad y = \sin x,$$

respectively, as shown in Figures 3.11 and 3.12 (see exercises 7 and 8).
Notice that the asymptotes of the graph of $y = \sec x$ are the graphs of

$$x = \frac{\pi}{2} + k\pi, \quad k \in J;$$

the x-intercepts of the asymptotes are the zeros of the function defined by $y = \cos x$. The asymptotes of the graph of $y = \csc x$ are the graphs of

$$x = k\pi, \quad k \in J;$$

the x-intercepts of the asymptotes are the zeros of the function defined by $y = \sin x$. The range of either the secant or the cosecant function is $\{y \mid y \leq -1 \text{ or } y \geq 1\}$. Thus, there are no zeros of these functions.

Figure 3.11

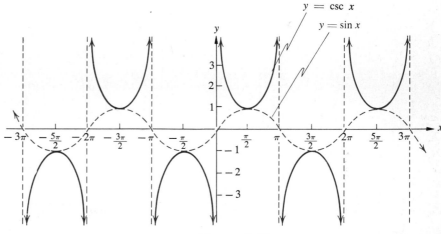

Figure 3.12

Variation of Constants

Variations in the constants A, B, and C in equations such as

$$y = A \tan B(x - C) \qquad (3)$$

affect their graphs in much the same way that these constants affect the graphs of $y = A \sin B(x - C)$ and $y = A \cos B(x - C)$. However, in the graphs of functions other than sine and cosine, the coefficient A is not referred to as the amplitude.

EXERCISE SET 3.4

A

Sketch the graph of the function defined by each equation over the interval $0 \leq x \leq 2\pi$ and specify (a) the range, (b) the equations of the asymptotes, and (c) the zeros of the function (if they exist).

Example $y = \tan 2x$

Solution   Since the period of $y = \tan x$ is π, the period of $y = \tan 2x$ is $\pi/2$. The graph is sketched as shown.

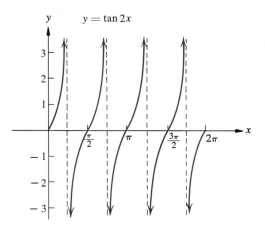

a. Range: $\{y \,|\, y \in R\}$

b. Since the equations of the asymptotes of $y = \tan x$ are

$$x = \frac{\pi}{2} + k\pi, \quad k \in J,$$

the equations of the asymptotes of $y = \tan 2x$ are

$$x = \frac{1}{2}\left(\frac{\pi}{2} + k\pi\right) = \frac{\pi}{4} + k \cdot \frac{\pi}{2}, \quad k \in J.$$

c. Zeros: $\left\{x \,|\, x = k \cdot \dfrac{\pi}{2}, k \in J\right\}$

1. $y = \tan 3x$

2. $y = \tan \dfrac{1}{2}x$

3. $y = \tan\left[x - \left(-\dfrac{\pi}{6}\right)\right]$

4. $y = \tan\left(x - \dfrac{\pi}{4}\right)$

5. $y = \cot \dfrac{1}{2} x$ **6.** $y = \cot 2x$

7. Without referring to Figure 3.11, graph $y = \sec x$, by using the fact that $\sec x = \dfrac{1}{\cos x}$, $x \neq \dfrac{\pi}{2} + k\pi$, $k \in J$.

8. Without referring to Figure 3.12, graph $y = \csc x$ by using the fact that $\csc x = \dfrac{1}{\sin x}$, $x \neq k\pi$, $k \in J$.

9. Use the graph obtained in Exercise 7 to write the equations of the asymptotes of the graph of $y = \sec x$.

10. Use the graph obtained in Exercise 8 to write the equations of the asymptotes of the graph of $y = \csc x$.

B

Graph each of the following over the interval $-\pi \leq x \leq 2\pi$.

11. $y = \dfrac{1}{2} \sec x$ **12.** $y = 2 \csc x$

13. $y = \csc 2x$ **14.** $y = \sec \dfrac{1}{2} x$

15. $y = \sec\left(x + \dfrac{\pi}{4}\right)$ **16.** $y = \csc\left(x - \dfrac{\pi}{6}\right)$

3.5 Graphing Systems of Equations

It is sometimes convenient to use systems of equations involving trigonometric function values to specify conditions on variables. As in algebraic equations, the solution of a system of equations is the intersection of the solution set of each equation in the system.

In this section we consider graphical methods of *approximating solutions* of such systems. In Chapter 4 we will consider algebraic methods to find such solutions.

Example   Solve the system

$$y = 2 \sin x$$

$$y = \dfrac{3}{2}$$

by graphical methods in the interval $0 \leq x \leq 2\pi$.   *(continued)*

Solution   Graph the solution set of each equation and approximate the coordinates of the points of intersection.

An approximation to the solution set is $\{(0.8, 1.5), (2.3, 1.5)\}$.

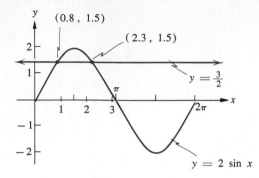

Intersection: $\{(0.8,\ 1.5), (2.3,\ 1.5)\}$

Example   Solve the system

$$y = 2 \sin x$$

$$y = \cos 2x$$

by graphical methods in the interval $0 \leq x \leq 2\pi$.

Solution   Graph each equation and approximate the coordinates of the points of intersection.

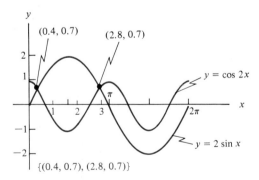

An approximation to the solution set is $\{(0.4, 0.7), (2.8, 0.7)\}$.

Parametric Equations

Sometimes it is convenient to define a relation (or function) in which the variable representing an element in the domain and the variable representing an element in the range are each related to a third variable, frequently a variable representing time or an angle. For example, the pair of equations

$$x = f(t) = t - 3 \quad \text{and} \quad y = g(t) = t + 1$$

express x in terms of t and y in terms of t. Such equations are called **parametric equations;** t is the **parameter.** To graph the relation between x and y as shown in the example below, we can choose arbitrary real number replacement values for t, and then obtain the corresponding values for x and y. Alternatively, by solving the system of equations, we can eliminate the parameter t to first obtain an equation in the variables x and y. Then we can graph the equation in the usual way.

Example
   a. Graph the function defined by the parametric equations

$$x = t - 3 \quad \text{and} \quad y = t + 1 \tag{1}$$

by assigning arbitrary values for t to obtain some ordered pairs (x, y) in the function.

   b. Eliminate t and find an equation in x and y.

Solution
   a. Choosing 0, 1, 2, and 3 as replacements for t, the corresponding ordered pairs (x, y) shown in the figure can be obtained. The construction of a table is sometimes helpful to obtain these ordered pairs.

t	x	y
0	-3	1
1	-2	2
2	-1	3
3	0	4

   b. Equations (1) can be written equivalently as

$$t = x + 3 \quad \text{and} \quad t = y - 1,$$

from which

$$x + 3 = y - 1 \quad \leftrightarrow \quad x - y = -4,*$$

the equation of a straight line in x and y.

* The symbol $\leftrightarrow$ should be read "is equivalent to."

The next example illustrates a case where the parameter is an angle.

Example
a. Graph the function defined by the parametric equations

$$x = 3 \cos \alpha \quad \text{and} \quad y = 4 \sin \alpha$$

by assigning arbitrary values for α to obtain some ordered pairs (x, y) in the function.

b. Eliminate α and find an equation in x and y.

Solution
a. Choosing selected elements of $\{\alpha \mid 0 \leq \alpha < 2\pi\}$ as replacements for α, we can obtain the corresponding ordered pairs (x, y) as shown in the figure. The tabulation of values is helpful.

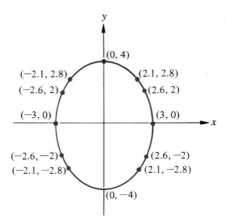

α	x $3 \cos \alpha$	y $4 \sin \alpha$
0	3	0
$\dfrac{\pi}{6}$	$\dfrac{3\sqrt{3}}{2} \approx 2.6$	2
$\dfrac{\pi}{4}$	$\dfrac{3}{\sqrt{2}} \approx 2.1$	$\dfrac{4}{\sqrt{2}} \approx 2.8$
$\dfrac{\pi}{2}$	0	4
$\dfrac{5\pi}{6}$	-2.6	2
$\vdots$	$\vdots$	$\vdots$
2π	3	0

b. Motivated by the fact that $\cos^2 \alpha + \sin^2 \alpha = 1$, we first solve the given equations for $\cos \alpha$ and $\sin \alpha$.

$$x = 3 \cos \alpha \quad \leftrightarrow \quad \cos \alpha = \frac{x}{3};$$

$$y = 4 \sin \alpha \quad \leftrightarrow \quad \sin \alpha = \frac{y}{4}.$$

Hence,

$$\cos^2 \alpha + \sin^2 \alpha = \frac{x^2}{9} + \frac{y^2}{16}.$$

Because $\cos^2 \alpha + \sin^2 \alpha = 1$,

$$\frac{x^2}{9} + \frac{y^2}{16} = 1,$$

from which

$$16x^2 + 9y^2 = 144.$$

EXERCISE SET 3.5

A

Solve each system by graphical methods over the interval $0 \le x \le 2\pi$.

Example $y = \sin x$

   $y = 2 \cos x$

Solution

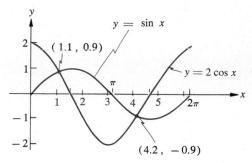

An approximation to the solution set is $\{(1.1, 0.9), (4.2, -0.9)\}$.

1. $y = \cos x$
   $y = 1$

2. $y = \sin 2x$
   $y = -1$

3. $y = \cos \frac{1}{2}x$
   $y = \frac{1}{2}$

4. $y = 2 \sin x$
   $y = -\frac{1}{2}$

5. $y = \tan x$
   $y = -1$

6. $y = \cot x$
   $y = \sqrt{3}$

7. $y = \sin 2x$
   $y = \frac{1}{2} \sin x$

8. $y = \cos 2x$
   $y = 2 \sin x$

9. $y = 2 \cos 2x$
   $y = \sin \frac{1}{2}x$

10. $y = \cos x$
   $y = 2 \cos \frac{1}{2}x$

In Exercises 11–18, (a) graph the curve represented by the parametric equations by assigning values to α, where $0 \le \alpha < 2\pi$, to obtain some ordered pairs (x, y), and (b) eliminate the parameter to find an equation in x and y.

Example $x = 2 \cos \alpha, \ y = \sin \alpha$

Solution

a.

α	x	y
0	2	0
$\dfrac{\pi}{4}$	1.4	0.7
$\dfrac{\pi}{2}$	0	1
$\dfrac{3\pi}{4}$	−1.4	0.7
π	−2	0
$\dfrac{5\pi}{4}$	−1.4	−0.7
$\dfrac{3\pi}{2}$	0	−1
$\dfrac{7\pi}{4}$	1.4	−0.7

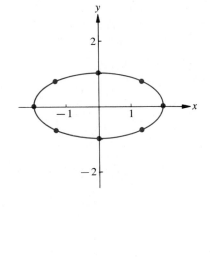

b. Solve for cos α and sin α:

$$\cos \alpha = \frac{x}{2} \quad \text{and} \quad \sin \alpha = y,$$

from which

$$\cos^2 \alpha = \frac{x^2}{4} \quad \text{and} \quad \sin^2 \alpha = y^2.$$

Since $\cos^2 \alpha + \sin^2 \alpha = 1$, we have

$$\frac{x^2}{4} + y^2 = 1,$$

from which

$$x^2 + 4y^2 = 4.$$

11. $x = \cos \alpha, \ y = \sin \alpha$ **12.** $x = 2 \cos \alpha, \ y = 2 \sin \alpha$

13. $x = 4 \cos \alpha, \ y = 3 \sin \alpha$ **14.** $x = 2 \cos \alpha, \ y = 5 \sin \alpha$

15. $x = 5 \cos \alpha, \ y = 2 \sin \alpha$ **16.** $x = 3 \cos \alpha, \ y = 6 \sin \alpha$

17. $x = -\cos \alpha, \ y = 3 \sin \alpha$ **18.** $x = 2 \cos \alpha, \ y = -\sin \alpha$

B

In Exercises 19 and 20 obtain one equation in x and y.

19. $\begin{cases} x = 2 \cot t \\ y = 2 \sin^2 t \end{cases}$ **20.** $\begin{cases} x = \sin t \\ y = (2 + \sin t) \cos^2 t \end{cases}$

21. A curve called the **cycloid** is determined by the equations

$$x = a(\alpha - \sin \alpha),$$
$$y = a(1 - \cos \alpha).$$

Graph these equations in an x, y coordinate system for $a = 1$ and $0 \le \alpha \le 2\pi$.

22. Graph the equations in Exercise 21 for $a = 2$ and for $a = 3$.

In Exercises 23 and 24, obtain one equation in x and y and sketch the graph over the interval $-\pi \le t \le \pi$.

23. $\begin{cases} x = \sec t \\ y = \cos t \end{cases}$ **24.** $\begin{cases} x = \sec t \\ y = \tan t \end{cases}$

In Exercises 25 and 26, obtain one equation that does not contain the parameter t.

25. $\begin{cases} x = a \cos t + h \\ y = b \sin t + k \end{cases}$ **26.** $\begin{cases} x = a \sec t + h \\ y = b \csc t + k \end{cases}$

3.6 Inverse Trigonometric Functions

We have observed from our work in Chapter 1 and the graphs of the trigonometric functions in this chapter that *for each element in the domain of each function there is exactly one element in the respective range.* However, since the trigonometric functions are periodic, *for each element in a range, there are an infinite number of elements in their respective domains.*

Figure 3.13

For example, consider the graph of $y = \cos x$ in Figure 3.13, and note that the *least positive x* for which

$$\cos x = \frac{1}{\sqrt{2}} \tag{1}$$

is $\pi/4$. However, each value of x such that

$$x = \frac{\pi}{4} + k \cdot 2\pi, \quad k \in J$$

is also a solution of Equation (1) because the cosine function has a period 2π.

Similar considerations are also applicable to the other trigonometric functions.

Recall that in Section 1.5, we restricted the values of x in the domains of each of the trigonometric functions

$$y = \text{trig } x$$

to be between 0 and $\pi/2$, so that only one value of x would be associated with each y in this interval and used the notation

$$x = \text{Trig}^{-1} y$$

to specify this unique value of x.

For example, $\text{Cos}^{-1}(1/\sqrt{2})$ represents the angle or real number between 0 and $\pi/2$ that satisfies Equation (1) above. Thus,

$$x = \text{Cos}^{-1} \frac{1}{\sqrt{2}} = 45°$$

or

$$x = \text{Cos}^{-1} \frac{1}{\sqrt{2}} = \frac{\pi}{4}.$$

In this section we are interested in obtaining unique elements in the domains of all the trigonometric functions for *all* elements in their ranges. We will do so by first considering the notion of "inverses of the trigonometric functions." You may find it helpful to first review Appendix A.6 for a discussion of inverses of algebraic functions.

Inverse of a Function

The inverse of a function is the relation obtained by interchanging the components of every ordered pair in the function. Hence, the defining equation of the inverse relation of a function can be obtained by interchanging the variables in the defining equation of the original function. Furthermore, the graph of the inverse of a function is the reflection of the graph of the function with respect to the graph of the equation $y = x$. Consider the graph of the function

$$y = \sin x,$$

with domain R and range $\{y \mid -1 \leq y \leq 1\}$, and its inverse

$$x = \sin y,$$

with domain $\{x \mid -1 \leq x \leq 1\}$ and range $y \in R$, in Figure 3.14.

Figure 3.14

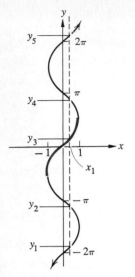

Figure 3.15

Notice that the graph of $x = \sin y$ is the reflection of the graph of $y = \sin x$ in the graph of the equation $y = x$. Further, notice that

$$x - \sin y$$

defines a relation that is not a function because for each element x in its domain, there is an unlimited number of elements y_1, y_2, etc., in its range as shown in Figure 3.15.

Inverse Functions

Each trigonometric function has an inverse relation. By suitably restricting the domain of the function or the range of the respective inverse relation, we can define an *inverse function* for each circular function. For example, if we restrict the domain of the sine function to $\{x \mid -\pi/2 \le x \le \pi/2\}$, then its inverse

$$x = \sin y, \quad -\frac{\pi}{2} \le y \le \frac{\pi}{2}, \tag{1}$$

defines an inverse function (there is only one y for each x in the domain). Its graph is shown in color in Figure 3.15.

Generally, we wish to write the defining Equation (1) in a form in which y is expressed explicitly in terms of x. This is done by using the inverse notation introduced in Section 1.5. For example,

$$x = \sin y, \quad -\frac{\pi}{2} \le y \le \frac{\pi}{2} \quad \leftrightarrow \quad y = \text{Arcsin } x \quad \leftrightarrow \quad y = \text{Sin}^{-1} x.$$

Similar notation is used for the other inverse circular functions, as shown in Definition 3.3 where the restrictions specified for y limit x to a single value.

Definition 3.3 *The inverse notation corresponding to the six inverse trigonometric functions is*

$$y = \text{Sin}^{-1} x \quad \leftrightarrow \quad x = \sin y, \quad -\frac{\pi}{2} \le y \le \frac{\pi}{2},$$

$$y = \text{Cos}^{-1} x \quad \leftrightarrow \quad x = \cos y, \quad 0 \le y \le \pi,$$

$$y = \text{Tan}^{-1} x \quad \leftrightarrow \quad x = \tan y, \quad -\frac{\pi}{2} < y < \frac{\pi}{2},$$

$$y = \text{Csc}^{-1} x \quad \leftrightarrow \quad x = \csc y, \quad -\frac{\pi}{2} \le y < 0 \quad \text{or} \quad 0 < y \le \frac{\pi}{2},$$

$$y = \mathrm{Sec}^{-1}\, x \quad \leftrightarrow \quad x = \sec y, \qquad 0 \le y < \frac{\pi}{2} \quad \text{or} \quad \frac{\pi}{2} < y \le \pi,$$

$$y = \mathrm{Cot}^{-1}\, x \quad \leftrightarrow \quad x = \cot y, \qquad -\frac{\pi}{2} \le y < 0 \quad \text{or} \quad 0 < y \le \frac{\pi}{2}.$$

Notice that the ranges (values of y) of the inverse functions are specified in the definition rather than following the usual procedure of stating the domain (values of x). Furthermore, the choices for the ranges are somewhat arbitrary. Those that have been chosen here involve small values of y and include values between 0 and $\pi/2$, have relatively simple graphs, and yield a one-to-one correspondence between elements in the domain and range. They are also the values that are obtained with most calculators.

For some purposes in the calculus, ranges for some of the inverse functions differ from those shown in Definition 3.3.

As we noted in Section 1.5, it is sometimes easier to interpret a relationship expressed in inverse notation if it is rewritten as an equation without using such notation.

Example   The equation $y = \mathrm{Arcsin}(-1/2)$ can be rewritten equivalently without inverse notation as

$$\sin y = -\frac{1}{2}, \quad -\frac{\pi}{2} \le y \le \frac{\pi}{2}.$$

This equation is satisfied for $y = -\pi/6$. Hence,

$$y = \mathrm{Arcsin}\left(-\frac{1}{2}\right) = -\frac{\pi}{6}.$$

The result could also be expressed in degree measure as $-30°$.

While a little practice you will be able to find an inverse function value directly from the expression involving the inverse notation.

Examples   Find the exact value of each of the following.

a. $\mathrm{Arccos}\, \dfrac{1}{\sqrt{2}}$ b. $\mathrm{Tan}^{-1}\, \dfrac{1}{\sqrt{3}}$ c. $\mathrm{Csc}^{-1}\, \dfrac{2}{\sqrt{3}}$

Solutions

a. From Definition 3.3, $\pi/4$ is the number between 0 and $\pi/2$ such that $\cos(\pi/4) = 1/\sqrt{2}$. Hence,

$$\mathrm{Arccos}\, \frac{1}{\sqrt{2}} = \frac{\pi}{4}.$$

(continued)

b. $\pi/6$ is the number between 0 and $\pi/2$ such that $\tan(\pi/6) = 1/\sqrt{3}$. Hence,

$$\text{Tan}^{-1}\frac{1}{\sqrt{3}} = \frac{\pi}{6}.$$

c. Because $\csc x = 1/\sin x$, we want a number x between 0 and $\pi/2$ such that $\csc x = 2/\sqrt{3}$ or $\sin x = \sqrt{3}/2$. Since $\pi/3$ is that number,

$$\text{Csc}^{-1}\frac{2}{\sqrt{3}} = \frac{\pi}{3}.$$

Careful attention should be given to the ranges specified in Definition 3.3 when finding values for $\text{Trig}^{-1} x$ for negative values of x.

Examples   Find the exact value of each of the following.

a. $\text{Arccos}\left(-\dfrac{1}{\sqrt{2}}\right)$ b. $\text{Tan}^{-1}\left(-\dfrac{1}{\sqrt{3}}\right)$ c. $\text{Csc}^{-1}\left(-\dfrac{2}{\sqrt{3}}\right)$

Solutions
a. Since $3\pi/4$ is the number between $\pi/2$ and π such that $\cos 3\pi/4 = -1/\sqrt{2}$,

$$\text{Arccos}\left(-\frac{1}{\sqrt{2}}\right) = \frac{3\pi}{4}.$$

b. Since $-\pi/6$ is the number between 0 and $-\pi/2$ such that $\tan(-\pi/6) = -1/\sqrt{3}$,

$$\text{Tan}^{-1}\left(-\frac{1}{\sqrt{3}}\right) = -\frac{\pi}{6}.$$

c. Since $-\pi/3$ is the number between 0 and $-\pi/2$ such that $\csc(-\pi/3) = -2/\sqrt{3}$,

$$\text{Csc}^{-1}\left(-\frac{2}{\sqrt{3}}\right) = -\frac{\pi}{3}.$$

A calculator (or tables) can be used to find approximations for values of the six trigonometric inverse functions for elements in their respective domains that are not listed in Table 1.1. Values for Arcsin x, Arccos x, and Arctan x can be found directly.

Examples Find an approximation for each of the following. Use a calculator in radian mode (or Table III in the appendix).

a. Arcsin 0.2377   b. $\text{Tan}^{-1}(-0.6989)$ c. $\sin(\text{Cos}^{-1} 0.8727)$

Solutions

a. Arcsin $0.2377 \approx 0.24$.

b. $\text{Tan}^{-1}(-0.6989) \approx -0.61$.

c. Since $\text{Cos}^{-1} 0.8727 \approx 0.51$,

$$\sin(\text{Cos}^{-1} 0.8727) \approx \sin 0.51 \approx 0.4882.$$

Values for Arcsec x, Arccsc x, and Arccot x cannot be obtained directly when using a calculator. However, we can obtain them by using the reciprocal relationships

$$\cos x = \frac{1}{\sec x}, \quad \sin x = \frac{1}{\csc x}, \quad \text{and} \quad \tan x = \frac{1}{\cot x}.$$

Examples Find an approximation for each of the following.

a. Arcsec 1.016   b. $\text{Cot}^{-1} 1.163$

Solutions

a. Arcsec $1.016 = \text{Arccos} \dfrac{1}{1.016}$ b. $\text{Cot}^{-1} 1.163 = \text{Tan}^{-1} \dfrac{1}{1.163}$

≈ 0.18. ≈ 0.71.

Writing Equivalent Equations

Inverse notation for the circular functions enables us to rewrite equations in one or more variables so that one of the variables can be expressed explicitly in terms of the other variables and constants.

Example Solve $4 \cos 3x = 1$ for x, using inverse notation. Assume restrictions specified in Definition 3.3.

Solution We can multiply each member of

$$4 \cos 3x = 1$$

by 1/4 to obtain

$$\cos 3x = \frac{1}{4}.$$

(*continued*)

By Definition 3.3,

$$\cos 3x = \frac{1}{4} \quad \leftrightarrow \quad 3x = \text{Cos}^{-1}\frac{1}{4}.$$

Then, multiplying each member by 1/3, we obtain

$$x = \frac{1}{3}\text{Cos}^{-1}\frac{1}{4}.$$

Of course, if we want to, we can now obtain a decimal approximation for x by first finding an approximation for $\text{Cos}^{-1}(1/4)$.

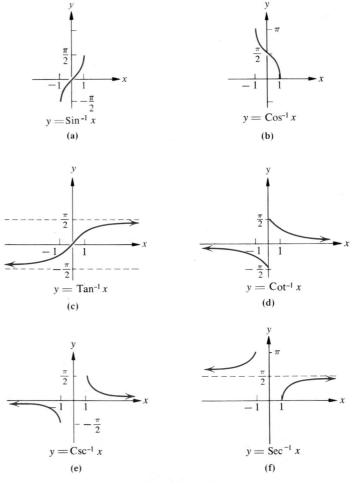

(a) $y = \text{Sin}^{-1} x$

(b) $y = \text{Cos}^{-1} x$

(c) $y = \text{Tan}^{-1} x$

(d) $y = \text{Cot}^{-1} x$

(e) $y = \text{Csc}^{-1} x$

(f) $y = \text{Sec}^{-1} x$

Figure 3.16

Example   Solve $y = \dfrac{1}{3} \tan \dfrac{x}{2}$ explicitly for x in terms of y using inverse notation. Assume restrictions specified in Definition 3.3.

Solution   We can multiply each member of

$$y = \frac{1}{3} \tan \frac{x}{2}$$

by 3 to obtain

$$3y = \tan \frac{x}{2} \quad \leftrightarrow \quad \frac{x}{2} = \text{Tan}^{-1} 3y.$$

Then, multiplying each member by 2, we obtain

$$x = 2\,\text{Tan}^{-1} 3y.$$

Graphs of the Inverse Trigonometric Functions

The graph of the Arcsine function (shown in color in Figure 3.15) and the graphs of the other inverse functions, which can be obtained by graphing ordered pairs or by reflecting the graphs of the respective functions over the appropriate domain, are shown in Figure 3.16.

EXERCISE SET 3.6

A

Rewrite each expression without using inverse notation and then specify the exact value of y for which the statement is true.

Example $y = \text{Arccot } 1$

Solution   From Definition 3.3,

$$y = \text{Arccot } 1 \quad \leftrightarrow \quad \cot y = 1, \quad 0 < y \le \frac{\pi}{2}.$$

Hence, $y = \pi/4$.

1. $y = \text{Arccos } \dfrac{\sqrt{3}}{2}$ **2.** $y = \text{Arctan } 1$ **3.** $y = \text{Sin}^{-1} 1$

4. $y = \text{Csc}^{-1} 2$ **5.** $y = \text{Cot}^{-1} \dfrac{1}{\sqrt{3}}$ **6.** $y = \text{Sec}^{-1} \sqrt{2}$

Find the exact value of each of the following.

Example $\mathrm{Cos}^{-1}\left(-\dfrac{\sqrt{3}}{2}\right)$

Solution   Let $y = \mathrm{Cos}^{-1}(-\sqrt{3}/2)$. Then, from Definition 3.3,

$$\cos y = -\frac{\sqrt{3}}{2}, \quad 0 \le y \le \pi,$$

from which

$$y = \mathrm{Cos}^{-1}\left(-\frac{\sqrt{3}}{2}\right) = \frac{5\pi}{6}.$$

7. $\mathrm{Arcsin}\ \dfrac{1}{\sqrt{2}}$ **8.** $\mathrm{Arccos}\ \dfrac{\sqrt{3}}{2}$ **9.** $\mathrm{Arctan}\ 1$

10. $\mathrm{Arcsin}\ 0$ **11.** $\mathrm{Tan}^{-1}\ \sqrt{3}$ **12.** $\mathrm{Cos}^{-1}\ 0$

13. $\mathrm{Cos}^{-1}\ 1$ **14.** $\mathrm{Tan}^{-1}\ 0$ **15.** $\mathrm{Sec}^{-1}\ \dfrac{2}{\sqrt{3}}$

16. $\mathrm{Csc}^{-1}\ 1$ **17.** $\mathrm{Cot}^{-1}\ 1$ **18.** $\mathrm{Sec}^{-1}\ 1$

19. $\mathrm{Sin}^{-1}\left(-\dfrac{1}{2}\right)$ **20.** $\mathrm{Cos}^{-1}(-1)$ **21.** $\mathrm{Tan}^{-1}(-\sqrt{3})$

22. $\mathrm{Tan}^{-1}(-1)$ **23.** $\mathrm{Cos}^{-1}\left(-\dfrac{1}{2}\right)$ **24.** $\mathrm{Sin}^{-1}(-1)$

Find an approximation to the nearest hundredth for each of the following. Use a calculator in radian mode (or Table III in the appendix).

Examples   a. $\mathrm{Sin}^{-1}\ 0.8772$ b. $\mathrm{Arcsec}\ 1.238$

Solutions   a. $\mathrm{Sin}^{-1}\ 0.8772 \approx 1.07$.   b. $\mathrm{Arcsec}\ 1.238 = \mathrm{Arccos}\ \dfrac{1}{1.238}$

$$\approx 0.63.$$

25. $\mathrm{Arctan}\ 0.5726$ **26.** $\mathrm{Arcsin}\ 0.6442$ **27.** $\mathrm{Arccos}\ 0.8961$

28. $\mathrm{Arctan}\ 0.1104$ **29.** $\mathrm{Sin}^{-1}\ 0.3523$ **30.** $\mathrm{Cos}^{-1}\ 0.6600$

31. $\mathrm{Tan}^{-1}\ 2.912$ **32.** $\mathrm{Sin}^{-1}\ 0.8724$ **33.** $\mathrm{Sec}^{-1}\ 8.299$

34. $\mathrm{Cot}^{-1}\ 0.6563$ **35.** $\mathrm{Csc}^{-1}\ 1.077$ **36.** $\mathrm{Sec}^{-1}\ 1.004$

37. $\text{Tan}^{-1}(-0.5463)$ **38.** $\text{Sin}^{-1}(-0.6518)$ **39.** $\text{Cos}^{-1}(-0.9759)$

40. $\text{Cos}^{-1}(-0.3153)$ **41.** $\text{Sin}^{-1}(-0.9927)$ **42.** $\text{Tan}^{-1}(-1.827)$

Rewrite each equation so that x is expressed explicitly in terms of y and/or the constants using inverse notation. Assume restrictions specified in Definition 3.3.

Example $3 \sin 2x = 1$

Solution   Multiplying each member by $1/3$ yields $\sin 2x = 1/3$. By Definition 3.3,

$$\sin 2x = \frac{1}{3} \quad \leftrightarrow \quad 2x = \text{Sin}^{-1}\frac{1}{3}.$$

Then, multiplying each member by $1/2$,

$$x = \frac{1}{2}\text{Sin}^{-1}\frac{1}{3}.$$

43. $4 \cos x = 3$ **44.** $2 \tan x = 3$ **45.** $\sin 3x = \dfrac{1}{4}$

46. $\cos 2x = \dfrac{1}{3}$ **47.** $2 \tan 2x = 5$ **48.** $3 \sin 4x = 2$

49. $2 \cos 3x = 1.24$ **50.** $5 \tan 2x = 2.25$ **51.** $y = 2 \tan 3x$

52. $y = 3 \cos 2x$ **53.** $y = 4 \sin \dfrac{x}{2}$ **54.** $y = 2 \tan \dfrac{x}{3}$

Find the exact value of each of the following.

Example $\cos\left(\text{Sin}^{-1}\dfrac{1}{2}\right)$

Solution $\text{Sin}^{-1}\left(\dfrac{1}{2}\right) = \dfrac{\pi}{6}$; therefore

$$\cos\left(\text{Sin}^{-1}\frac{1}{2}\right) = \cos\frac{\pi}{6} = \frac{\sqrt{3}}{2}.$$

55. $\cos\left(\text{Cos}^{-1}\dfrac{\sqrt{3}}{2}\right)$ **56.** $\tan(\text{Tan}^{-1} 1)$

57. $\tan(\text{Sin}^{-1} 0)$ **58.** $\cos(\text{Sin}^{-1} 1)$

59. $\sin[\text{Arctan}(-1)]$

60. $\sin\left[\text{Cos}^{-1}\left(-\dfrac{1}{2}\right)\right]$

61. $\text{Sin}^{-1}\left(\sin\dfrac{\pi}{4}\right)$

62. $\text{Cos}^{-1}\left(\cos\dfrac{\pi}{3}\right)$

63. $\text{Cos}^{-1}\left[\sin\left(\dfrac{-\pi}{3}\right)\right]$

64. $\text{Sin}^{-1}\left[\cos\left(\dfrac{-\pi}{6}\right)\right]$

65. $\text{Tan}^{-1}\left(\cos\dfrac{\pi}{2}\right)$

66. $\text{Csc}^{-1}(\cos \pi)$

67. $\text{Arcsec}\left[\tan\left(\dfrac{-3\pi}{4}\right)\right]$

68. $\text{Arccot}\left[\sin\left(\dfrac{-3\pi}{2}\right)\right]$

Find an approximation for each of the following.

69. $\cot(\text{Sin}^{-1} 0.8134)$

70. $\csc(\text{Arccot } 4.933)$

71. $\sec(\text{Cot}^{-1} 0.4475)$

72. $\cot(\text{Arcsec } 1.005)$

73. $\csc(\text{Arcsec } 2.760)$

74. $\sec(\text{Csc}^{-1} 2.452)$

B

Write each of the following as a numerical, or an algebraic, expression without using inverse notation. Assume values for the variables for which each function value exists.

Example $\sin\left(\text{Arccos}\dfrac{3}{5}\right)$

Solution   Because the value of Arccos 3/5 is not evident by inspection, we let $\beta = \text{Arccos } 3/5$, from which $\cos \beta = 3/5$. The sketch shows a right triangle in which $\cos \beta = 3/5$. From the Pythagorean theorem, we have

$$y^2 + 3^2 = 5^2$$
$$y^2 = 25 - 9 = 16$$
$$y = 4.$$

Hence,

$$\sin \beta = \sin\left(\text{Arccos}\dfrac{3}{5}\right)$$

$$= \dfrac{y}{5} = \dfrac{4}{5}.$$

75. $\cos\left(\text{Arcsin }\dfrac{3}{5}\right)$

76. $\sin\left(\text{Arccos }\dfrac{5}{13}\right)$

77. $\sin\left(\text{Arcsin }\dfrac{2}{5}\right)$

78. $\cos\left(\text{Arccos }\dfrac{1}{3}\right)$

79. $\tan\left(\text{Arccos }\dfrac{4}{5}\right)$

80. $\tan\left(\text{Arcsin }\dfrac{5}{13}\right)$

81. $\cos\left(\text{Arcsin }\dfrac{y}{3}\right)$

82. $\tan\left(\text{Arccos }\dfrac{x}{4}\right)$

83. $\sin(\text{Arccos } x)$

84. $\sin(\text{Arctan } y)$

Chapter Summary

3.1 Graphs of the form of the graph of the sine function are called **sine waves** or **sinusoidal waves.** The part of the graph over one period is called a **cycle** of the wave. The absolute value of half the difference of the maximum and the minimum ordinates is called the **amplitude** of the wave.

Zeros of the sine function are $k\pi$, $k \in J$; zeros of the cosine function are $(\pi/2) + k\pi$, $k \in J$. These values are associated with the points where the graphs of $y = \sin x$ and $y = \cos x$ intersect the x-axis. The range of both functions is $\{y \mid -1 \le y \le 1\}$.

3.2 Functions defined by equations of the form

$$y = A \sin B(x - C) \qquad \text{or} \qquad y = A \cos B(x - C),$$

where $A, B, C, x, y \in R$ and $A, B \ne 0$, always have sine waves for their graphs. The **amplitude** of the wave or function is $|A|$, the **period** is $2\pi/|B|$, and the **phase shift** is $|C|$.

3.3 Functions defined by equations of the form

$$y = A_1 \sin B_1 x + A_2 \cos B_2 x$$

can be graphed by first graphing $y_1 = A_1 \sin B_1 x$ and $y_2 = A_2 \cos B_2 x$ and then "adding" their ordinates graphically for arbitrary values of x.

3.4 Graphs of trigonometric functions defined by $y = \tan x$, $y = \cot x$, $y = \sec x$, and $y = \csc x$ exhibit the periodic nature of these functions. The table on page 122 shows the zeros and/or the equations for the asymptotes of these functions.

Function	Zeros	Equations for Asymptotes	Range
tangent	$k\pi,\ k \in J$	$x = \dfrac{\pi}{2} + k\pi,\ k \in J$	$\{y \mid y \in R\}$
cotangent	$\dfrac{\pi}{2} + k\pi,\ k \in J$	$x = k\pi,\ k \in J$	$\{y \mid y \in R\}$
secant	none	$x = \dfrac{\pi}{2} + k\pi,\ k \in J$	$\{y \mid y \leq -1 \text{ or } y \geq 1\}$
cosecant	none	$x = k\pi,\ k \in J$	$\{y \mid y \leq -1 \text{ or } y \geq 1\}$

3.5 Solutions of systems of equations involving trigonometric function values can be approximated by graphical methods.

A pair of equations of the form

$$x = f(\alpha) \qquad \text{and} \qquad y = g(\alpha),$$

in which two variables, in this case x and y, are each expressed in terms of a third variable, are called **parametric equations**. The relation between x and y can be graphed by arbitrarily choosing real number replacements for the third variable, in this case α, and then obtaining the corresponding values for x and y.

3.6 The **inverse of a function** is the relation obtained by interchanging the components of every ordered pair in the function.

By suitably restricting the domains of the trigonometric functions, or the ranges of the respective inverse relations, an inverse function is defined for each trigonometric function (see Definition 3.3).

Review Exercises

3.1–3.2 *Sketch the graph of the function defined by each equation over the interval* $-2\pi \leq x \leq 2\pi$. *Specify* (**a**) *the amplitude,* (**b**) *the period, and* (**c**) *the phase shift* (*if applicable*).

1. $y = 4 \sin x$

2. $y = \dfrac{1}{2} \cos 2x$

3. $y = \sin\left(x + \dfrac{\pi}{2}\right)$

4. $y = 2 \cos\left(\dfrac{1}{2}x - \dfrac{\pi}{6}\right)$

Graph two cycles of each of the following.

5. $y = \cos \dfrac{\pi}{4} x$

6. $y = 2 \sin \dfrac{\pi}{3} x$

3.3 *Using the method of graphical addition of ordinates, graph the equations over the interval* $0 \le x \le 2\pi$.

7. $y = \sin 2x - \dfrac{1}{2} \cos x$ **8.** $y = \dfrac{1}{3} x - \sin x$

3.4 *Sketch the graph of the function defined by each equation over the interval* $0 \le x \le 2\pi$. *Specify* **(a)** *the range,* **(b)** *the equations of the asymptotes, and* **(c)** *the set of all zeros of the function (if applicable).*

9. $y = \tan\left(x - \dfrac{\pi}{3}\right)$ **10.** $y = \cot 3x$

11. $y = \csc \dfrac{1}{2} x$ **12.** $y = 2 \sec x$

3.5 *Find approximations to solutions of each system over the interval* $0 \le x \le 2\pi$ *by graphical methods.*

13. $y = \cos x$ **14.** $y = 2 \sin x$

$y = \dfrac{1}{3}$ $y = \dfrac{3}{4}$

15. $y = \sin 2x$ **16.** $y = \cos 2x$

$y = \dfrac{1}{2} \cos x$ $y = \dfrac{1}{2} \sin x$

17. Graph $x = 3 \cos \alpha$ and $y = 3 \sin \alpha$ in an xy system.

18. Graph $x = 3 \cos \alpha$ and $y = 4 \sin \alpha$ in an xy system.

19. Eliminate the parameter in the equations in Exercise 17 to find an equation in x and y.

20. Eliminate the parameter in the equations in Exercise 18 to find an equation in x and y.

3.6 *Rewrite each expression without using inverse notation and then specify the exact value of y for which the statement is true.*

21. $y = \text{Arccos} \dfrac{1}{\sqrt{2}}$ **22.** $y = \text{Arccot}(-1)$

23. $y = \text{Csc}^{-1}(-2)$ **24.** $y = \text{Sec}^{-1} \dfrac{2}{\sqrt{3}}$

Find the exact value of each expression.

25. Arcsin 1

26. $\text{Arccot}(-\sqrt{3})$

27. $\text{Cos}^{-1}\left(-\dfrac{1}{\sqrt{2}}\right)$

28. $\text{Tan}^{-1}\dfrac{1}{\sqrt{3}}$

Find an approximation to the nearest hundredth for each of the following.

29. Arccos 0.1205

30. Arctan 1.398

31. $\text{Sin}^{-1}\,0.4529$

32. $\text{Cot}^{-1}(-0.6142)$

Rewrite each equation using inverse notation so that x is expressed explicitly in terms of y and/or the constants. Assume restrictions specified in Definition 3.3.

33. $5\cos x = 3$

34. $\sin 2x = \dfrac{1}{5}$

35. $y = 3\tan 2x$

36. $y = 2\cos\dfrac{x}{2}$

Find the exact value of each of the following.

37. $\tan\left(\text{Cos}^{-1}\dfrac{1}{2}\right)$

38. $\cos(\text{Arccot}\,\sqrt{3})$

39. $\text{Sin}^{-1}\left(\tan\dfrac{\pi}{4}\right)$

40. $\text{Tan}^{-1}(\cos 0)$

4 IDENTITIES AND CONDITIONAL EQUATIONS

Equations such as

$$a \cdot 1 = a, \quad a + b = b + a, \quad \text{and} \quad a(b + c) = ab + ac,$$

which are true for all real numbers a, b, and c, are called **identities;** *the left-hand member and the right-hand member of each equation are equivalent expressions.** In algebra, we have used such identities to simplify algebraic expressions whenever we substituted one expression for an equivalent expression.

In this chapter we will review some basic identities that involve trigonometric (or circular) function values that have been introduced in previous chapters and we will introduce additional ones that will be useful in our work. We will then use these identities to rewrite expressions that involve trigonometric function values as equivalent expressions.

In our work with identities we shall not explicitly specify the replacement set of the variable in each case. The variables α, β, γ, and θ will represent elements of the set of all angles, and the variables x and y will be considered as representing elements either of the set of all angles or of the set of real numbers. Furthermore, it should be understood that *an equation that is specified to be an identity without noting the restrictions is in fact an identity only for all replacements of the variables for which both members of the equation are defined.*

---

* Also see Appendix A.3.

4.1 Basic Trigonometric Identities

The trigonometric identities

$$\csc \alpha = \frac{1}{\sin \alpha}, \quad \sec \alpha = \frac{1}{\cos \alpha}, \quad \cot \alpha = \frac{1}{\tan \alpha},$$

and

$$\cos^2 \alpha + \sin^2 \alpha = 1,$$

stated in Chapter 1, have been shown to be consequences of the definition of the trigonometric ratios. See Definition 1.3 on page 9.

Two other basic identities also follow directly from Definition 1.3. Since $\sin \alpha = y/r$ and $\cos \alpha = x/r$ $(r \neq 0)$,

$$\frac{\sin \alpha}{\cos \alpha} = \frac{\dfrac{y}{r}}{\dfrac{x}{r}} = \frac{y}{x}, \quad (x \neq 0).$$

By Definition 1.3, $\tan \alpha = y/x$. Hence,

$$\tan \alpha = \frac{\sin \alpha}{\cos \alpha}$$

for those values of α for which $\cos \alpha \neq 0$. Because $\cot \alpha = x/y$,

$$\frac{\cos \alpha}{\sin \alpha} = \frac{\dfrac{x}{r}}{\dfrac{y}{r}} = \frac{x}{y}, \quad (y \neq 0),$$

$$\cot \alpha = \frac{\cos \alpha}{\sin \alpha},$$

for those values of α for which $\sin \alpha \neq 0$.

Two other identities follow from the identity $\sin^2 \alpha + \cos^2 \alpha = 1$. For any α for which $\cos \alpha \neq 0$, we multiply each member by $1/\cos^2 \alpha$ to obtain the equivalent equations

$$\frac{\sin^2 \alpha}{\cos^2 \alpha} + \frac{\cos^2 \alpha}{\cos^2 \alpha} = \frac{1}{\cos^2 \alpha},$$

$$\tan^2 \alpha + 1 = \sec^2 \alpha.$$

Multiplying each member of $\sin^2 \alpha + \cos^2 \alpha = 1$ by $1/\sin^2 \alpha$ for $\sin \alpha \neq 0$, we obtain

$$\frac{\sin^2 \alpha}{\sin^2 \alpha} + \frac{\cos^2 \alpha}{\sin^2 \alpha} = \frac{1}{\sin^2 \alpha},$$

$$\mathbf{1 + \cot^2 \alpha = \csc^2 \alpha.}$$

The basic identities that we have stated above should be recognized in equivalent forms. For example, using the symbol $\leftrightarrow$ we have

$$\sec \alpha = \frac{1}{\cos \alpha} \quad \leftrightarrow \quad (\cos \alpha)(\sec \alpha) = 1 \quad \leftrightarrow \quad \cos \alpha = \frac{1}{\sec \alpha};$$

$$\csc \alpha = \frac{1}{\sin \alpha} \quad \leftrightarrow \quad (\sin \alpha)(\csc \alpha) = 1 \quad \leftrightarrow \quad \sin \alpha = \frac{1}{\csc \alpha};$$

$$\cot \alpha = \frac{1}{\tan \alpha} \quad \leftrightarrow \quad (\tan \alpha)(\cot \alpha) = 1 \quad \leftrightarrow \quad \tan \alpha = \frac{1}{\cot \alpha};$$

$$\sin^2 \alpha + \cos^2 \alpha = 1 \quad \leftrightarrow \quad \sin^2 \alpha = 1 - \cos^2 \alpha \quad \leftrightarrow \quad \cos^2 \alpha = 1 - \sin^2 \alpha;$$

$$\tan^2 \alpha + 1 = \sec^2 \alpha \quad \leftrightarrow \quad \tan^2 \alpha = \sec^2 \alpha - 1;$$

$$\cot^2 \alpha + 1 = \csc^2 \alpha \quad \leftrightarrow \quad \cot^2 \alpha = \csc^2 \alpha - 1.$$

Identities for $\sin(-\alpha)$, $\cos(-\alpha)$, and $\tan(-\alpha)$

Several other useful trigonometric relationships can be obtained from geometric considerations. In Figure 4.1a,

$$\angle AOB = \angle COB, \quad OB = OB, \quad \text{and} \quad \angle OBA = \angle OBC.$$

Hence,

$$\triangle AOB \cong \triangle COB,$$

from which we have that each pair of corresponding sides are congruent and the absolute values of the respective components of the ordered pairs corresponding to points A and C are equal. Because

$$\sin(-\alpha) = \frac{-y}{r} \quad \text{and} \quad \sin \alpha = \frac{y}{r},$$

$$\mathbf{\sin(-\alpha) = -\sin \alpha.} \tag{1}$$

Also, $\cos(-\alpha) = x/r$ and $\cos \alpha = x/r$, from which

$$\mathbf{\cos(-\alpha) = \cos \alpha,} \tag{2}$$

(a)

Figure 4.1

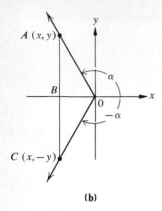

(b)

Figure 4.1

and $\tan(-\alpha) = -y/x$ and $\tan \alpha = y/x$, from which

$$\tan(-\alpha) = -\tan \alpha. \tag{3}$$

A similar argument for angles α and $-\alpha$ that are in Quadrants II and III, as shown in Figure 4.1b, also leads to the results stated by Equations (1), (2), and (3).

Because identities we have introduced are simply equations in which the left member and right member are equivalent expressions and the function values in these expressions are real numbers, we can sometimes simplify expressions that involve trigonometric function values by using these identities and applying the familiar algebraic properties for real numbers.

Examples   Write each expression in terms of sine and cosine and simplify.

a. $(\tan \alpha)(\cos^2 \alpha)$ b. $\dfrac{\cos \beta}{\cot^2 \beta}$ c. $\tan(-\gamma) + \sec(-\gamma)$

Solutions

   a. Substitute $\sin \alpha/\cos \alpha$ for $\tan \alpha$ and simplify.

$$(\tan \alpha)(\cos^2 \alpha) = \frac{\sin \alpha}{\cos \alpha} \cdot \cos^2 \alpha$$

$$= \sin \alpha \cdot \cos \alpha$$

   b. Substitute $\cos^2 \beta/\sin^2 \beta$ for $\cot^2 \beta$ and simplify.

$$\frac{\cos \beta}{\cot^2 \beta} = \frac{\cos \beta}{\dfrac{\cos^2 \beta}{\sin^2 \beta}}$$

$$= \cos \beta \cdot \frac{\sin^2 \beta}{\cos^2 \beta} = \frac{\sin^2 \beta}{\cos \beta}$$

   c. Substitute $\sin(-\gamma)/\cos(-\gamma)$ for $\tan(-\gamma)$ and $1/\cos(-\gamma)$ for $\sec(-\gamma)$ and simplify.

$$\tan(-\gamma) + \sec(-\gamma) = \frac{\sin(-\gamma)}{\cos(-\gamma)} + \frac{1}{\cos(-\gamma)}$$

$$= \frac{-\sin \gamma}{\cos \gamma} + \frac{1}{\cos \gamma}$$

$$= \frac{1 - \sin \gamma}{\cos \gamma}$$

EXERCISE SET 4.1

A

Use the identities of this section to express each of the following in terms of a single function value.

Examples   a. $1 - \sin^2 \alpha$ b. $\dfrac{\cos^2 \alpha}{\sin^2 \alpha}$

Solutions

a. Since $\sin^2 \alpha + \cos^2 \alpha = 1,$

$$1 - \sin^2 \alpha = \cos^2 \alpha.$$

b. Since $\dfrac{\cos \alpha}{\sin \alpha} = \cot \alpha,$

$$\frac{\cos^2 \alpha}{\sin^2 \alpha} = \cot^2 \alpha.$$

1. $1 + \tan^2 \alpha$ **2.** $1 + \cot^2 \alpha$ **3.** $1 - \cos^2 \alpha$

4. $\sec^2 \alpha - 1$ **5.** $\cos^2 \alpha - 1$ **6.** $1 - \sec^2 \alpha$

7. $\dfrac{\sin^3 \alpha}{\cos^3 \alpha}$ **8.** $\dfrac{\cos^4 \alpha}{\sin^4 \alpha}$

Use the identities of this section to write each expression in terms of sine and cosine and simplify.

Examples   a. $(\sin \alpha)(\cot^2 \alpha)(\sec^2 \alpha)$ b. $\dfrac{\cot^2 x}{\cos x}$

Solutions

a. Substituting $\cos \alpha / \sin \alpha$ for $\cot \alpha$ and $1/\cos \alpha$ for $\sec \alpha$ yields

$$(\sin \alpha)(\cot^2 \alpha)(\sec^2 \alpha) = (\sin \alpha)\left(\frac{\cos^2 \alpha}{\sin^2 \alpha}\right)\left(\frac{1}{\cos^2 \alpha}\right)$$

$$= \frac{1}{\sin \alpha}.$$

b. Since $\cot x = \cos x / \sin x,$

$$\frac{\cot^2 x}{\cos x} = \frac{\cos^2 x / \sin^2 x}{\cos x}$$

$$= \frac{\cos^2 x}{\sin^2 x} \cdot \frac{1}{\cos x} = \frac{\cos x}{\sin^2 x}.$$

9. $(\sec \alpha)(\cot \alpha)$

10. $(\csc \alpha)(\tan \alpha)$

11. $(\cot \alpha)(\csc \alpha)$

12. $(\tan \alpha)(\sec \alpha)$

13. $1 - \tan^2 \alpha$

14. $1 - \cot^2 \alpha$

15. $1 - \sec^2 \alpha$

16. $1 - \csc^2 \alpha$

17. $(\csc^2 \beta)(\tan \beta)$

18. $(\sec^2 \beta)(\cot \beta)$

19. $(\csc^2 \beta)(\cot \beta)$

20. $(\sec^2 \beta)(\tan \beta)$

21. $\dfrac{\sec \gamma}{\tan \gamma}$

22. $\dfrac{\csc \gamma}{\cot \gamma}$

23. $\dfrac{\cot \gamma}{\sec \gamma}$

24. $\dfrac{\tan \gamma}{\csc \gamma}$

25. $\dfrac{\cot^2 x}{\csc^2 x}$

26. $\dfrac{\tan^2 x}{\sec^2 x}$

27. $\dfrac{\sec^2 x}{\cot^2 x}$

28. $\dfrac{\csc^2 x}{\tan^2 x}$

29. $\tan y + \cot y$

30. $\sec y + \csc y$

31. $\tan^2 \theta + \cot^2 \theta$

32. $\sec^2 \theta + \csc^2 \theta$

33. $\sec \theta + \dfrac{\csc^2 \theta}{\sec \theta}$

34. $\csc \theta + \dfrac{\sec^2 \theta}{\csc \theta}$

35. $(\sin \alpha)(\cot^2 \alpha)(\csc^2 \alpha)$

36. $(\cos \alpha)(\tan^2 \alpha)(\sec^2 \alpha)$

37. $\tan \alpha - \dfrac{\cot^2 \alpha}{\tan \alpha}$

38. $\sec \alpha - \dfrac{\csc^2 \alpha}{\sec \alpha}$

Examples   a. $\cos(-\alpha) \cdot \tan(-\alpha)$ b. $\cot(-\alpha) \cdot \sin \alpha$

Solutions

a. Since $\cos(-\alpha) = \cos \alpha$ and $\tan(-\alpha) = -\tan \alpha$,

$$\cos(-\alpha) \cdot \tan(-\alpha) = \cos \alpha \cdot (-\tan \alpha)$$

$$= \cos \alpha \cdot \left(\frac{-\sin \alpha}{\cos \alpha} \right)$$

$$= -\sin \alpha.$$

b. Since $\cot(-\alpha) = \dfrac{\cos(-\alpha)}{\sin(-\alpha)}$, $\cos(-\alpha) = \cos \alpha$, and $\sin(-\alpha) = -\sin \alpha$,

$$\cot(-\alpha) \cdot \sin \alpha = \frac{\cos \alpha}{-\sin \alpha} \cdot \sin \alpha$$

$$= -\cos \alpha.$$

39. $\csc(-\alpha) \cdot \tan \alpha$

40. $\sec(-\alpha) \cdot \cot(-\alpha)$

41. $\sec^2(-\beta) \cdot \cot(-\beta)$

42. $\csc^2 \beta \cdot \tan(-\beta)$

43. $\dfrac{\csc x}{\cot(-x)}$

44. $\dfrac{\sec(-x)}{\tan(-x)}$

45. $\cot(-y) + \tan(-y)$

46. $\csc(-y) + \sec y$

B

Simplify each expression.

47. $\dfrac{(\cos \alpha + \sin \alpha)^2}{\cos \alpha - \sin \alpha} - \dfrac{1}{\dfrac{\cos \alpha - \sin \alpha}{2 \cos \alpha \sin \alpha}}$

48. $\dfrac{\sin \alpha}{\dfrac{\sin \alpha}{\sin \alpha + \cos \alpha}} - \dfrac{2 \sin \alpha \cos \alpha}{\sin \alpha + \cos \alpha}$

49. $\dfrac{\dfrac{1}{\dfrac{\sin \alpha - \cos \alpha}{\sin \alpha}}}{\sin^2 \alpha - \cos^2 \alpha}$

50. $\dfrac{\dfrac{1}{\sin^2 \alpha} + \dfrac{1}{\cos^2 \alpha}}{\dfrac{1}{\sin \alpha \cos \alpha}} - \dfrac{\cos \alpha}{\sin \alpha}$

4.2 Proving Identities

In Section 4.1 we introduced some basic identities that we used to rewrite and simplify expressions that involved trigonometric function values. In Section 4.1 we always expressed our results in terms of sine and cosine function values. Sometimes we may also want to rewrite trigonometric expressions in terms of other function values or constants.

Example   Show that $\csc \alpha \cos \alpha$ is equivalent to $\cot \alpha$.

Solution   Since $\csc \alpha = 1/\sin \alpha$,

$$\csc \alpha \cos \alpha = \frac{1}{\sin \alpha} \cdot \cos \alpha$$

$$= \frac{\cos \alpha}{\sin \alpha}.$$

Since $\cos \alpha / \sin \alpha = \cot \alpha$, we have that

$$\csc \alpha \cos \alpha = \cot \alpha.$$

It is customary to express the statement in the above example as follows:

"Show that $\csc \alpha \cos \alpha = \cot \alpha$ is an identity."

Of course for this identity the proof is exactly as given in the above solution.

Let us consider another example.

Example   Show that

$$\frac{\cos \alpha - \sin \alpha}{\cos \alpha} = 1 - \tan \alpha \qquad (1)$$

is an identity.

Solution   We shall rewrite the right-hand member in terms of sines and cosines. Substituting $\sin \alpha / \cos \alpha$ for $\tan \alpha$ in the right-hand member,

$$\frac{\cos \alpha - \sin \alpha}{\cos \alpha} = 1 - \frac{\sin \alpha}{\cos \alpha}$$

Writing the right-hand member as a single term yields

$$\frac{\cos \alpha - \sin \alpha}{\cos \alpha} = \frac{\cos \alpha - \sin \alpha}{\cos \alpha},$$

which is clearly an identity.

Example   Show that

$$1 + \tan^2 \alpha = \frac{1}{\cos^2 \alpha} \qquad (2)$$

is an identity.

Solution   We shall rewrite the left-hand member in terms of sines and cosines. Substituting $\sin^2 \alpha / \cos^2 \alpha$ for $\tan^2 \alpha$ in the left-hand member, we obtain

$$1 + \frac{\sin^2 \alpha}{\cos^2 \alpha} = \frac{1}{\cos^2 \alpha}.$$

Writing the left-hand member as a single term yields

$$\frac{\cos^2 \alpha + \sin^2 \alpha}{\cos^2 \alpha} = \frac{1}{\cos^2 \alpha}.$$

Since $\cos^2 \alpha + \sin^2 \alpha$ equals 1, we obtain

$$\frac{1}{\cos^2 \alpha} = \frac{1}{\cos^2 \alpha}$$

which is clearly an identity.

Sometimes it is helpful to express *both* members of an identity in terms of sine and cosines.

Example   Show that

$$\cot \alpha + \tan \alpha = \csc \alpha \sec \alpha \tag{3}$$

is an identity.

Solution   Substituting $\cos \alpha / \sin \alpha$ for $\cot \alpha$ and $\sin \alpha / \cos \alpha$ for $\tan \alpha$ in the left-hand member,

$$\frac{\cos \alpha}{\sin \alpha} + \frac{\sin \alpha}{\cos \alpha} = \csc \alpha \sec \alpha,$$

$$\frac{\cos^2 \alpha + \sin^2 \alpha}{\sin \alpha \cos \alpha} = \csc \alpha \sec \alpha,$$

$$\frac{1}{\sin \alpha \cos \alpha} = \csc \alpha \sec \alpha,$$

Substituting $1/\sin \alpha$ for $\csc \alpha$ and $1/\cos \alpha$ for $\sec \alpha$ in the right-hand member,

$$\frac{1}{\sin \alpha \cos \alpha} = \frac{1}{\sin \alpha} \cdot \frac{1}{\cos \alpha},$$

$$\frac{1}{\sin \alpha \cos \alpha} = \frac{1}{\sin \alpha \cos \alpha}.$$

The two members are identical, and the original equation is an identity.

In the above examples members of the equations were written in terms of sine and cosine function values to establish that the equations were indeed identities. Although there is no general method to prove an identity, as shown in the above examples there are three possible ways to proceed using algebraic properties and the basic identities that have been established:

1. Rewrite the right-hand member so that it is identical to the left-hand member. (See Equation (1) above.)
2. Rewrite the left-hand member so that it is identical to the right-hand member. (See Equation 2 above.)
3. Rewrite both members *independently* so that they are identical. (See Equation (3) above.)

Proving identities will provide you with the opportunity to use the basic identities and hence become familiar with these important relationships, and also to use the trigonometric function values (which simply represent real numbers) in conjunction with properties of algebra.

Following is a list of suggestions that may help in determining the best method to use to prove an identity.

1. If one member of an equation appears to be the more complicated, it should be transformed so that it reduces to the simpler member of the equation.
2. In each step of a proof, look for any possible application of algebraic procedures such as combining fractions, writing a single fraction as two or more fractions, using the distributive law, factoring, or multiplying the numerator and denominator of a fraction by the conjugate of one or the other.
3. If an elegant approach in which an identity can be established in one or two steps is not obvious, as a last resort the identity can usually be proven by changing all the terms to expressions involving only sines and/or cosines, and then simplifying both members.

In the following examples, we illustrate several of these approaches.

Example   Show that $\dfrac{\sin^2 x}{1 + \cos x} = \dfrac{\tan x - \sin x}{\tan x}$ is an identity.

Solution   In this case we will simplify both members by appropriate substitutions. Expressing the right-hand member in terms of sines and cosines,

$$\frac{\sin^2 x}{1 + \cos x} = \frac{\tan x - \sin x}{\tan x}$$

$$= \frac{\tan x}{\tan x} - \frac{\sin x}{\tan x}$$

$$= 1 - \frac{\sin x}{\dfrac{\sin x}{\cos x}}$$

$$= 1 - \cos x.$$

Rewriting the left-hand member by substituting $1 - \cos^2 x$ for $\sin^2 x$,

$$\frac{1 - \cos^2 x}{1 + \cos x} = 1 - \cos x$$

$$\frac{(1 - \cos x)(1 + \cos x)}{1 + \cos x} = 1 - \cos x$$

$$1 - \cos x = 1 - \cos x.$$

Hence, the equation is an identity.

Example Show that $\dfrac{\cos x}{1 + \sin x} = \dfrac{1 - \sin x}{\cos x}$ is an identity.

Solution We first note that the numerator of the right-hand member is $1 - \sin x$. Hence, we try to obtain this expression in the numerator of the left-hand member by using an algebraic property. In this case, we can multiply the left-hand member by $\dfrac{1 - \sin x}{1 - \sin x}$ to obtain

$$\frac{\cos x(1 - \sin x)}{(1 + \sin x)(1 - \sin x)} = \frac{1 - \sin x}{\cos x},$$

from which

$$\frac{\cos x(1 - \sin x)}{1 - \sin^2 x} = \frac{1 - \sin x}{\cos x}.$$

Substituting $\cos^2 x$ for $1 - \sin^2 x$ in the left-hand member yields

$$\frac{\cos x(1 - \sin x)}{\cos^2 x} = \frac{1 - \sin x}{\cos x}.$$

Then, dividing the numerator and denominator of the left-hand member by $\cos x$,

$$\frac{1 - \sin x}{\cos x} = \frac{1 - \sin x}{\cos x},$$

which is clearly an identity.

Example Show that $\dfrac{\sec^2 \theta - \tan^2 \theta}{\sec \theta + \tan \theta} = \dfrac{1 - \sin \theta}{\cos \theta}$ is an identity.

Solution If we note that $\sec^2 \theta - \tan^2 \theta$ (of the form $a^2 - b^2$) can be factored as $(\sec \theta - \tan \theta)(\sec \theta + \tan \theta)$, we can make a substitution to obtain

$$\frac{(\sec \theta - \tan \theta)(\sec \theta + \tan \theta)}{\sec \theta + \tan \theta} = \frac{1 - \sin \theta}{\cos \theta}$$

from which

$$\sec \theta - \tan \theta = \frac{1 - \sin \theta}{\cos \theta}.$$

(*continued*)

Now substituting $1/\cos\theta$ for $\sec\theta$ and $\sin\theta/\cos\theta$ for $\tan\theta$ and writing the left-hand member as a single fraction, we have that

$$\frac{1}{\cos\theta} - \frac{\sin\theta}{\cos\theta} = \frac{1 - \sin\theta}{\cos\theta}$$

$$\frac{1 - \sin\theta}{\cos\theta} = \frac{1 - \sin\theta}{\cos\theta}.$$

Hence, the equation is an identity.

Counterexamples

In general, it is a relatively simple matter to show that an equation is not an identity, if such is the case, by finding a *counterexample*. For example, substituting 0 for α in the equation

$$\sin\alpha = \cos\alpha$$

yields

$$\sin 0 = \cos 0,$$

$$0 = 1,$$

which is a false statement. Thus, the equation $\sin\alpha = \cos\alpha$ is not an identity.

Of course, in demonstrating a counterexample, we should select a replacement for the variable so that each member of the equation is defined.

EXERCISE SET 4.2

A

Simplify each expression on the left to show that it is equivalent to the expression on the right for all values of the variable for which both expressions are defined.

Example $(1 - \cos^2\beta)(1 + \cot^2\beta); \; 1$

Solution   Substituting $\sin^2\beta$ for $1 - \cos^2\beta$ and $\csc^2\beta$ for $1 + \cot^2\beta$ yields

$$(1 - \cos^2\beta)(1 + \cot^2\beta) = (\sin^2\beta)(\csc^2\beta).$$

Since $\csc^2\beta = 1/\sin^2\beta$,

$$(1 - \cos^2\beta)(1 + \cot^2\beta) = (\sin^2\beta)\left(\frac{1}{\sin^2\beta}\right) = 1.$$

1. $\sec \alpha \sin \alpha$; $\tan \alpha$

2. $\tan \alpha \cos \alpha$; $\sin \alpha$

3. $\sin \beta \cot \beta$; $\cos \beta$

4. $\csc \beta \tan \beta$; $\sec \beta$

5. $\dfrac{\sin \gamma}{\tan \gamma}$; $\cos(-\gamma)$

6. $\dfrac{\sin(-\gamma) \sec \gamma}{\tan(-\gamma)}$; 1

7. $\cot^2 \gamma \sin^2 \gamma$; $\cos^2 \gamma$

8. $\tan^2 \alpha \cos^2 \alpha$; $\sin^2 \alpha$

9. $\cos^2 \alpha(1 + \tan^2 \alpha)$; 1

10. $(\csc^2 \beta - 1)(\sin^2 \beta)$; $\cos^2 \beta$

11. $\dfrac{(1 + \sin \beta)(1 - \sin \beta)}{\cos \beta}$; $\cos \beta$

12. $\dfrac{(1 + \cos \gamma)(1 - \cos \gamma)}{\sin \gamma}$; $\sin \gamma$

13. $\dfrac{\cos^2 \gamma}{\cot^2 \gamma}$; $\sin^2 \gamma$

14. $\dfrac{1 + \tan^2 \gamma}{\csc^2 \gamma}$; $\tan^2 \gamma$

15. $\dfrac{\sin^2 \alpha}{\tan^2 \alpha - \sin^2 \alpha}$; $\cot^2 \alpha$

16. $\dfrac{\cos^2 \alpha}{\cot^2 \alpha - \cos^2 \alpha}$; $\tan^2 \alpha$

17. $\dfrac{1}{\tan \alpha + \cot \alpha}$; $\dfrac{\cos \alpha}{\csc \alpha}$

18. $\dfrac{1}{\csc \alpha - \sin \alpha}$; $\dfrac{\tan \alpha}{\cos \alpha}$

Verify that each equation is an identity.

19. $\cos x \tan x \csc x = 1$

20. $\cot x \sec x = \csc x$

21. $\sec y - \cos y = \tan y \sin y$

22. $\sec y - \tan y \sin y = \cos y$

23. $\cos^2 x - \sin^2 x = 2 \cos^2 x - 1$

24. $(\sec^2 y - 1)(\csc^2 y - 1) = 1$

25. $\tan^2 y + \sec^2 y = \dfrac{2}{\cos^2 y} - 1$

26. $(\tan x + \cot x)\dfrac{\sin x}{\sec x} = 1$

27. $\dfrac{\cos x + \sin x}{\sin x} = 1 + \cot x$

28. $\dfrac{1 - \tan^2 y}{\tan y} = \cot y - \tan y$

29. $\tan^2 y - \sin^2 y = \tan^2 y \sin^2 y$

30. $\dfrac{1}{1 + \sin x} + \dfrac{1}{1 - \sin x} = \dfrac{2}{\cos^2 y}$

31. $\tan y + \cot y = \sec y \csc y$

32. $\dfrac{1 + \tan^2 y}{\tan^2 y} = \csc^2 y$

33. $\cos^4 x - \sin^4 x = \cos^2 x - \sin^2 x$ *Hint:* Factor left-hand member.

34. $\dfrac{1 + \sec x}{\sec x} = \dfrac{\sin^2 x}{1 - \cos x}$ *Hint:* $\sin^2 x = (1 - \cos x)(1 + \cos x)$.

35. $\dfrac{1 - \cos x}{\sin x} = \dfrac{\sin x}{1 + \cos x}$ *Hint:* $\dfrac{\sin x}{1 + \cos x} = \dfrac{\sin x(1 - \cos x)}{(1 + \cos x)(1 - \cos x)}$.

36. $\dfrac{1 + \sin x}{\cos x} = \dfrac{\cos x}{1 - \sin x}$ *Hint:* $\dfrac{1 + \sin x}{\cos x} = \dfrac{(1 + \sin x)(1 - \sin x)}{\cos x(1 - \sin x)}$

37. $\csc y = \dfrac{\sec y + \csc y}{1 + \tan y}$

38. $\dfrac{1}{\sec y - \tan y} - \sec y + \tan y$

39. $\dfrac{\sec y - 1}{\sec y + 1} = \dfrac{1 - \cos y}{1 + \cos y}$

40. $\dfrac{\cos y}{\sec y - \tan y} = \dfrac{\cos^2 y}{1 - \sin y}$

41. $\sec^2 y + \csc^2 y = \sec^2 y \csc^2 y$

42. $\sec^4 y - \sec^2 y = \tan^4 y + \tan^2 y$

43. $\cos(-\alpha) \tan(-\alpha) \csc \alpha = -1$

44. $\cot(-\alpha) \sec(-\alpha) = \csc(-\alpha)$

45. $\dfrac{\cos(-\alpha) + \sin \alpha}{\sin(-\alpha)} = -(1 + \cot \alpha)$

46. $\dfrac{\cos(-\alpha) - \sin(-\alpha)}{\cos \alpha} = 1 + \tan \alpha$

47. $\sec(-\alpha) \sin(-\alpha) = -\tan \alpha$

48. $\dfrac{\sin(-\alpha)}{\tan(-\alpha)} = \cos \alpha$

49. $\dfrac{\sin(-\alpha) \sec \alpha}{\tan(-\alpha)} = 1$

50. $\dfrac{1 + \tan^2(-\alpha)}{\csc^2(-\alpha)} = \tan^2 \alpha$

By using counterexamples show that each equation is not an identity.

51. $\sin x + \cos x \tan x = 2$

52. $\sin^2 x + 2 \cos x - \cos^2 x = 1$

53. $2 \sin y \cos y + \sin y = 0$

54. $\tan^2 y - 2 \tan y = 0$

B

Verify that each equation is an identity.

55. $\sec \alpha = \dfrac{\cot \alpha + \tan \alpha}{\csc \alpha}$

56. $\dfrac{\cot \alpha + \tan \alpha}{\sec \alpha} = \csc \alpha$

57. $\sin^2 \alpha = \dfrac{\cos^2 \alpha + \tan^2 \alpha - 1}{\tan^2 \alpha}$

58. $\dfrac{\sin^2 \alpha + \cot^2 \alpha - 1}{\cot^2 \alpha} = \cos^2 \alpha$

59. $\sec \beta - 1 = \dfrac{\sec \beta + 1}{(\csc \beta + \cot \beta)^2}$

60. $\tan \beta + \sec \beta = \dfrac{1}{\sec \beta - \tan \beta}$

61. $\csc \gamma - \cot \gamma = \dfrac{1}{\cot \gamma + \csc \gamma}$

62. $\sec^2 \gamma - \csc^2 \gamma = \dfrac{\tan \gamma}{\cot \gamma} - \dfrac{\cot \gamma}{\tan \gamma}$

63. $\sec^2 x - \tan x = \dfrac{1 + \tan^3 x}{1 + \tan x}$ **64.** $\sec x = \dfrac{\tan^2 x - 2 \tan x}{\cos x \tan x - 2 \cos x}$

65. $\dfrac{\cos y + \sin y}{\cos^2 y + \cos y \sin y} = \dfrac{\cos y + 1}{\cos y + \cos^2 y}$

66. $\sin y + \cos y - 1 = \dfrac{2 \sin y \cos y}{1 + \sin y + \cos y}$

4.3 Sum and Difference Formulas for Cosine

In this section we shall establish several trigonometric identities involving the sum or difference of two angles or two real numbers. We first derive a formula for $\cos(\alpha + \beta)$. In Figure 4.2, angles α, $\alpha + \beta$, and $-\beta$ are in

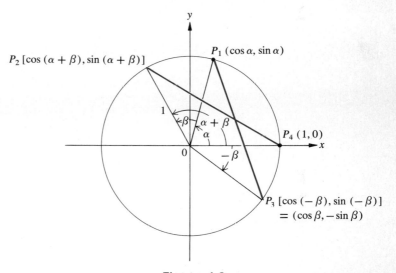

Figure 4.2

standard position, with their sides intersecting the circle of radius 1 at points P_1, P_2, P_3, and P_4. If P_1 has coordinates (x_1, y_1), then, since the radius r of the circle equals 1,

$$\cos \alpha = \frac{x_1}{1} \quad \text{and} \quad \sin \alpha = \frac{y_1}{1}.$$

Hence, the coordinates at P_1 are $x_1 = \cos \alpha$ and $y_1 = \sin \alpha$. We obtain the coordinates at P_2 and P_3 in a similar way. Note that from Equations (1) and (2) in Section 4.1, we have $\sin(-\beta) = -\sin \beta$ and $\cos(-\beta) = \cos \beta$. Since all radii are equal and the central angles $\angle P_4 O P_2$ and $\angle P_3 O P_1$ are equal,

$$\triangle P_4 O P_2 \cong \triangle P_3 O P_1$$

and

$$P_4 P_2 = P_3 P_1.$$

From the distance formula (see Appendix A.4), we have

$$(P_3 P_1)^2 = [\cos \alpha - \cos \beta]^2 + [\sin \alpha - (-\sin \beta)]^2$$
$$= (\cos \alpha - \cos \beta)^2 + (\sin \alpha + \sin \beta)^2$$
$$= \cos^2 \alpha - 2 \cos \alpha \cos \beta + \cos^2 \beta + \sin^2 \alpha + 2 \sin \alpha \sin \beta + \sin^2 \beta.$$

Since $\cos^2 \alpha + \sin^2 \alpha = 1$ and $\cos^2 \beta + \sin^2 \beta = 1$,

$$(P_3 P_1)^2 = 2 - 2 \cos \alpha \cos \beta + 2 \sin \alpha \sin \beta.$$

Again using the distance formula, we have

$$(P_4 P_2)^2 = [\cos(\alpha + \beta) - 1]^2 + [\sin(\alpha + \beta) - 0]^2$$
$$= \cos^2(\alpha + \beta) - 2 \cos(\alpha + \beta) + 1 + \sin^2(\alpha + \beta).$$

Since $\cos^2(\alpha + \beta) + \sin^2(\alpha + \beta) = 1$,

$$(P_4 P_2)^2 = 2 - 2 \cos(\alpha + \beta).$$

Since $P_4 P_2 = P_3 P_1$, we have

$$2 - 2 \cos(\alpha + \beta) = 2 - 2 \cos \alpha \cos \beta + 2 \sin \alpha \sin \beta,$$

which gives

$$\mathbf{\cos(\alpha + \beta) = \cos \alpha \cos \beta - \sin \alpha \sin \beta.} \qquad (1)$$

Substituting $-\beta$ for β in Equation (1), we have

$$\cos[\alpha + (-\beta)] = \cos \alpha \cos(-\beta) - \sin \alpha \sin(-\beta).$$

Since

$$\cos(-\beta) = \cos \beta \qquad \text{and} \qquad \sin(-\beta) = -\sin \beta,$$

we obtain

$$\cos(\alpha - \beta) = \cos \alpha \cos \beta + \sin \alpha \sin \beta. \tag{2}$$

These very important relationships for $\cos(\alpha + \beta)$ and $\cos(\alpha - \beta)$ are called the **sum formula** and the **difference formula,** respectively, for the cosine function.

We can, as illustrated in the following examples, find some function values using the sum and difference formulas. Although using Table II, Table III, or a calculator is a more efficient way of finding these function values, such examples give us practice in applying the formulas.

Example   Compute $\cos 15°$.

Solution   Because we know the exact function values of $45°$ and $30°$, we replace $15°$ with $45° - 30°$ in Equation (2) and obtain

$$\cos 15° = \cos(45° - 30°)$$

$$= \cos 45° \cos 30° + \sin 45° \sin 30°$$

$$= \frac{1}{\sqrt{2}} \cdot \frac{\sqrt{3}}{2} + \frac{1}{\sqrt{2}} \cdot \frac{1}{2} = \frac{\sqrt{3} + 1}{2\sqrt{2}}.$$

Example   Compute $\cos \dfrac{7\pi}{12}$.

Solution   Since $\dfrac{1}{3} + \dfrac{1}{4} = \dfrac{7}{12}$, we have $\dfrac{7\pi}{12} = \dfrac{\pi}{3} + \dfrac{\pi}{4}$. By Equation (1),

$$\cos \frac{7\pi}{12} = \cos\left(\frac{\pi}{3} + \frac{\pi}{4}\right) = \cos \frac{\pi}{3} \cos \frac{\pi}{4} - \sin \frac{\pi}{3} \sin \frac{\pi}{4}.$$

The appropriate substitutions for the function values yields

$$\cos \frac{7\pi}{12} = \frac{1}{2} \cdot \frac{1}{\sqrt{2}} - \frac{\sqrt{3}}{2} \cdot \frac{1}{\sqrt{2}} = \frac{1 - \sqrt{3}}{2\sqrt{2}}.$$

In general we shall leave such results in radical form. However, you can, if you wish, obtain a rational number approximation. For example, since $\sqrt{2} \approx 1.414$ and $\sqrt{3} \approx 1.732$,

$$\cos \frac{7\pi}{12} \approx \frac{1 - 1.732}{2(1.414)} \approx -0.2588.$$

The sum and difference formulas for the cosine function can be used to establish other identities. Three identities that we will be using are

$$\cos(180° − α) = −\cos α. \tag{3}$$

$$\cos(180° + α) = −\cos α, \tag{4}$$

$$\cos(360° − α) = \cos α. \tag{5}$$

Identities such as these are known as **reduction formulas.** The proof that Equation (3) is an identity is shown here; the proofs of (4) and (5) are left as exercises.

Using the difference formula [Equation (2)] and substituting $180°$ for $α$ and $α$ for $β$, we obtain

$$\cos(180° − α) = (\cos 180°)(\cos α) + (\sin 180°)(\sin α).$$

Since $\cos 180° = −1$ and $\sin 180° = 0$, we have

$$\cos(180° − α) = (−1)\cos α + (0)\sin α = −\cos α.$$

Reduction formulas can be used, instead of relying on sketches of reference angles, to rewrite a function value of an angle greater than $90°$ or less than $0°$ as a function value of an acute angle.

Examples   Write each expression as the cosine of an acute angle.

a. $\cos 150°$ b. $\cos 225°$

Solutions

a. Since $150° = 180° − 30°$, we use Equation (3) to obtain

$$\cos 150° = \cos(180° − 30°) = −\cos 30°.$$

b. Since $225° = 180° + 45°$, we use Equation (4) to obtain

$$\cos 225° = \cos(180° + 45°) = −\cos 45°.$$

EXERCISE SET 4.3

A

Find the exact value of each expression using Equation (1) or Equation (2) of this section.

1. $\cos 75°$ (Use: $75° = 120° − 45°$)

2. $\cos 105°$ (Use: $105° = 60° + 45°$)

3. $\cos 165°$ (Use: $165° = 120° + 45°$)

4. $\cos 195°$ (Use: $195° = 135° + 60°$)

5. $\cos \dfrac{\pi}{12}$ $\left(\text{Use: } \dfrac{\pi}{12} = \dfrac{\pi}{3} - \dfrac{\pi}{4}\right)$ **6.** $\cos \dfrac{5\pi}{12}$ $\left(\text{Use: } \dfrac{5\pi}{12} = \dfrac{\pi}{6} + \dfrac{\pi}{4}\right)$

Verify that each equation is correct.

Example $\cos 110° \cos 70° - \sin 110° \sin 70° = -1$

Solution   From Equation (1), if we replace α with $110°$ and β with $70°$, we obtain

$$\cos 110° \cos 70° - \sin 110° \sin 70° = \cos(110° + 70°)$$
$$= \cos 180° = -1.$$

7. $\cos 40° \cos 140° - \sin 40° \sin 140° = -1$

8. $\cos 20° \cos 160° - \sin 20° \sin 160° = -1$

9. $\cos 35° \cos 55° - \sin 35° \sin 55° = 0$

10. $\cos 50° \cos 40° - \sin 50° \sin 40° = 0$

11. $\cos 100° \cos 55° + \sin 100° \sin 55° = \dfrac{1}{\sqrt{2}}$

12. $\cos 130° \cos 70° + \sin 130° \sin 70° = \dfrac{1}{2}$

Write each expression as a single function of kx, k an integer.

Example $\cos 2x \cos 3x - \sin 2x \sin 3x$

Solution   From Equation (1),

$$\cos 2x \cos 3x - \sin 2x \sin 3x = \cos(2x + 3x) = \cos 5x.$$

13. $\cos 4x \cos 5x - \sin 4x \sin 5x$ **14.** $\cos x \cos 2x - \sin x \sin 2x$

15. $\cos 8x \cos 6x + \sin 8x \sin 6x$ **16.** $\cos 5x \cos x + \sin 5x \sin x$

17. $\cos \dfrac{2}{3} x \cos \dfrac{1}{3} x - \sin \dfrac{2}{3} x \sin \dfrac{1}{3} x$

18. $\cos \dfrac{3}{4} x \cos \dfrac{5}{4} x - \sin \dfrac{3}{4} x \sin \dfrac{5}{4} x$

19. $\cos(x + 70°) \cos(x - 70°) - \sin(x + 70°) \sin(x - 70°)$

20. $\cos(65° + x) \cos(65° - x) + \sin(65° + x) \sin(65° - x)$

If angles α and β have the given function values and are in the indicated quadrant, use Equation (1) to find $\cos(α + β)$.

Example $\cos α = -\dfrac{1}{\sqrt{5}}$ and $\sin β = \dfrac{2}{\sqrt{13}}$; α in II and β in I

Solution   Since α is in Quadrant II, as shown in the sketch, using the Pythagorean theorem,

$$y_1{}^2 + (-1)^2 = (\sqrt{5})^2,$$
$$y_1{}^2 + 1 = 5,$$
$$y_1{}^2 = 4,$$
$$y_1 = 2 \quad (-2 \text{ does not apply}).$$

Hence, $\sin α = \dfrac{y_1}{\sqrt{5}} = \dfrac{2}{\sqrt{5}}.$

Since β is in Quadrant I, from the sketch,

$$x_2{}^2 + 2^2 = (\sqrt{13})^2,$$
$$x_2{}^2 + 4 = 13,$$
$$x_2{}^2 = 9,$$
$$x_2 = 3 \quad (-3 \text{ does not apply}).$$

Hence, $\cos β = \dfrac{x_2}{\sqrt{13}} = \dfrac{3}{\sqrt{13}}.$ Therefore,

$$\cos(α + β) = \cos α \cos β - \sin α \sin β$$

$$= -\frac{1}{\sqrt{5}} \cdot \frac{3}{\sqrt{13}} - \frac{2}{\sqrt{5}} \cdot \frac{2}{\sqrt{13}}$$

$$= -\frac{3}{\sqrt{65}} - \frac{4}{\sqrt{65}} = -\frac{7}{\sqrt{65}}.$$

21. $\cos α = \dfrac{3}{5}$ and $\sin β = \dfrac{5}{13}$; α and β in Quadrant I

22. $\cos α = -\dfrac{4}{5}$ and $\sin β = -\dfrac{7}{25}$; α in Quadrant III and β in IV

23. $\sin \alpha = \dfrac{1}{\sqrt{10}}$ and $\cos \beta = -\dfrac{4}{\sqrt{17}}$; α in Quadrant II and β in III

24. $\sin \alpha = -\dfrac{5}{\sqrt{29}}$ and $\cos \beta = \dfrac{1}{\sqrt{26}}$; α in Quadrant IV and β in I

Use one of the reduction formulas [Equation (3), (4), or (5)] of this section to write each expression as a function value of an acute angle.

Example $\cos 135°$

Solution   Using Equation (3), we find that

$$\cos 135° = \cos(180° - 45°) = -\cos 45°.$$

25. $\cos 120°$ **26.** $\cos 240°$ **27.** $\cos 210°$

28. $\cos 300°$ **29.** $\cos 315°$ **30.** $\cos 330°$

Verify that each equation in Exercises 31–35 is an identity.

31. $\cos(180° + \alpha) = -\cos \alpha$ **32.** $\cos(360° - \alpha) = \cos \alpha$

33. $\cos(90° - \alpha) = \sin \alpha$ **34.** $\cos(90° + \alpha) = -\sin \alpha$

35. $\cos 2x = \cos^2 x - \sin^2 x.$ *Hint:* $\cos 2x = \cos(x + x).$

36. Use the results of Exercise 35 to show that

$$\cos 2x = 2\cos^2 x - 1 = 1 - 2\sin^2 x.$$

37. Show by counterexample that $\cos(\alpha + \beta) = \cos \alpha + \cos \beta$ is not an identity.

38. Show by counterexample that $\cos(\alpha - \beta) = \cos \alpha - \cos \beta$ is not an identity.

39. Show that $\cos\left(\dfrac{\pi}{2} - \alpha\right) = \sin \alpha.$

40. Use the results of Exercise 39 to show that $\sin\left(\dfrac{\pi}{2} - \alpha\right) = \cos \alpha.$

B

Evaluate each of the following.

Example $\cos\left(\text{Arccot } \dfrac{1}{\sqrt{3}} + \text{Arcsin } \dfrac{1}{\sqrt{2}}\right)$

(continued)

Solution Let

$$\alpha = \text{Arccot}\,\frac{1}{\sqrt{3}} \quad \text{and} \quad \beta = \text{Arcsin}\,\frac{1}{\sqrt{2}}. \tag{1}$$

Then, using symbols α and β and the sum formula on page 140, we have

$$\cos(\alpha + \beta) = \cos \alpha \cos \beta - \sin \alpha \sin \beta.$$

By inspection of Equations (1), $\alpha = (\pi/3)^{\text{R}}$. Hence,

$$\cos \alpha = \frac{1}{2} \quad \text{and} \quad \sin \alpha = \frac{\sqrt{3}}{2}.$$

Also by inspection of (1), $\beta = (\pi/4)^{\text{R}}$. Thus, we have that

$$\sin \beta = \frac{1}{\sqrt{2}} \quad \text{and} \quad \cos \beta = \frac{1}{\sqrt{2}}.$$

Therefore,

$$\cos\left(\text{Arccot}\,\frac{1}{\sqrt{3}} + \text{Arcsin}\,\frac{1}{\sqrt{2}}\right) = \frac{1}{2}\cdot\frac{1}{\sqrt{2}} - \frac{\sqrt{3}}{2}\cdot\frac{1}{\sqrt{2}}$$

$$= \frac{1 - \sqrt{3}}{2\sqrt{2}}.$$

41. $\cos\left(\text{Arcsin}\,\frac{1}{2} + \text{Arcsec}\,2\right)$ **42.** $\cos\left(\text{Arccos}\,\frac{1}{2} + \text{Arccsc}\,2\right)$

43. $\cos\left(\text{Arctan}\,1 - \text{Arccos}\,\frac{\sqrt{3}}{2}\right)$ **44.** $\cos\left(\text{Arcsin}\,\frac{1}{\sqrt{2}} - \text{Arcsec}\,1\right)$

45. $\cos[\text{Csc}^{-1}(-2) - \text{Sec}^{-1}\sqrt{2}]$

46. $\cos\left[\text{Cos}^{-1}\left(-\frac{1}{\sqrt{2}}\right) - \text{Sin}^{-1}(-1)\right]$

4.4 Sum and Difference Formulas for Sine

Sum and difference formulas for the sine function can be obtained from the sum and difference formulas for the cosine function. First, however, we consider two relationships between function values for the cosine and sine functions. From the difference formula for the cosine function we have

$$\cos\left(\frac{\pi}{2} - \theta\right) = \cos\frac{\pi}{2}\cos\theta + \sin\frac{\pi}{2}\sin\theta.$$

Replacing $\cos(\pi/2)$ with 0 and $\sin(\pi/2)$ with 1, we have

$$\cos\left(\frac{\pi}{2} - \theta\right) = 0\cdot\cos\theta + 1\cdot\sin\theta,$$

$$\cos\left(\frac{\pi}{2} - \theta\right) = \sin\theta. \tag{1}$$

Replacing θ with $\pi/2 - \theta$ in Equation (1), we obtain

$$\cos\left[\frac{\pi}{2} - \left(\frac{\pi}{2} - \theta\right)\right] = \sin\left(\frac{\pi}{2} - \theta\right),$$

$$\cos\left[\left(\frac{\pi}{2} - \frac{\pi}{2}\right) + \theta\right] = \sin\left(\frac{\pi}{2} - \theta\right),$$

$$\cos\theta = \sin\left(\frac{\pi}{2} - \theta\right),$$

$$\sin\left(\frac{\pi}{2} - \theta\right) = \cos\theta. \tag{2}$$

Examples

a. $\sin\dfrac{\pi}{3} = \cos\left(\dfrac{\pi}{2} - \dfrac{\pi}{3}\right)$

$\qquad = \cos\dfrac{\pi}{6}$

b. $\cos 46° = \sin(90° - 46°)$

$\qquad\qquad = \sin 44°$

In Section 1.3 we noted that the sine and cosine are cofunctions. Equations (1) and (2) also establish this fact because in each case $(\pi/2 - \theta) + \theta = \pi/2$.

Now we consider the sum and difference formulas for the sine function. If we replace θ with $\alpha + \beta$ in Equation (1), we have

$$\sin(\alpha + \beta) = \cos\left[\frac{\pi}{2} - (\alpha + \beta)\right]$$

$$= \cos\left[\left(\frac{\pi}{2} - \alpha\right) - \beta\right],$$

which by the difference formula for the cosine can be written as

$$\sin(\alpha + \beta) = \cos\left(\frac{\pi}{2} - \alpha\right)\cos\beta + \sin\left(\frac{\pi}{2} - \alpha\right)\sin\beta.$$

By Equations (1) and (2),

$$\cos\left(\frac{\pi}{2} - \alpha\right) = \sin\alpha \quad \text{and} \quad \sin\left(\frac{\pi}{2} - \alpha\right) = \cos\alpha.$$

Hence, we have

$$\sin(\alpha + \beta) = \sin\alpha\cos\beta + \cos\alpha\sin\beta. \tag{3}$$

Replacing β with $-\beta$ in Equation (3) yields

$$\sin[\alpha + (-\beta)] = \sin\alpha\cos(-\beta) + \cos\alpha\sin(-\beta).$$

Substituting $\cos\beta$ for $\cos(-\beta)$ and $(-\sin\beta)$ for $\sin(-\beta)$, we have

$$\sin(\alpha - \beta) = \sin\alpha\cos\beta + \cos\alpha(-\sin\beta),$$

from which

$$\sin(\alpha - \beta) = \sin\alpha\cos\beta - \cos\alpha\sin\beta. \tag{4}$$

Example   Compute $\sin 15°$.

Solution   Since $15° = 45° - 30°$, by Equation (4),

$$\sin 15° = \sin(45° - 30°)$$
$$= \sin 45° \cos 30° - \cos 45° \sin 30°$$
$$= \frac{1}{\sqrt{2}} \cdot \frac{\sqrt{3}}{2} - \frac{1}{\sqrt{2}} \cdot \frac{1}{2} = \frac{\sqrt{3} - 1}{2\sqrt{2}}.$$

Example   Compute $\sin\dfrac{\pi}{12}$.

Solution   Observe that $\dfrac{\pi}{12} = \dfrac{\pi}{3} - \dfrac{\pi}{4}$. By using Equation (4),

$$\sin\frac{\pi}{12} = \sin\left(\frac{\pi}{3} - \frac{\pi}{4}\right) = \sin\frac{\pi}{3}\cos\frac{\pi}{4} - \cos\frac{\pi}{3}\sin\frac{\pi}{4}.$$

Making the appropriate substitutions for the function values yields

$$\sin\frac{\pi}{12} = \frac{\sqrt{3}}{2}\cdot\frac{1}{\sqrt{2}} - \frac{1}{2}\cdot\frac{1}{\sqrt{2}} = \frac{\sqrt{3}-1}{2\sqrt{2}}.$$

The same results are obtained if $\pi/12$ is written as $(\pi/4) - (\pi/6)$.

Again we note that in practical applications, approximations for function values such as $\sin(\pi/12)$ can be found readily by using Table II or a calculator.

The sum and difference formulas for the sine function can be used to establish the following reduction formulas.

$$\sin(180° - \alpha) = \sin\alpha, \tag{5}$$

$$\sin(180° + \alpha) = -\sin\alpha, \tag{6}$$

$$\sin(360° - \alpha) = -\sin\alpha. \tag{7}$$

The proof that Equation (5) is an identity is presented here; the proofs of (6) and (7) are left as exercises.

Using the difference formula, Equation (4), and substituting 180° for α and α for β, we obtain

$$\sin(180° - \alpha) = \sin 180° \cos\alpha - \cos 180° \sin\alpha.$$

Since

$$\sin 180° = 0 \quad\text{and}\quad \cos 180° = -1,$$

we have

$$\sin(180° - \alpha) = 0\cdot\cos\alpha - (-1)\sin\alpha = \sin\alpha.$$

EXERCISE SET 4.4

A

Find the value of each expression using Equation (3) or (4).

1. $\sin 105°$ (Use: $105° = 60° + 45°$)

2. $\sin 75°$ (Use: $75° = 120° - 45°$)

3. $\sin 165°$ (Use: $165° = 210° - 45°$)

4. $\sin 195°$ (Use: $195° = 135° + 60°$)

5. $\sin\dfrac{5\pi}{12}$ $\left(\text{Use: } \dfrac{5\pi}{12} = \dfrac{\pi}{6} + \dfrac{\pi}{4}\right)$ 6. $\sin\dfrac{\pi}{12}$ $\left(\text{Use: } \dfrac{\pi}{12} = \dfrac{\pi}{3} - \dfrac{\pi}{4}\right)$

Verify that each equation is correct.

Example $\sin 100° \cos 80° + \cos 100° \sin 80° = 0$

Solution   From Equation (3) of this section, if we replace α with $100°$ and β with $80°$,

$$\sin 100° \cos 80° + \cos 100° \sin 80° = \sin(100° + 80°)$$
$$= \sin 180° = 0.$$

7. $\sin 20° \cos 160° + \cos 20° \sin 160° = 0$

8. $\sin 40° \cos 140° + \cos 40° \sin 140° = 0$

9. $\sin 50° \cos 40° + \cos 50° \sin 40° = 1$

10. $\sin 35° \cos 55° + \cos 35° \sin 55° = 1$

11. $\sin 110° \cos 80° - \cos 110° \sin 80° = \dfrac{1}{2}$

12. $\sin 95° \cos 50° - \cos 95° \sin 50° = \dfrac{1}{\sqrt{2}}$

Write each expression as a single function of kx, where k is an integer.

Example $\sin 2x \cos 3x + \cos 2x \sin 3x$

Solution   Using Equation (3),

$$\sin 2x \cos 3x + \cos 2x \sin 3x = \sin(2x + 3x)$$
$$= \sin 5x.$$

13. $\sin x \cos 2x + \cos x \sin 2x$ **14.** $\sin 4x \cos 5x + \cos 4x \sin 5x$

15. $\sin 5x \cos x - \cos 5x \sin x$ **16.** $\sin 8x \cos 6x - \cos 8x \sin 6x$

17. $\sin \dfrac{3}{4} x \cos \dfrac{5}{4} x + \cos \dfrac{3}{4} x \sin \dfrac{5}{4} x$

18. $\sin \dfrac{4}{3} x \cos \dfrac{1}{3} x - \cos \dfrac{4}{3} x \sin \dfrac{1}{3} x$

19. $\sin(50° + x) \cos(50° - x) - \cos(50° + x) \sin(50° - x)$

20. $\sin(x + 60°) \cos(x - 60°) + \cos(x + 60°) \sin(x - 60°)$

If angles α and β have the given function values and are in the indicated quadrants, use Equation (3) to find $\sin(\alpha + \beta)$.

Example $\cos \alpha = -\dfrac{1}{\sqrt{5}}$, $\sin \beta = \dfrac{2}{\sqrt{13}}$, α in Quadrant II, β in I

Solution   From the example on page 144, we have $\sin \alpha = 2/\sqrt{5}$ and $\cos \beta = 3/\sqrt{13}$. Therefore, from Equation (3),

$$\sin(\alpha + \beta) = \sin \alpha \cos \beta + \cos \alpha \sin \beta$$

$$= \frac{2}{\sqrt{5}} \cdot \frac{3}{\sqrt{13}} + \left(-\frac{1}{\sqrt{5}} \right) \cdot \frac{2}{\sqrt{13}}$$

$$= \frac{6}{\sqrt{65}} - \frac{2}{\sqrt{65}} = \frac{4}{\sqrt{65}}.$$

21. $\cos \alpha = -\dfrac{4}{5}$, $\sin \beta = -\dfrac{7}{25}$, α in Quadrant III, β in IV

22. $\cos \alpha = \dfrac{3}{5}$, $\sin \beta = \dfrac{5}{13}$, α and β in Quadrant I

23. $\sin \alpha = -\dfrac{5}{\sqrt{29}}$, $\cos \beta = \dfrac{1}{\sqrt{26}}$, α in Quadrant IV, β in I

24. $\sin \alpha = \dfrac{1}{\sqrt{10}}$, $\cos \beta = -\dfrac{4}{\sqrt{17}}$, α in Quadrant II, β in III

Use one of the reduction formulas [Equation (5), (6), or (7)] of this section to write each expression as a function of an acute angle.

Example $\sin 135°$

Solution   By Equation (5),

$$\sin 135° = \sin(180° - 45°) = \sin 45°.$$

25. $\sin 240°$ **26.** $\sin 120°$ **27.** $\sin 150°$

28. $\sin 210°$ **29.** $\sin 330°$ **30.** $\sin 315°$

Verify that each equation is an identity.

31. $\sin(180° + \alpha) = -\sin \alpha$ **32.** $\sin(360° - \alpha) = -\sin \alpha$

33. $\sin(270° - \alpha) = -\cos \alpha$ **34.** $\sin(270° + \alpha) = -\cos \alpha$

35. $\sin 2x = 2 \sin x \cos x$.   *Hint:* $\sin 2x = \sin(x + x)$.

36. $\sin(\alpha - 30°) = \dfrac{\sqrt{3}\sin\alpha - \cos\alpha}{2}$

37. Show by a counterexample that $\sin(\alpha - \beta) = \sin\alpha - \sin\beta$ is not an identity.

38. Show by a counterexample that $\sin(\alpha + \beta) = \sin\alpha + \sin\beta$ is not an identity.

B

Evaluate each of the following.

Example $\sin\left(\text{Arcsin}\,\dfrac{1}{\sqrt{2}} + \text{Arccos}\,\dfrac{1}{2}\right)$

Solution   Let

$$\alpha = \text{Arcsin}\,\dfrac{1}{\sqrt{2}} \quad \text{and} \quad \beta = \text{Arccos}\,\dfrac{1}{2}. \tag{8}$$

Then, by using the symbols α and β, and by the sum formula,

$$\sin\left(\text{Arcsin}\,\dfrac{1}{\sqrt{2}} + \text{Arccos}\,\dfrac{1}{2}\right) = \sin(\alpha + \beta)$$

$$= \sin\alpha\cos\beta + \cos\alpha\sin\beta.$$

By inspection of Equations (1), $\alpha = (\pi/4)^R$, and hence

$$\sin\alpha = \dfrac{1}{\sqrt{2}} \quad \text{and} \quad \cos\alpha = \dfrac{1}{\sqrt{2}}.$$

Also, by inspection of (8), $\beta = (\pi/3)^R$, and hence $\cos\beta = 1/2$ and $\sin\beta = \sqrt{3}/2$. Therefore,

$$\sin\left(\text{Arcsin}\,\dfrac{1}{\sqrt{2}} + \text{Arccos}\,\dfrac{1}{2}\right) = \dfrac{1}{\sqrt{2}}\cdot\dfrac{1}{2} + \dfrac{1}{\sqrt{2}}\cdot\dfrac{\sqrt{3}}{2}$$

$$= \dfrac{1}{2\sqrt{2}} + \dfrac{\sqrt{3}}{2\sqrt{2}} = \dfrac{1+\sqrt{3}}{2\sqrt{2}}.$$

39. $\sin(\text{Arcsin } 0 + \text{Arccos } 1)$ **40.** $\sin(\text{Arccos } 0 + \text{Arcsin } 1)$

41. $\sin\left(\text{Arcsin } 0 - \text{Arcsec}\,\dfrac{2}{\sqrt{3}}\right)$ **42.** $\sin\left(\text{Arccos } 1 - \text{Arcsin}\,\dfrac{1}{\sqrt{2}}\right)$

43. $\sin\left(\text{Arctan } 1 - \text{Arccos } \dfrac{\sqrt{3}}{2}\right)$ **44.** $\sin\left(\text{Arcsin } \dfrac{1}{2} - \text{Arcsec } 2\right)$

4.5 Sum and Difference Formulas for Tangent

Sum and difference formulas for the tangent function follow from the sum and difference formulas for the sine and cosine functions. Since

$$\tan(\alpha + \beta) = \frac{\sin(\alpha + \beta)}{\cos(\alpha + \beta)},$$

by substituting appropriately, we have

$$\tan(\alpha + \beta) = \frac{\sin \alpha \cos \beta + \cos \alpha \sin \beta}{\cos \alpha \cos \beta - \sin \alpha \sin \beta}.$$

Multiplying each term of the numerator and denominator of the right-hand member by $1/\cos \alpha \cos \beta$ yields

$$\tan(\alpha + \beta) = \frac{\dfrac{\sin \alpha \cos \beta}{\cos \alpha \cos \beta} + \dfrac{\cos \alpha \sin \beta}{\cos \alpha \cos \beta}}{\dfrac{\cos \alpha \cos \beta}{\cos \alpha \cos \beta} - \dfrac{\sin \alpha \sin \beta}{\cos \alpha \cos \beta}},$$

from which, by simplifying the right-hand member, we obtain

$$\tan(\alpha + \beta) = \frac{\tan \alpha + \tan \beta}{1 - \tan \alpha \tan \beta}. \tag{1}$$

By substituting $-\beta$ for β in Equation (1), we have

$$\tan[\alpha + (-\beta)] = \frac{\tan \alpha + \tan(-\beta)}{1 - \tan \alpha \tan(-\beta)}.$$

Substituting $-\tan \beta$ for $\tan(-\beta)$ yields

$$\tan[\alpha + (-\beta)] = \frac{\tan \alpha + (-\tan \beta)}{1 - \tan \alpha(-\tan \beta)},$$

$$\tan(\alpha - \beta) = \frac{\tan \alpha - \tan \beta}{1 + \tan \alpha \tan \beta}. \tag{2}$$

Example Compute tan 75°.

Solution Observe that 75° = 45° + 30°. Then, using Equation (1) and replacing α with 45° and β with 30°,

$$\tan 75° = \tan(45° + 30°) = \frac{\tan 45° + \tan 30°}{1 - \tan 45° \tan 30°}$$

$$= \frac{1 + \dfrac{1}{\sqrt{3}}}{1 - 1\left(\dfrac{1}{\sqrt{3}}\right)} = \frac{\left(1 + \dfrac{1}{\sqrt{3}}\right) \cdot \sqrt{3}}{\left(1 - \dfrac{1}{\sqrt{3}}\right) \cdot \sqrt{3}} = \frac{\sqrt{3} + 1}{\sqrt{3} - 1}.$$

The sum and difference formulas established in this section can be used to establish the following reduction formulas.

$$\tan(180° - \alpha) = -\tan \alpha, \tag{3}$$

$$\tan(180° + \alpha) = \tan \alpha, \tag{4}$$

$$\tan(360° - \alpha) = -\tan \alpha. \tag{5}$$

The proof that Equation (3) is an identity is presented here; the proofs of (4) and (5) are left as exercises.

Using the difference formula [Equation (2)] and substituting 180° for α and α for β, we obtain

$$\tan(180° - \alpha) = \frac{\tan 180° - \tan \alpha}{1 + \tan 180° \tan \alpha}$$

$$= \frac{0 - \tan \alpha}{1 + (0)\tan \alpha} = -\tan \alpha.$$

EXERCISE SET 4.5

A

Use Equation (1) or (2) to find the value of each expression.

1. tan 105° (Use: 105° = 45° + 60°)

2. tan 15° (Use: 15° = 45° − 30°)

3. tan 195° (Use: 195° = 240° − 45°)

4. tan 165° (Use: 165° = 120° + 45°)

5. $\tan \dfrac{\pi}{12}$ $\left(\text{Use: } \dfrac{\pi}{12} = \dfrac{\pi}{3} - \dfrac{\pi}{4}\right)$ **6.** $\tan \dfrac{3\pi}{4}$ $\left(\text{Use: } \dfrac{3\pi}{4} = \pi - \dfrac{\pi}{4}\right)$

Verify that each equation is correct.

Example $\dfrac{\tan 20° + \tan 25°}{1 - \tan 20° \tan 25°} = 1$

Solution From Equation (1), if we replace α with 20° and β with 25°,

$$\dfrac{\tan 20° + \tan 25°}{1 - \tan 20° \tan 25°} = \tan(20° + 25°) = \tan 45° = 1.$$

7. $\dfrac{\tan 50° + \tan 130°}{1 - \tan 50° \tan 130°} = 0$ **8.** $\dfrac{\tan 100° + \tan 50°}{1 - \tan 100° \tan 50°} = -\dfrac{1}{\sqrt{3}}$

9. $\dfrac{\tan 140° + \tan 100°}{1 - \tan 140° \tan 100°} = \sqrt{3}$ **10.** $\dfrac{\tan 95° + \tan 40°}{1 - \tan 95° \tan 40°} = -1$

11. $\dfrac{\tan 110° - \tan 50°}{1 + \tan 110° \tan 50°} = \sqrt{3}$ **12.** $\dfrac{\tan 115° - \tan 70°}{1 + \tan 115° \tan 70°} = 1$

Write each expression as a single function of kx where k is an integer.

Example $\dfrac{\tan 2x + \tan 3x}{1 - \tan 2x \tan 3x}$

Solution Using Equation (1), we obtain

$$\dfrac{\tan 2x + \tan 3x}{1 - \tan 2x \tan 3x} = \tan(2x + 3x) = \tan 5x.$$

13. $\dfrac{\tan 4x + \tan 5x}{1 - \tan 4x \tan 5x}$ **14.** $\dfrac{\tan 2x + \tan x}{1 - \tan 2x \tan x}$

15. $\dfrac{\tan 8x - \tan 6x}{1 + \tan 8x \tan 6x}$ **16.** $\dfrac{\tan 5x - \tan x}{1 + \tan 5x \tan x}$

17. $\dfrac{\tan \frac{4}{3}x - \tan \frac{1}{3}x}{1 + \tan \frac{4}{3}x \tan \frac{1}{3}x}$ **18.** $\dfrac{\tan \frac{5}{4}x + \tan \frac{3}{4}x}{1 - \tan \frac{5}{4}x \tan \frac{3}{4}x}$

Use one of the reduction formulas [Equation (3), (4), or (5)] of this section to write each expression as a function of an acute angle.

Example $\tan 135°$

Solution Using Equation (3),

$$\tan 135° = \tan(180° - 45°) = -45°.$$

19. $\tan 120°$ **20.** $\tan 240°$ **21.** $\tan 210°$

22. $\tan 150°$ **23.** $\tan 315°$ **24.** $\tan 330°$

Verify that each equation is an identity.

25. $\tan(180° + \alpha) = \tan \alpha$ **26.** $\tan(360° - \alpha) = -\tan \alpha$

27. $\tan(90° - \alpha) = \cot \alpha$ **28.** $\cot(90° - \alpha) = \tan \alpha$

29. $\tan 2x = \dfrac{2 \tan x}{1 - \tan^2 x}$.   *Hint:* $\tan 2x = \tan(x + x)$.

30. $\tan\left(x + \dfrac{\pi}{4}\right) = \dfrac{1 + \tan x}{1 - \tan x}$

B

31. $\dfrac{\tan(45° - \alpha)}{\tan(45° + \alpha)} = \dfrac{(1 - \tan \alpha)^2}{(1 + \tan \alpha)^2}$ **32.** $\tan \beta = \dfrac{\tan(45° + \beta) - 1}{1 + \tan(45° + \beta)}$

33. $\cot(\alpha + \beta) = \dfrac{\cot \alpha \cot \beta - 1}{\cot \beta + \cot \alpha}$ **34.** $\cot(\alpha - \beta) = \dfrac{\cot \alpha \cot \beta + 1}{\cot \beta - \cot \alpha}$

4.6 Double-Angle Formulas

Additional identities follow from the sum and difference formulas developed in Sections 4.2 and 4.3. For example, if α is substituted for β in

$$\cos(\alpha + \beta) = \cos \alpha \cos \beta - \sin \alpha \sin \beta,$$

we obtain

$$\cos(\alpha + \alpha) = \cos \alpha \cos \alpha - \sin \alpha \sin \alpha,$$

$$\cos 2\alpha = \cos^2 \alpha - \sin^2 \alpha. \tag{1}$$

The right-hand member can be written in a variety of convenient forms. Thus, if $1 - \sin^2 \alpha$ is substituted for $\cos^2 \alpha$, we have

$$\cos 2\alpha = (1 - \sin^2 \alpha) - \sin^2 \alpha,$$

$$\cos 2\alpha = 1 - 2 \sin^2 \alpha. \tag{2}$$

If $1 - \cos^2 \alpha$ is substituted for $\sin^2 \alpha$ in Equation (1), we have

$$\cos 2\alpha = \cos^2 \alpha - (1 - \cos^2 \alpha),$$

$$\cos 2\alpha = 2 \cos^2 \alpha - 1. \tag{3}$$

Identities (1), (2), and (3), and similar identities that follow for the sine and tangent functions, called **double-angle formulas,** are often helpful for rewriting expressions in more useful forms. You will prove additional identities in the exercises to become familiar with these important relationships.

Example   Show that $\dfrac{1 - \cos 2x}{2 \sin x \cos x} = \tan x$ is a identity.

Solution   From Equation (2), we can substitute $1 - 2 \sin^2 x$ for $\cos 2x$ in the left-hand member to obtain

$$\frac{1 - (1 - 2 \sin^2 x)}{2 \sin x \cos x} = \tan x$$

$$\frac{2 \sin^2 x}{2 \sin x \cos x} = \tan x$$

$$\frac{\sin x}{\cos x} = \tan x$$

$$\tan x = \tan x.$$

Hence, the original equation is an identity.

Example   Show that $2 \sec^2 \beta = \dfrac{5 + \cos 2\beta}{1 + \cos 2\beta} - 1$ is an identity.

Solution   From Equation (3), we substitute $2 \cos^2 \beta - 1$ for $\cos 2\beta$ in the right-hand member to obtain

$$2 \sec^2 \beta = \frac{5 + (2 \cos^2 \beta - 1)}{1 + (2 \cos^2 \beta - 1)} - 1 = \frac{4 + 2 \cos^2 \beta}{2 \cos^2 \beta} - 1$$

$$= \frac{4 + 2 \cos^2 \beta - 2 \cos^2 \beta}{2 \cos^2 \beta}$$

$$= \frac{4}{2 \cos^2 \beta};$$

$$2 \sec^2 \beta = 2 \sec^2 \beta.$$

Hence, the original equation is an identity.

If α is substituted for β in

$$\sin(\alpha + \beta) = \sin \alpha \cos \beta + \cos \alpha \sin \beta,$$

we obtain

$$\sin(\alpha + \alpha) = \sin \alpha \cos \alpha + \cos \alpha \sin \alpha,$$

$$\mathbf{\sin 2\alpha = 2 \sin \alpha \cos \alpha.} \tag{4}$$

Example   Show that $\dfrac{2}{\sin 2\theta} = \tan \theta + \cot \theta$ is an identity.

Solution   Substituting $2 \sin \theta \cos \theta$ for $\sin 2\theta$ in the left-hand member, and $\sin \theta / \cos \theta$ for $\tan \theta$ and $\cos \theta / \sin \theta$ for $\cot \theta$ in the right-hand member yields

$$\frac{2}{2 \sin \theta \cos \theta} = \frac{\sin \theta}{\cos \theta} + \frac{\cos \theta}{\sin \theta}$$

$$= \frac{\sin^2 \theta + \cos^2 \theta}{\sin \theta \cos \theta} = \frac{1}{\sin \theta \cos \theta}$$

$$= \frac{2}{2 \sin \theta \cos \theta}.$$

Hence, the original equation is an identity.

If α is substituted for β in

$$\tan(\alpha + \beta) = \frac{\tan \alpha + \tan \beta}{1 - \tan \alpha \tan \beta},$$

we obtain

$$\tan(\alpha + \alpha) = \frac{\tan \alpha + \tan \alpha}{1 - \tan \alpha \tan \alpha},$$

$$\mathbf{\tan 2\alpha = \frac{2 \tan \alpha}{1 - \tan^2 \alpha}.} \tag{5}$$

Example   Show that $\tan 2x = \dfrac{2 \tan x}{2 - \sec^2 x}$ is an identity.

Solution   Substituting $\tan^2 x + 1$ for $\sec^2 x$ yields

$$\tan 2x = \frac{2 \tan x}{2 - (\tan^2 x + 1)}$$

$$= \frac{2 \tan x}{1 - \tan^2 x},$$

which is an identity by Equation (5).

Relationships similar to those developed in this section for sine, cosine, and tangent can be derived for cotangent, secant, and cosecant. However, they are of less importance and will not be considered.

EXERCISE SET 4.6

A

Express each of the following as a single function of kx, or kα, where k is a positive integer. Assume that x or α takes on no value for which the expression is undefined.

Examples

a. $1 - 2 \sin^2 x$

b. $\dfrac{\tan 3x}{1 - \tan^2 3x}$

Solutions

a. From Equation (2),

$$1 - 2 \sin^2 x = \cos 2x.$$

b. From Equation (5),

$$\frac{\tan 3x}{1 - \tan^2 3x} = \frac{1}{2} \cdot \frac{2 \tan 3x}{1 - \tan^2 3x}$$

$$= \frac{1}{2} \tan 2(3x) = \frac{1}{2} \tan 6x.$$

1. $2 \sin x \cos x$

2. $2 \cos^2 x - 1$

3. $\cos^2 x - \sin^2 x$

4. $\dfrac{2 \tan x}{1 - \tan^2 x}$

5. $2 \sin 2x \cos 2x$

6. $\dfrac{4 \tan 2x}{1 - \tan^2 2x}$

7. $\sin \dfrac{x}{2} \cos \dfrac{x}{2}$

8. $4 \sin 3x \cos 3x$

9. $\cos^2 5\alpha - \sin^2 5\alpha$

10. $1 - 2 \sin^2 4\alpha$

11. $4 \cos^2 6\alpha - 2$

12. $2 \sin 5\alpha \cos 5\alpha$

13. $\sin 2\alpha \cos \alpha - \cos 2\alpha \sin \alpha$

14. $\sin 3\alpha \cos 2\alpha + \cos 3\alpha \sin 2\alpha$

15. $\cos 3x \cos 5x + \sin 3x \sin 5x$

16. $\cos 4x \cos 2x - \sin 4x \sin 2x$

Verify that each equation is an identity.

17. $\cos^2 \alpha = \dfrac{1 + \cos 2\alpha}{2}$

18. $\sin^2 \alpha = \dfrac{1 - \cos 2\alpha}{2}$

19. $\cos 2x + 2 \sin^2 x = 1$

20. $\sin 2x \csc x = 2 \cos x$

21. $\dfrac{\sin 2x}{1 + \cos 2x} = \tan x$

22. $\sin 2x = \dfrac{2 \tan x}{1 + \tan^2 x}$

23. $\sec^2 x = \dfrac{2}{1 + \cos 2x}$

24. $\cot x = \dfrac{1 + \cos 2x}{\sin 2x}$

25. $\cos 2\beta = \dfrac{1 - \tan^2 \beta}{1 + \tan^2 \beta}$

26. $\cot^2 \beta = \dfrac{1 + \cos 2\beta}{1 - \cos 2\beta}$

27. $\cot 4x = \dfrac{1 - \tan^2 2x}{2 \tan 2x}$

28. $\cos^4 x - \sin^4 x = \cos 2x$

29. $\sin 3x = 3 \sin x - 4 \sin^3 x$
Hint: First rewrite $\sin 3x$ as $\sin(2x + x)$.

30. $\cos 3x = 4 \cos^3 x - 3 \cos x$

31. $\sin 4\alpha = (\cos \alpha)(4 \sin \alpha - 8 \sin^3 \alpha)$

32. $\cos 4\alpha = 1 - 8 \sin^2 \alpha \cos^2 \alpha$

Example If $\sin x = \dfrac{3}{5}$ and $\cos x < 0$, find $\sin 2x$.

Solution To evaluate

$$\sin 2x = 2 \sin x \cos x, \qquad (1)$$

the values for $\cos x$ for those values of x for which $\sin x = 3/5$ and $\cos x < 0$ are needed. Substituting $3/5$ for $\sin x$ in

$$\sin^2 x + \cos^2 x = 1$$

gives

$$\left(\frac{3}{5}\right)^2 + \cos^2 x = 1,$$

$$\cos^2 x = 1 - \frac{9}{25} = \frac{16}{25}.$$

Because $\cos x < 0$, $\cos x = -4/5$. Substituting $3/5$ for $\sin x$ and $-4/5$ for

$\cos x$ in $\sin 2x = 2 \sin x \cos x$ yields

$$\sin 2x = 2\left(\frac{3}{5}\right)\left(-\frac{4}{5}\right) = -\frac{24}{25}.$$

33. If $\sin x = \dfrac{4}{5}$ and $\cot x > 0$, find $\sin 2x$.

34. If $\sin x = \dfrac{12}{13}$ and $\sec x < 0$, find $\cos 2x$.

35. If $\cos 2\alpha = -\dfrac{8}{17}$ and $180° \leq 2\alpha \leq 270°$, find $\sin \alpha$.

36. If $\cos 2\alpha = \dfrac{5}{13}$ and $270° \leq 2\alpha \leq 360°$, find $\cos \alpha$.

37. If $\tan \beta = -\dfrac{3}{4}$, find $\tan 2\beta$.   **38.** If $\tan \beta = \dfrac{5}{12}$, find $\tan 2\beta$.

4.7 Additional Identities

The identities in this section are also useful in rewriting expressions involving trigonometric function values.

Identities for $\cos^2 \alpha/2$ and $\sin^2 \alpha/2$

In Section 4.6 we observed that

$$\cos 2\alpha = 2 \cos^2 \alpha - 1.$$

Solving this for $\cos^2 \alpha$, we have

$$1 + \cos 2\alpha = 2 \cos^2 \alpha,$$

from which

$$\mathbf{\cos^2 \alpha = \frac{1 + \cos 2\alpha}{2}}. \tag{1}$$

We also noted in Section 4.6 that

$$\cos 2\alpha = 1 - 2 \sin^2 \alpha.$$

Solving this for $\sin^2 \alpha$, we have

$$2 \sin^2 \alpha = 1 - \cos 2\alpha,$$

from which

$$\sin^2 \alpha = \frac{1 - \cos 2\alpha}{2}. \tag{2}$$

These identities enable us to write the second-degree terms, $\cos^2 \alpha$ and $\sin^2 \alpha$, in terms of first-degree terms. They also enable us to write additional useful identities.

Half-Angle Formulas

Substituting $\alpha/2$ for α in Equation (1) yields

$$\cos^2 \frac{\alpha}{2} = \frac{1 + \cos 2(\alpha/2)}{2},$$

from which

$$\cos^2 \frac{\alpha}{2} = \frac{1 + \cos \alpha}{2}. \tag{3}$$

Substituting $\alpha/2$ for α in Equation (2) yields

$$\sin^2 \frac{\alpha}{2} = \frac{1 - \cos 2(\alpha/2)}{2},$$

from which

$$\sin^2 \frac{\alpha}{2} = \frac{1 - \cos \alpha}{2}. \tag{4}$$

Values for $\cos(\alpha/2)$ and $\sin(\alpha/2)$ are positive or negative, depending on the value of $\alpha/2$. Table 1.2 (page 24) can be used to help you select the appropriate sign.

Example   Use Equation (4) to find $\sin \frac{\pi}{8}$.

Solution   With $\alpha/2 = \pi/8$ and $\alpha = 2(\pi/8) = \pi/4$, we obtain

$$\sin^2 \frac{\pi}{8} = \frac{1 - \cos \frac{\pi}{4}}{2}.$$

Because $0 < \pi/8 < \pi/2$, we use the positive square root for $\sin(\pi/8)$. Then, since $\cos(\pi/4) = 1/\sqrt{2}$,

$$\sin\frac{\pi}{8} = \sqrt{\frac{1 - \dfrac{1}{\sqrt{2}}}{2}} = \sqrt{\frac{\sqrt{2} - 1}{2\sqrt{2}}} = \frac{1}{2}\sqrt{2 - \sqrt{2}}.$$

Formulas for $\tan(\alpha/2)$ also follow directly from identities we have developed. Substituting $\alpha/2$ for α in $\tan\alpha = (\sin\alpha)/(\cos\alpha)$, we have

$$\tan\frac{\alpha}{2} = \frac{\sin\dfrac{\alpha}{2}}{\cos\dfrac{\alpha}{2}}.$$

Multiplying both the numerator and the denominator of the right-hand member by $2\sin(\alpha/2)$ yields

$$\tan\frac{\alpha}{2} = \frac{2\sin^2\dfrac{\alpha}{2}}{2\sin\dfrac{\alpha}{2}\cos\dfrac{\alpha}{2}}.$$

Substituting $(1 - \cos\alpha)/2$ for $\sin^2(\alpha/2)$ and substituting $\sin 2(\alpha/2)$ for $2\sin(\alpha/2)\cos(\alpha/2)$, we obtain

$$\tan\frac{\alpha}{2} = \frac{2\left(\dfrac{1 - \cos\alpha}{2}\right)}{\sin 2\left(\dfrac{\alpha}{2}\right)},$$

$$\tan\frac{\alpha}{2} = \frac{1 - \cos\alpha}{\sin\alpha}. \tag{5}$$

By multiplying the numerator and the denominator of the right-hand member of Equation (5) by $1 + \cos\alpha$, we obtain

$$\tan\frac{\alpha}{2} = \frac{(1 - \cos\alpha)(1 + \cos\alpha)}{\sin\alpha(1 + \cos\alpha)}$$

$$= \frac{\sin^2\alpha}{\sin\alpha(1 + \cos\alpha)},$$

from which

$$\tan\frac{\alpha}{2} = \frac{\sin\alpha}{1 + \cos\alpha}. \tag{6}$$

Example Use Equation (5) to find $\tan \dfrac{\pi}{8}$.

Solution If $\alpha/2 = \pi/8$, then $\alpha = \pi/4$ and

$$\tan \frac{\pi}{8} = \frac{1 - \cos \dfrac{\pi}{4}}{\sin \dfrac{\pi}{4}} = \frac{1 - \dfrac{1}{\sqrt{2}}}{\dfrac{1}{\sqrt{2}}}$$

$$= \sqrt{2} - 1.$$

Just as identities such as $\sin 2\alpha = 2 \sin \alpha \cos \alpha$ are sometimes called double-angle formulas, the identities developed above are sometimes called **half-angle formulas.**

Relationships similar to those developed above for sine, cosine, and tangent can be derived for cotangent, secant, and cosecant. They are of less importance and will not be considered.

Product, Sum, and Difference Formulas

The following identities are derived from the sum and difference formulas for the cosine and sine that were developed in Sections 4.3 and 4.4. The proofs are left as exercises.

$$\cos \alpha \cos \beta = \tfrac{1}{2}[\cos(\alpha + \beta) + \cos(\alpha - \beta)], \tag{7}$$

$$\sin \alpha \sin \beta = -\tfrac{1}{2}[\cos(\alpha + \beta) - \cos(\alpha - \beta)], \tag{8}$$

$$\sin \alpha \cos \beta = \tfrac{1}{2}[\sin(\alpha + \beta) + \sin(\alpha - \beta)], \tag{9}$$

$$\cos \alpha \sin \beta = \tfrac{1}{2}[\sin(\alpha + \beta) - \sin(\alpha - \beta)]. \tag{10}$$

Example Express $\cos 5x \sin 2x$ as a sum or difference.

Solution From Equation (10),

$$\cos 5x \sin 2x = \tfrac{1}{2}[\sin(5x + 2x) - \sin(5x - 2x)]$$
$$= \tfrac{1}{2}(\sin 7x - \sin 3x).$$

The following four identities enable us to express sums and differences of function values in terms of products.

$$\cos \alpha + \cos \beta = 2 \cos\left(\frac{\alpha + \beta}{2}\right) \cos\left(\frac{\alpha - \beta}{2}\right), \tag{11}$$

$$\cos \alpha - \cos \beta = -2 \sin\left(\frac{\alpha + \beta}{2}\right) \sin\left(\frac{\alpha - \beta}{2}\right), \tag{12}$$

$$\sin \alpha + \sin \beta = 2 \sin\left(\frac{\alpha + \beta}{2}\right) \cos\left(\frac{\alpha - \beta}{2}\right), \tag{13}$$

$$\sin \alpha - \sin \beta = 2 \cos\left(\frac{\alpha + \beta}{2}\right) \sin\left(\frac{\alpha - \beta}{2}\right). \tag{14}$$

Example   Express $\sin 4\alpha + \sin 2\alpha$ as a product.

Solution   From Equation (13),

$$\sin 4\alpha + \sin 2\alpha = 2 \sin\left(\frac{4\alpha + 2\alpha}{2}\right) \cos\left(\frac{4\alpha - 2\alpha}{2}\right)$$

$$= 2 \sin 3\alpha \cos \alpha.$$

EXERCISE SET 4.6

A

Use the half-angle formulas to find the exact value for each of the following.

Example $\sin 105°$

Solution   From Equation (4), with $\alpha/2 = 105°$ and $\alpha = 210°$,

$$\sin 105° = \sqrt{\frac{1 - \cos 210°}{2}},$$

where the positive square root is chosen because $105°$ is in Quadrant II and $\sin 105°$ is therefore positive. Since $\cos 210° = -\sqrt{3}/2$,

$$\sin 105° = \sqrt{\frac{1 - (-\sqrt{3}/2)}{2}}$$

$$= \frac{\sqrt{2 + \sqrt{3}}}{2}.$$

1. $\sin 15°$ **2.** $\cos 15°$ **3.** $\tan 15°$ **4.** $\sin 75°$

5. $\cos 75°$ **6.** $\tan 75°$ **7.** $\cos \dfrac{\pi}{8}$ **8.** $\sin \dfrac{\pi}{12}$

9. $\tan \dfrac{5\pi}{12}$ **10.** $\cos \dfrac{\pi}{12}$ **11.** $\cos \dfrac{7\pi}{12}$ **12.** $\sin \dfrac{7\pi}{12}$

Verify each identity.

13. $\tan \dfrac{x}{2} = \csc x - \cot x$

14. $\sin \dfrac{\theta}{2} \cos \dfrac{\theta}{2} = \dfrac{\sin \theta}{2}$

15. $\left(\cos \dfrac{\alpha}{2} - \sin \dfrac{\alpha}{2} \right)^2 = 1 - \sin \alpha$

16. $\sin \beta - \cos \beta \tan \dfrac{\beta}{2} = \tan \dfrac{\beta}{2}$

17. $\sin^2 \dfrac{\gamma}{2} = \dfrac{\sec \gamma - 1}{2 \sec \gamma}$

18. $\tan \dfrac{x}{2} \sin x = \dfrac{\tan x - \sin x}{\sin x \sec x}$

In Exercises 19–24, assume $0° \le x < 360°$.

Example   If $\tan x = -\dfrac{3}{4}$ and $\cos x$ is negative, find $\sin \dfrac{x}{2}$.

Solution   Since $\cos x$ and $\tan x$ are both negative, x is an angle in Quadrant II, and because $\tan x = -3/4$, a point on the terminal ray has the x-coordinate -4 and the y-coordinate 3. From the Pythagorean formula (see Appendix A.8),

$$r = \sqrt{x^2 + y^2}$$
$$= \sqrt{(-4)^2 + 3^2} = 5.$$

From Equation (4) with $\cos x = -4/5$, we obtain

$$\sin \dfrac{x}{2} = \sqrt{\dfrac{1 - (-4/5)}{2}} = \dfrac{3}{\sqrt{10}},$$

where the positive square root is taken because x is in Quadrant II and therefore $x/2$ must be in Quadrant I.

19. If $\sin x = \dfrac{\sqrt{3}}{2}$ and $\cos x > 0$, find $\tan \dfrac{x}{2}$.

20. If $\sin x = \dfrac{1}{2}$ and $\sec x > 0$, find $\sin \dfrac{x}{2}$.

21. If $\cos x = -\dfrac{1}{3}$ and $\tan x > 0$, find $\sin \dfrac{x}{2}$.

22. If $\tan x = 3$ and $\sin x > 0$, find $\cos \dfrac{x}{2}$.

23. If $\tan x = -1$ and $\sin x < 0$, find $\cos \dfrac{x}{2}$.

24. If $\sin x = \dfrac{1}{\sqrt{2}}$ and $\cos x < 0$, find $\tan \dfrac{x}{2}$.

Write each expression as a sum or difference of function values.

Example $\sin 3x \sin x$

Solution   From Equation (8),

$$\sin 3x \sin x = -\tfrac{1}{2}[\cos(3x + x) - \cos(3x - x)]$$
$$= -\tfrac{1}{2}(\cos 4x - \cos 2x).$$

25. $\sin 5° \cos 12°$ **26.** $\cos 7° \cos 20°$ **27.** $\cos 2x \sin x$

28. $\sin 2x \sin 5x$ **29.** $\cos 2x \cos 5x$ **30.** $\sin 6x \cos x$

Write each expression as a product.

Example $\cos 3\alpha - \cos \alpha$

Solution   From Equation (12),

$$\cos 3\alpha - \cos \alpha = -2 \sin\left(\frac{3\alpha + \alpha}{2}\right) \sin\left(\frac{3\alpha - \alpha}{2}\right)$$
$$= -2 \sin 2\alpha \sin \alpha.$$

31. $\cos 20° + \cos 4°$ **32.** $\sin 18° - \sin 6°$ **33.** $\sin 3\alpha + \sin 5\alpha$

34. $\cos \alpha + \cos 5\alpha$ **35.** $\sin 5\alpha - \sin \alpha$ **36.** $\cos 6\alpha - \cos 2\alpha$

B

Verify each of the following identities that are listed in this section.

37. Equation (7)   **38.** Equation (8)   **39.** Equation (9)

40. Equation (10)   **41.** Equation (11)   **42.** Equation (12)

43. Equation (13)   **44.** Equation (14)

4.8 Conditional Equations

In Sections 4.1–4.7 we considered identities—equations that are satisfied by all angle or all real number replacements for which both of the members are defined. We now consider equations that are conditional—that is, they are satisfied only for particular replacements for the variable.

Finding Values of Angles α, $0° \leq \alpha < 360°$

We have already solved many simple conditional equations, such as

$$\sin \alpha = \frac{1}{2}, \tag{1}$$

in Chapter 1 over the interval $0° \leq \alpha \leq 90°$. In this case

$$\alpha = \text{Arcsin} \frac{1}{2} = 30°.$$

However, over the interval $0°$ to $360°$, there are two elements in the domain of the sine function corresponding to $1/2$ in the range; either

$$\alpha_1 = 30° \quad \text{or} \quad \alpha_2 = 150°$$

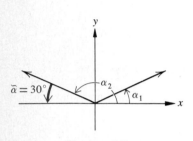

Figure 4.3

satisfies Equation (1), as shown in Figure 4.3, where $\tilde{\alpha} = 30°$.

In general, for most elements in the range of a trigonometric function, there are two corresponding elements in its domain over the interval $0°$ to $360°$ or 0^R to $2\pi^R$. A reference angle can be used to find all elements in the domain of a trigonometric function over those intervals that satisfy a given condition. For any positive function value, the angle obtained from a table or a calculator is an angle in the first quadrant and is the reference angle that can be used to find a possible second value that satisfies the specified condition.

Example   Given that $\tan \alpha = 1$, find the following for which the statement is true:

a. The least nonnegative angle α in degrees;

b. $\{\alpha \,|\, 0° \leq \alpha < 360°\}$.

Solution   The reference angle $\tilde{\alpha}$, for which

$$\tan \tilde{\alpha} = \tan \alpha = 1,$$

is

$$\tilde{\alpha} = \text{Tan}^{-1} 1 = 45°$$

Since $\tan \alpha$ (equal to 1) is positive, α is in Quadrant I or Quadrant III. A sketch showing a reference angle is helpful.

a. The least nonnegative angle α_1 is

$$\alpha = \tilde{\alpha} = 45°$$

b. Because the period of the tangent function is $180°$,

$$\alpha + 180° = 45° + 180° = 225°.$$

Hence,

$$\{\alpha \mid \tan \alpha = 1, \quad 0° \leq \alpha < 360°\} = \{45°, 225°\}.$$

A procedure similar to the one used in the above examples can also be used in cases in which the function value is negative. In these cases it is helpful to work with the absolute value of the function value to obtain the reference angle.

Example Given that $\sin \alpha = -0.6494$, find the following for which the statement is true:

    a. The least nonnegative angle α in degrees;

    b. $\{\alpha \mid 0° \leq \alpha < 360°\}$.

Solution The reference angle $\tilde{\alpha}$, for which

$$\sin \tilde{\alpha} = |\sin \alpha| = |-0.6494| = 0.6494,$$

is

$$\tilde{\alpha} = \text{Sin}^{-1} 0.6494 \approx 40.5°.$$

Since $\sin \alpha$ is negative, α is in Quadrant III or IV. A sketch showing the reference angle in each of these quadrants is helpful.

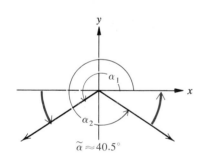

    a. The least nonnegative angle is

$$\alpha_1 = 180° + \tilde{\alpha}$$

$$\approx 180° + 40.5° = 220.5°.$$

    b. In Quadrant IV,

$$\alpha_2 = 360° - \tilde{\alpha}$$

$$\approx 360° - 40.5° = 319.5°.$$

Hence,

$$\{\alpha \mid \sin \alpha = -0.6494, \quad 0° \leq \alpha < 360°\} = \{220.5°, 319.5°\}.$$

Example Given that $\cos \alpha = -0.7518$, find the following for which the statement is true:

    a. The least nonnegative angle α in radians;

    b. $\{\alpha \mid 0^R \leq \alpha < 2\pi^R\}$.

(continued)

Solution   The reference angle $\tilde{\alpha}$, for which

$$\cos \tilde{\alpha} = |\cos \alpha| = |-0.7518| = 0.7518,$$

is

$$\tilde{\alpha} = \text{Cos}^{-1} \, 0.7518 \approx 0.72^R$$

Since $\cos \alpha$ is negative, α is in Quadrant II or Quadrant III. A sketch showing the reference angle in each of these quadrants is helpful.

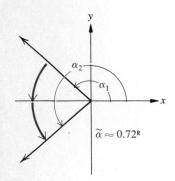

a. The least nonnegative angle is

$$\alpha_1 = \pi^R - \tilde{\alpha}$$
$$\approx (3.14 - 0.72)^R = 2.42^R.$$

b. In Quadrant III,

$$\alpha_2 = \pi^R + \tilde{\alpha}$$
$$\approx (3.14 + 0.72)^R = 3.86^R.$$

Hence,

$$\{\alpha \,|\, \cos \alpha = -0.7518, \quad 0^R \le \alpha < 2\pi^R\} = \{2.42^R, 3.86^R\}.$$

Finding All Values of Angles

The trigonometric functions are periodic. Hence, we can obtain the infinite set of values for angles α for a given trigonometric function value by first finding the value of α in the interval $0° \le \alpha < 360°$ (or $0° \le x < 2\pi^R$) by the methods used in the above examples.

Example   Find the set of all angles α (in degree measure) for which $\cos \alpha = 1/\sqrt{2}$.

Solution   The reference angle $\tilde{\alpha}$, for which

$$\cos \tilde{\alpha} = |\cos \alpha| = \frac{1}{\sqrt{2}}$$

is

$$\tilde{\alpha} = \text{Cos}^{-1} \frac{1}{\sqrt{2}} = 45°.$$

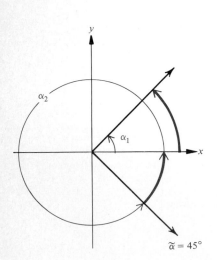

Since $\cos \alpha$ is positive, α is in Quadrants I or IV. The figure shows the reference angles in each of these quadrants. The least nonnegative angle is

$$\alpha_1 = 45°.$$

In Quadrant IV,

$$\alpha_2 = 360° - \tilde{\alpha}$$
$$= 360° - 45° = 315°.$$

Hence,

$$\left\{\alpha \,\middle|\, \cos \alpha = \frac{1}{\sqrt{2}}\right\} = \{\alpha \,|\, \alpha = 45° + k \cdot 360° \quad \text{or} \quad \alpha = 315° + k \cdot 360°\}.$$

Finding Values of x, x ∈ R

In the above discussion and examples we viewed the elements in the domain associated with a given element in the range as angles with degree or radian measures. Similar procedures are used if the elements in the domain are viewed as real numbers.

Example   Find the set of all real numbers for which $\tan x = -\sqrt{3}$.

Solution   The reference arc $\tilde{x}$, for which

$$\tan \tilde{x} = |\tan x| = \sqrt{3}$$

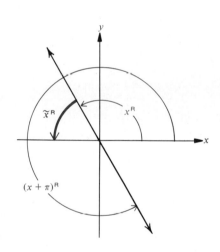

is

$$\tilde{x} = \text{Tan}^{-1} \sqrt{3} = \frac{\pi}{3}.$$

Since $\tan x$ is negative, x is a real number that can be associated with an angle in Quadrants II or IV. The figure shows $\tilde{x}$. The least nonnegative x is

$$x = \pi - \frac{\pi}{3} = \frac{2\pi}{3}.$$

Since the tangent function has period π,

$$\{x \,|\, \tan x = -\sqrt{3}\} = \left\{x \,\middle|\, x = \frac{2\pi}{3} + k \cdot \pi\right\}.$$

EXERCISE SET 4.8

A

For each function value, find **(a)** *the exact value of the least nonnegative angle* α *in degrees and* **(b)** $\{\alpha \,|\, 0° \leq \alpha < 360°\}$ *for which the statement is true.*

Example $\sin \alpha = -\dfrac{\sqrt{3}}{2}$

(continued)

Solution   The reference angle $\tilde{\alpha}$, for which

$$\sin \tilde{\alpha} = |\sin \alpha| = \left| -\frac{\sqrt{3}}{2} \right| = \frac{\sqrt{3}}{2},$$

is

$$\tilde{\alpha} = \text{Sin}^{-1} \frac{\sqrt{3}}{2} = 60°.$$

Since $\sin \alpha$ (equal to $-\sqrt{3}/2$) is negative, α is in Quadrant III or Quadrant IV. A sketch showing the reference angle in each of these quadrants is helpful.

a. The least nonnegative angle in Quadrant III is

$$\alpha_1 = 180° + \tilde{\alpha} = 180° + 60° = 240°.$$

b. The angle in Quadrant IV is

$$\alpha_2 = 360° - 60° = 300°.$$

Hence,

$$\left\{ \alpha \,\middle|\, \sin \alpha = -\frac{\sqrt{3}}{2}, \quad 0° \le \alpha < 360° \right\} = \{240°, 300°\}.$$

1. $\cos \alpha = \dfrac{\sqrt{3}}{2}$ 2. $\sin \alpha = \dfrac{1}{\sqrt{2}}$ 3. $\tan \alpha = -\dfrac{1}{\sqrt{3}}$

4. $\cos \alpha = \dfrac{1}{2}$ 5. $\cot \alpha = \sqrt{3}$ 6. $\tan \alpha = 1$

7. $\sec \alpha = \sqrt{2}$ 8. $\cot \alpha = -1$ 9. $\csc \alpha = -2$

10. $\sec \alpha = 2$ 11. $\sin \alpha = -\dfrac{1}{2}$ 12. $\csc \alpha = \dfrac{2}{\sqrt{3}}$

13. $\cos \alpha = -\dfrac{1}{\sqrt{2}}$ 14. $\sin \alpha = -\dfrac{\sqrt{3}}{2}$ 15. $\csc \alpha = \sqrt{2}$

16. $\cos \alpha = -1$ 17. $\sin \alpha = 0$ 18. $\tan \alpha = -\sqrt{3}$

19. $\sec \alpha = -\dfrac{2}{\sqrt{3}}$ 20. $\cot \alpha = 0$ 21. $\csc \alpha$ undefined

22. $\sec \alpha$ undefined   23. $\cot \alpha$ undefined   24. $\tan \alpha$ undefined

For each function value, find an approximation in degrees for (a) *the least nonnegative angle* α *and* (a) $\{\alpha \mid 0° \leq \alpha < 360°\}$ *for which each statement is true.*

Example $\tan \alpha = -0.8693$

Solution   The reference angle $\tilde{\alpha}$, for which

$$\tan \tilde{\alpha} = |\tan \alpha| = |-0.8693| = 0.8693,$$

is

$$\tilde{\alpha} = \text{Tan}^{-1}\, 0.8693 \approx 41°.$$

Since $\tan \alpha$ (equal to -0.8693) is negative, α is in Quadrant II or Quadrant IV. A sketch showing a reference angle is helpful.

a. The least nonnegative angle in Quadrant II is

$$\alpha = 180° - \tilde{\alpha}$$
$$\approx 180° - 41° = 139°.$$

b. Because the period of the tangent function is 180°,

$$\alpha + 180° \approx 139° + 180° = 319°$$

Hence,

$$\{\alpha \mid \tan \alpha = -0.8693, \quad 0° \leq \alpha < 360°\} = \{139°, 319°\}.$$

25. $\sin \alpha = 0.2419$ 26. $\cos \alpha = 0.5150$ 27. $\tan \alpha = -2.6051$

28. $\cot \alpha = -2.1445$ 29. $\sec \alpha = 1.0320$ 30. $\csc \alpha = 1.1570$

31. $\cos \alpha = \div 0.9085$ 32. $\sin \alpha = -0.3535$ 33. $\cot \alpha = 2.1060$

34. $\tan \alpha = 0.3134$ 35. $\csc \alpha = -1.2796$ 36. $\sec \alpha = -3.8140$

37. $\sin \alpha = -0.5476$ 38. $\cot \alpha = 0.6032$ 39. $\cos \alpha = 0.3955$

40. $\csc \alpha = -2.3400$ 41. $\tan \alpha = 1.0212$ 42. $\sec \alpha = 1.6316$

For each function value, find an approximation in radians for (a) *the least nonnegative angle* α *and* (b) $\{\alpha \mid 0^R \leq \alpha < 2\pi^R\}$ *for which each statement is true.*

Example $\cos \alpha = -0.9131$

(continued)

Solution   The reference angle $\tilde{\alpha}$, for which

$$\cos \tilde{\alpha} = |\cos \alpha| = |-0.9131| = 0.9131,$$

is

$$\tilde{\alpha} = \text{Cos}^{-1}\, 0.9131 \approx 0.42^{\text{R}}.$$

Since $\cos \alpha$ (equal to -0.9131) is negative, α is in Quadrant II or Quadrant III. A sketch showing the reference angle in each of these quadrants is helpful.

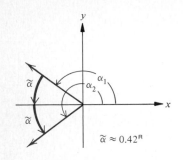

a. The least nonnegative angle in Quadrant II is

$$\alpha_1 = \pi^{\text{R}} - \tilde{\alpha} \approx \pi^{\text{R}} - 0.42^{\text{R}} \approx 2.72^{\text{R}}.$$

b. The angle in Quadrant III is

$$\alpha_2 \approx \pi^{\text{R}} + 0.42^{\text{R}} \approx 3.56^{\text{R}}.$$

43. $\sin \alpha = 0.6442$ **44.** $\cos \alpha = 0.5898$ **45.** $\tan \alpha = 0.2341$

46. $\sin \alpha = 0.2280$ **47.** $\sec \alpha = 1.795$ **48.** $\csc \alpha = 1.467$

49. $\cos \alpha = -0.9689$ **50.** $\sin \alpha = -0.5480$ **51.** $\sin \alpha = -0.1197$

52. $\tan \alpha = -1.050$ **53.** $\csc \alpha = -2.125$ **54.** $\sec \alpha = -5.273$

For each function value, find an approximation to the nearest tenth of a degree for (**a**) *the least nonnegative angle* α *and* (**b**) $\{\alpha \mid 0° \le \alpha < 360°\}$ *for which each statement is true.*

Example $\tan \alpha = -0.2162$

Solution   The reference angle $\tilde{\alpha}$, for which

$$\tan \tilde{\alpha} = |\tan \alpha| = |-0.2162| = 0.2162,$$

is

$$\tilde{\alpha} = \text{Tan}^{-1}\, 0.2162 \approx 12.2°.$$

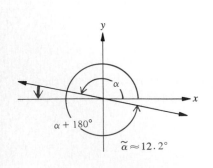

Since $\tan \alpha$ (equal to -0.2162) is negative, α is in Quadrant II or IV. The figure shows a reference angle. The least nonnegative angle is

$$\alpha = 180° - \tilde{\alpha}$$
$$\approx 180° - 12.2° = 167.8°.$$

Because the period of the tangent function is 180°,

$$\{\alpha \mid \tan \alpha = -0.2162\}$$

$$= \{\alpha \mid \alpha \approx 167.8 + k \cdot 180°\}, \quad k \in J.$$

55. $\sin \alpha = 0.2790$ **56.** $\cos \alpha = 0.9311$

57. $\cot \alpha = -1.2799$ **58.** $\tan \alpha = -1.4882$

59. $\csc \alpha = -1.2813$ **60.** $\sec \alpha = -1.1164$

61. $\tan \alpha = 0.3019$ **62.** $\sin \alpha = -0.9426$

4.9 Additional Conditional Equations

In Section 4.8 we obtained the solution set over various domains of simple conditional equations such as

$$\sin x = \frac{1}{2}, \quad \cos x = -0.3824, \quad \text{and} \quad \tan x = 1.752$$

in which the left-hand member contained only one function value and that function value was of first degree. In the case $\sin x = 1/2$, we obtained solutions by inspection.

Perhaps you can determine the solution of

$$\sin x = \cos x, \quad 0 < x < \frac{\pi}{2}, \tag{1}$$

by inspection—perhaps not. However, you can generate equivalent equations and hopefully you will generate one from which the solution(s) will be evident by inspection. Generally, solutions can be determined most easily from equations that involve *values of one function only*. The identities developed in the preceding sections are useful in obtaining such equivalent equations from equations involving values of more than one function. For example, in (1) above, for $0 < x < \pi/2$, $\cos x \neq 0$. Thus, we can multiply each member by $1/\cos x$ to obtain

$$\frac{\sin x}{\cos x} = 1.$$

Substituting $\tan x$ for $\sin x/\cos x$, we obtain

$$\tan x = 1.$$

Since, in this case, values of x are restricted to $0 < x < \pi/2$,

$$x = \mathrm{Tan}^{-1}\, 1 = \frac{\pi}{4}.$$

Example   Solve $2 \sin x \cos x - \sin x = 0$ over the interval $0 \leq x \leq \pi/2$.

Solution   Factoring the left-hand member we obtain the equivalent equation

$$\sin x (2 \cos x - 1) = 0.$$

The left-hand member equals zero if

$$\sin x = 0 \qquad \text{or} \qquad \cos x = \frac{1}{2},$$

from which

$$x = \mathrm{Sin}^{-1}\, 0 = 0 \qquad \text{or} \qquad x = \mathrm{Cos}^{-1}\, \frac{1}{2} = \frac{\pi}{3},$$

and the solution set over the interval $0 \leq x \leq \pi/2$ is $\left\{ 0, \dfrac{\pi}{3} \right\}$.

Because function values ($\cos x$, $\sin x$, etc.) are real numbers, equations that are quadratic in a given trigonometric function value, such as

$$\cos^2 x + \cos x - 2 = 0, \quad x \in A, \tag{2}$$

can be solved by first solving the quadratic equation for the function value involved. To emphasize that we are using a standard algebraic method for solving a quadratic equation, we also show each step of the process, where we let $\cos x = u$ and write Equation (2) as

$$u^2 + u - 2 = 0. \tag{3}$$

Since the left-hand members of Equations (2) and (3) are factorable, the equations can be written equivalently as

$$(\cos x + 2)(\cos x - 1) = 0 \qquad \text{or} \qquad (u + 2)(u - 1) = 0.$$

Since $\cos x + 2$ cannot equal 0 (there is no angle x such that $\cos x = -2$), we seek values of x for which

$$\cos x - 1 = 0 \qquad \text{or} \qquad u - 1 = 0.$$

From either equation, we have

$$\cos x = 1.$$

The least nonnegative value of x is given by

$$x = \text{Cos}^{-1}\, 1 = 0.$$

Hence the solution set of Equation (2) is

$$\{x \,|\, x = k \cdot 360°, \quad k \in J\} \qquad \text{or} \qquad \{x \,|\, x = k \cdot 2\pi^R, \quad k \in J\}.$$

EXERCISE SET 4.9

A

Solve each equation. Find all solutions in the interval $0 \leq x \leq \pi/2$.

Examples

a. $(2 \cos x - \sqrt{3})(\sin x - 1) = 0$ b. $\sqrt{2} \sin x \cos x - \cos x = 0$

Solutions

a. The left-hand member equals 0 if

$$2 \cos x - \sqrt{3} = 0 \qquad \text{or} \qquad \sin x - 1 = 0.$$

From these equations

$$\cos x = \frac{\sqrt{3}}{2} \qquad \text{or} \qquad \sin x = 1,$$

and the solution set is $\{\pi/6, \pi/2\}$.

b. Factoring the left-hand member

$$\cos x(\sqrt{2} \sin x - 1) = 0.$$

The left-hand member equals 0 if

$$\cos x = 0 \qquad \text{or} \qquad \sin x = \frac{1}{\sqrt{2}},$$

and the solution set is $\{\pi/2, \pi/4\}$.

1. $2 \sin x - 1 = 0$ **2.** $2 \cos x - 1 = 0$

3. $2 \cos^2 x = 1$ **4.** $4 \sin^2 x = 3$

5. $3 \csc^2 x - 3 = 0$ **6.** $\sec^2 x = 4$

7. $(2 \cos^2 x - 1)(\sin x - 1) = 0$ **8.** $(\sin^2 x - 1)(\sec x - 2) = 0$

9. $\cos x \sin x = 0$ **10.** $3 \tan x \sec x = 0$

11. $2 \cos^3 x - \cos x = 0$ **12.** $2 \sin^3 x - \sin x = 0$

13. $2 \sin x \cos x + \sin x = 0$ **14.** $2 \sin x \cos x - \cos x = 0$

Solve each equation. Find all solutions in the interval $0 \le x < 2\pi$.

Example $2 \cos^2 x + \cos x - 1 = 0$

Solution   Factoring the left-hand member,

$$(2 \cos x - 1)(\cos x + 1) = 0,$$

$$\cos x = \frac{1}{2} \quad \text{or} \quad \cos x = -1,$$

and the solution set is $\{\pi/3, 5\pi/3, \pi\}$.

15. $2 \cos^2 x - \cos x - 1 = 0$ **16.** $2 \cos^2 x + 3 \cos x - 2 = 0$

17. $9 \cos^2 x - 21 \cos x - 8 = 0$ **18.** $25 \sin^2 x - 20 \sin x + 4 = 0$

Solve each equation **(a)** *in the interval* $0° \le x < 360°$ *and* **(b)** *for all* $x \in A$.

Example $\sec x - 2 \cos x = 1$

Solution

    a. First write each term using the same function value. Substitute $1/\cos x$ for $\sec x$ to obtain

$$\frac{1}{\cos x} - 2 \cos x = 1.$$

Multiplying each member by $\cos x$ ($\cos x \ne 0$),

$$1 - 2 \cos^2 x = \cos x,$$

from which

$$2 \cos^2 x + \cos x - 1 = 0.$$

This equation is the same as in the preceding example. Thus, the solution set over the interval $0° \le x < 360°$ is $\{60°, 300°, 180°\}$.

b. For all $x \in A$, the solution set is

$$\{x \mid x = 60° + k \cdot 360°\} \cup \{x \mid x = 180° + k \cdot 360°\}$$
$$\cup \{x \mid x = 300° + k \cdot 360°\}, \quad k \in J,$$

19. $\csc x + \sin x + 2 = 0$ 20. $2 \cos x + \sec x - 3 = 0$

21. $2 \sin^2 x - \cos x - 1 = 0$ 22. $2 \cos^2 x - 3 \sin x - 3 = 0$

23. $2 \sin x + \csc x - 3 = 0$ 24. $\tan x + \cot x + 2 = 0$

B

Solve each equation (**a**) *in the interval* $0° \leq x < 360°$ *and* (**b**) *for all $x \in A$. Approximate answers to the nearest tenth of a degree.*   Hint: *Use the quadratic formula.*

25. $\sin^2 x - \sin x - 1 = 0$ 26. $\cos^2 x + 5 \cos x + 2 = 0$

27. $3 \cos x - 1 = \sec x$ 28. $4 \sin x \cos x - 2 \sin x = \tan x$

Solve each equation in the interval $0 \leq x \leq 2\pi$.   Hint: *Square each member.*

29. $1 + \cos x = \sqrt{3} \sin x$ 30. $\cos x + \sin x = 1$

31. $\cot x = 1 + \csc x$ 32. $1 + \tan x = -\sqrt{2} \sec x$

Solve each equation for all $x \in R$.

33. $2 \cot x \sec x + \cot x + 2 \sec x = -1$

34. $\cos x \tan x - \cos x + \sin x \tan x - \sin x = 0$

4.10 Conditional Equations for Multiples

Conditional equations such as

$$2 \sin x \cos x = \frac{\sqrt{3}}{2}, \tag{1}$$

$$\tan 3x = 1, \tag{2}$$

and

$$\cos 2x = \cos^2 x - 1 \tag{3}$$

need further discussion.

The identities involving multiples of the variable are often useful to help solve certain kinds of equations. For example, the solution set of

$$2 \sin x \cos x = \frac{\sqrt{3}}{2}, \tag{1}$$

over the interval $0 \leq x \leq \pi/2$, can readily be found if $\sin 2x$ is substituted for the left-hand member, $2 \sin x \cos x$, to obtain

$$\sin 2x = \frac{\sqrt{3}}{2}.$$

By inspection we have that

$$2x = \frac{\pi}{3} + k \cdot 2\pi \qquad \text{or} \qquad 2x = \frac{2\pi}{3} + k \cdot 2\pi,$$

$$x = \frac{\pi}{6} + k \cdot \pi \qquad\qquad x = \frac{\pi}{3} + k \cdot \pi.$$

If $k = 0$, $x = \pi/6$ or $x = \pi/3$; if $k \geq 1$, then $x > \pi$. Hence, the solution set of Equation (1) over the interval from 0 to $\pi/2$ is $\{\pi/6, \pi/3\}$.

Now consider the solution set of

$$\tan 3x = 1 \tag{2}$$

over the interval $0 \leq x < 2\pi$. By inspection we have that

$$3x = \frac{\pi}{4} + k \cdot \pi,$$

and, dividing each term by 3,

$$x = \frac{\pi}{12} + k \cdot \frac{\pi}{3}, \quad k \in J.$$

To find the solution set over the interval $0 \leq x < 2\pi$, we take 0, 1, 2, 3, 4, and 5 as replacements for k and obtain

$$\left\{ \frac{\pi}{12}, \frac{5\pi}{12}, \frac{3\pi}{4}, \frac{13\pi}{12}, \frac{17\pi}{12}, \frac{7\pi}{4} \right\}.$$

Note that if $k > 5$, then $x > 2\pi$.

Now consider the equation

$$\cos 2x = \cos^2 x - 1. \tag{3}$$

One method of solving this equation for $x \in R$ consists of substituting $2\cos^2 x - 1$ for $\cos 2x$ to obtain the equivalent equations

$$2\cos^2 x - 1 = \cos^2 x - 1,$$

$$\cos^2 x = 0.$$

Then, we have as the solution set

$$\left\{ x \,\middle|\, x = \frac{\pi}{2} + k\pi, \quad k \in J \right\}.$$

Alternatively, we could have substituted $(1 + \cos 2x)/2$ for $\cos^2 x$ in the right-hand member of Equation (3) to obtain the equivalent equations

$$\cos 2x = \frac{1 + \cos 2x}{2} - 1,$$

$$2\cos 2x = 1 + \cos 2x - 2,$$

$$\cos 2x = -1.$$

By inspection, we have that

$$2x = \pi + k \cdot 2\pi, \quad k \in J,$$

from which the solution set is given by

$$\left\{ x \,\middle|\, x = \frac{\pi}{2} + k\pi, \quad k \in J \right\},$$

as we found above using the other method.

EXERCISE SET 4.10

A

Solve each equation for all x, such that $0 \le x \le \pi/2$.

Example $\cos 3x = \dfrac{1}{2}$

Solution   Using Table 1.2 and the fact that the cosine of an angle in Quadrant I or IV is positive, we first consider all possible solutions and

(continued)

obtain

$$3x = \frac{\pi}{3} + k \cdot 2\pi \qquad \text{or} \qquad 3x = \frac{5\pi}{3} + k \cdot 2\pi.$$

From which, dividing each term by 3,

$$x = \frac{\pi}{9} + k \cdot \frac{2\pi}{3} \qquad \text{or} \qquad x = \frac{5\pi}{9} + k \cdot \frac{2\pi}{3}.$$

Since the problem restricts x to the interval $0 \le x \le \pi/2$, we need only the value 0 for k. If $k = 0$, $x = \pi/9$ or $x = 5\pi/9$. Furthermore, since $5\pi/9$ is not in the required interval, the solution set is $\{\pi/9\}$.

1. $\sin 3x = 0$ 2. $\cos 3x = 1$

3. $\tan 4x = -\sqrt{3}$ 4. $\cot 4x = -1$

5. $\sec 5x = 2$ 6. $\csc 5x = \sqrt{2}$

7. $\cos 2x - \sin x = 0$ 8. $\cos 2x + \cos x = 0$

9. $\sin 2x + \sin x = 0$ 10. $\sin 2x - \cos x = 0$

11. $\sin 2x = \cos 2x$ 12. $\sin \dfrac{x}{2} = \tan \dfrac{x}{2}$

13. $\tan \dfrac{x}{2} - \cos x = -1$ 14. $\cos x = \sin \dfrac{x}{2}$

Solve each equation in the interval $0° \le x < 360°$.

Example $\sin 4x = \dfrac{\sqrt{3}}{2}$

Solution   By inspection we have $4x = 60°$. However, since $\sin \theta$ is also positive for angles in Quadrant II, we know that $4x = 120°$ will also satisfy the equation. Thus for $k \in J$,

$$4x = 60° + k \cdot 360° \qquad \text{or} \qquad 4x = 120° + k \cdot 360°.$$

Dividing each member of both equations by 4, we obtain

$$x = 15° + k \cdot 90° \qquad \text{or} \qquad x = 30° + k \cdot 90°.$$

Since the problem restricts x to the interval $0° \le x < 360°$, we need only the values of 0, 1, 2, and 3 for k.

If $k = 0$, $x = 15°$ or $x = 30°$,

if $k = 1$, $x = 105°$ or $x = 120°$,

if $k = 2$, $x = 195°$ or $x = 210°$,

if $k = 3$, $x = 285°$ or $x = 300°$,

and the solution set is $\{15°, 30°, 105°, 120°, 195°, 210°, 285°, 300°\}$.

15. $\sin 3x = \dfrac{1}{2}$ **16.** $\cos 4x = -1$

17. $\sin 3x = \cos 3x$ **18.** $\tan 2x = \cot 2x$

19. $\sin^2 2x = 1$ **20.** $\cos^2 2x = 1$

21. $\tan^2 3x = 1$ **22.** $\cot^2 3x = 3$

23. $\left(1 + \cos \dfrac{x}{2}\right) \cos \dfrac{x}{2} = 0$ **24.** $\sin^2 \dfrac{x}{4} - \dfrac{1}{2} \sin \dfrac{x}{4} = 0$

25. $5 \sec 3x = 2 \cos 3x - 3$ **26.** $\tan x \tan 2x = 1$

B

Solve each equation for all x, such that $x \in R$.

27. $\sin 2x = \cos x$ **28.** $2 \cos 2x - 1 = -3 \sin^2 x$

29. $\sin x \cos 3x = 0$ **30.** $(\sin 2x - 1)(\cos 3x + 1) = 0$

31. $\tan 2x - 3 \sec^2 2x = -5$ **32.** $\cos^2 2x + 3 \sin 2x - 3 = 0$

33. $\sin 2x \cos x + \cos 2x \sin x = \dfrac{1}{\sqrt{2}}$

34. $\sin 3x \cos x - \cos 3x \sin x = \dfrac{\sqrt{3}}{2}$

35. Use graphical methods to approximate a positive solution for

$$\cos 2x = x, \quad x \in R.$$

Hint: Graph $y = \cos 2x$ and $y = x$ on the same axes.

36. Use graphical methods to approximate a positive solution for

$$x^2 - 3 \sin \dfrac{x}{2} = 0, \quad x \in R.$$

Chapter Summary

4.1–4.7 Equations are **identities** if they are satisfied for all real number replacements of the variable for which both members of the equation are defined.

For convenience, the basic identities that have been established are listed inside the back cover.

An equation can be shown *not* to be an identity by replacing the variable with a real number for which the resulting statement is false.

4.8–4.10 **Conditional equations** involving circular function values can be solved by inspection or by generating equivalent equations from which the solutions are evident by inspection. The identities introduced in this chapter are useful in obtaining such equivalent equations.

A list of symbols introduced in this chapter is shown inside the front cover, and important properties are shown inside the back cover.

Review Exercises

4.1–4.5 *Express each of the following in terms of the function values* $\sin \alpha$ *and/or* $\cos \alpha$ *and simplify the expression.*

1. $\csc \alpha \tan \alpha$

2. $\cot^2 \alpha \sec^2 \alpha$

3. $\sin^2 \alpha (\csc^2 \alpha - 1)$

4. $\cot \alpha \cos \alpha \sec^2 \alpha$

5. $\sin(-\alpha) \cot \alpha$

6. $\sec^2(-\alpha) \cos(-\alpha)$

Verify that each equation is an identity.

7. $\sin^2 \alpha(1 + \cot^2 \alpha) = 1$

8. $\dfrac{\cos \alpha \csc \alpha}{\cot \alpha} = 1$

9. $\tan \alpha = \dfrac{1}{\cos \alpha \csc \alpha}$

10. $1 + \tan^2 \alpha = \csc^2 \alpha \tan^2 \alpha$

11. $\sin(-\alpha) = -\dfrac{\tan \alpha}{\sec \alpha}$

12. $\tan(-\alpha) = \dfrac{\cos(-\alpha) \sin(-\alpha)}{1 - \sin^2 \alpha}$

13. $\cos(270° + x) = \sin x$

14. $\cos(270° - x) = -\sin x$

15. $\sin(90° + x) = \cos x$

16. $\sin(\alpha - 45°) = \dfrac{\sin \alpha - \cos \alpha}{\sqrt{2}}$

17. $\cos(\alpha - 45°) = \dfrac{\cos \alpha + \sin \alpha}{\sqrt{2}}$

18. $\tan(\alpha - 45°) = \dfrac{\tan \alpha - 1}{1 + \tan \alpha}$

4.6 **19.** $\cos 2x + 1 = 2 \cos^2 x$ **20.** $\sin 2x = 2 \sin^2 x \cos x \csc x$

21. $\csc x = 2 \cos x \csc 2x$ **22.** $\cos 2x = \cos^4 x - \sin^4 x$

By counterexamples, show that each equation is not an identity.

23. $\sin \alpha \sec \alpha = 2 \tan \alpha$ **24.** $\dfrac{1 - \cos \alpha}{\sin \alpha} = \dfrac{1 + \sin \alpha}{\cos \alpha}$

4.7 *Use half-angle formulas to find the exact value for each of the following.*

25. $\cos 105°$ **26.** $\tan 165°$

Verify the identities in Exercises 27 and 28.

27. $\cos^2 \dfrac{x}{2} = \dfrac{\sin^2 x}{2(1 - \cos x)}$ **28.** $\sin^2 \dfrac{x}{2} = \dfrac{\csc x - \cot x}{2 \csc x}$

29. Express $\cos x \cos 5x$ in terms of sums of function values.

30. Express $\sin 6x + \sin 4x$ in terms of products of function values.

4.8 *For each function value, find (a) the exact value of the least nonnegative angle α in degrees and (b) $\{\alpha \mid 0° \leq \alpha < 360°\}$ for which the statement is true.*

31. $\sin \alpha = \dfrac{\sqrt{3}}{2}$ **32.** $\tan \alpha = -1$

For each function value, find an approximation in degrees for (a) the least nonnegative angle α and (b) $\{\alpha \mid 0° \leq \alpha < 360°\}$ for which each statement is true.

33. $\cos \alpha = -0.4695$ **34.** $\csc \alpha = 1.6972$

For each function value, find an approximation in radians for (a) the least nonnegative angle α and (b) $\{\alpha \mid 0^R \leq \alpha < 2\pi^R\}$ for which each statement is true.

35. $\sec \alpha = 1.0720$ **36.** $\cot \alpha = -0.2439$

37. $\sin \alpha = 0.3107$ **38.** $\cos \alpha = 0.8965$

4.9 *Solve each equation in the interval $0 \leq x < \pi/2$.*

39. $2 \sin x - \sqrt{3} = 0$ **40.** $\csc^2 x = 4$

41. $\sec x - 2 \cos x = 0$ **42.** $\cot x \csc x = 0$

Solve each equation in the interval $0° \leq x < 360°$.

43. $\cot^2 x - 2 \csc^2 x + 5 = 0$ **44.** $\sec^2 x - 4 \cos^2 x = 0$

4.10 *Solve each equation in the interval* $0° \leq x < 360°$.

45. $\cos 3x = \dfrac{1}{\sqrt{2}}$ **46.** $\sin 4x = \dfrac{\sqrt{3}}{2}$

47. $\sin 2x - \tan x = 0$ **48.** $\tan 2x = \tan x$

5 VECTORS

5.1 Geometric Vectors

Some physical quantities, such as length and mass, are characterized by *magnitude alone*. If units of measure are specified, the quantities can be represented by real numbers and associated with points in a number line. In the description of other physical quantities, such as force, velocity, and acceleration, *direction as well as magnitude* is required. Such a quantity can be represented by a pair of real numbers and associated with a line segment in the form of an arrow with length representing the magnitude and pointing in an assigned direction.

Definition 5.1 *A two-dimensional* **geometric vector*** *designated by* $\vec{v}$ *is a line segment with a specified direction.*

If the endpoints of a geometric vector are named, it is sometimes convenient to specify the vector by using these names. For example, in Figure 5.1 the geometric vector can also be designated by $\overrightarrow{PQ}$, where the point named P is called the **initial point** and the point named Q is called the **terminal point.** For discussion purposes, the geometric vector in Figure 5.1 is shown in the framework of a rectangular coordinate system.

Sometimes a geometric vector is simply represented by **v** without using the arrow above the symbol. We are using the arrow to differentiate this symbol for a geometric vector from the symbol v as a variable for a real

Figure 5.1

---

* The term *geometric vector* is used to differentiate this notion of a directed line segment from a vector as an ordered pair of real numbers introduced in Section 5.3.

number. Also the arrow emphasizes the fact that a geometric vector has *magnitude* and *direction*.

Magnitude and Direction

We shall assume that each geometric vector in the plane has a length that is a real number that depends on the unit of measurement.

Definition 5.2 *For all geometric vectors $\vec{\mathbf{v}}$, the* **norm** *(or* **magnitude***) of $\vec{\mathbf{v}}$ is the length of $\vec{\mathbf{v}}$ and is designated by $\|\vec{\mathbf{v}}\|$.*

When a geometric vector is considered in relation to a rectangular coordinate system, the following concept is useful.

Definition 5.3 *The* **direction angle** *of a geometric vector $\vec{\mathbf{v}}$ is an angle α, $-180° < \alpha \le 180°$, such that the initial side of α is the ray from the initial point of $\vec{\mathbf{v}}$ parallel to the x-axis and directed in the positive x direction; the terminal side of α is the ray from the initial point of $\vec{\mathbf{v}}$ and containing $\vec{\mathbf{v}}$.*

Figure 5.2 shows $\|\vec{\mathbf{v}}\|$ and the direction angle α associated with a vector $\vec{\mathbf{v}}$. From a vector viewpoint, the real number $\|\vec{\mathbf{v}}\|$ is called a **scalar.**

Definition 5.4 *For all geometric vectors $\vec{\mathbf{v}}$, the geometric vector $-\vec{\mathbf{v}}$ has the same magnitude as $\vec{\mathbf{v}}$ and has a direction angle whose measure differs from the measure of the direction angle of $\vec{\mathbf{v}}$ by $180°$.*

Figure 5.3 shows two vectors $\vec{\mathbf{v}}$ and $-\vec{\mathbf{v}}$.

Definition 5.5 *Two geometric vectors $\vec{\mathbf{v}}_1$ and $\vec{\mathbf{v}}_2$ are* **equivalent** *if they have the same magnitude and the same direction angle.*

Figure 5.4 illustrates three equivalent vectors. We use the symbol $=$ to indicate that two vectors are equivalent.

If we can imagine "sliding" a geometric vector so that it remains parallel to its original position in the plane, then any geometric vector can "slide" onto any vector equivalent to it.

Operations

Geometric vectors are very useful in setting up mathematical models of practical problems. However, before this can be done, certain operations must be defined.

Definition 5.6 *For all geometric vectors $\vec{\mathbf{v}}_1$ and $\vec{\mathbf{v}}_2$, such that the terminal point of $\vec{\mathbf{v}}_1$ is the initial point of $\vec{\mathbf{v}}_2$, $\vec{\mathbf{v}}_1 + \vec{\mathbf{v}}_2$ is the geometric vector having the same initial point as $\vec{\mathbf{v}}_1$ and the same terminal point as $\vec{\mathbf{v}}_2$. $\vec{\mathbf{v}}_1 + \vec{\mathbf{v}}_2$ is called the* **sum** *(or* **resultant***) of $\vec{\mathbf{v}}_1$ and $\vec{\mathbf{v}}_2$.*

Figure 5.2

Figure 5.3

Figure 5.4

Figure 5.5

As indicated in Figure 5.5a, if $\vec{v}_1$ and $\vec{v}_2$ are not in the relative locations required by Definition 5.6, we "slide" one of them, say $\vec{v}_2$, parallel to its original position into the correct location and then find the sum $\vec{v}_1 + \vec{v}_2$.

Because a triangle is formed by $\vec{v}_1, \vec{v}_2$, and $\vec{v}_1 + \vec{v}_2$, we sometimes say that geometric vectors are added according to "the triangle law." From the properties of a parallelogram, we have that, if $\vec{v}_1$ and $\vec{v}_2$ have the same initial point, as shown in Figure 5.5b, the sum $\vec{v}_1 + \vec{v}_2$ is a diagonal of the parallelogram with adjacent sides $\vec{v}_1$ and $\vec{v}_2$. Therefore, we sometimes also say that geometric vectors are added according to "the parallelogram law." Figure 5.5c shows the sum $\vec{v}_3 + \vec{v}_4$ if $\vec{v}_3$ and $\vec{v}_4$ are collinear (vectors lie on the same line).

Note that the symbol "+" in a vector sum is being used in a different sense from that of its ordinary usage to pair two real numbers in a sum, although the operation is commonly referred to as the "addition" of geometric vectors.

Example   Construct the sum $\vec{v}_1 + \vec{v}_2$.

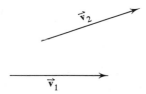

Solution   "Slide" the initial point of $\vec{v}_2$ onto the terminal point of $\vec{v}_1$. The sum $\vec{v}_1 + \vec{v}_2$ is the geometric vector from the initial point of $\vec{v}_1$ to the terminal point of $\vec{v}_2$ (figure a).

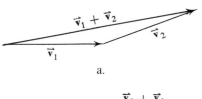

Alternate Solution   "Slide" the initial point of $\vec{v}_2$ onto the initial point of $\vec{v}_1$ and form a parallelogram. The sum $\vec{v}_1 + \vec{v}_2$ is the diagonal of the parallelogram (figure b).

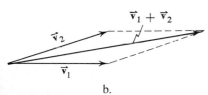

Definition 5.7   *For all geometric vectors $\vec{v}$ and any real number c, $c\vec{v}$ is a geometric vector with magnitude $\|c\vec{v}\| = |c| \cdot \|\vec{v}\|$, having the same direction angle as $\vec{v}$ if $c > 0$ and the same direction angle as $-\vec{v}$ if $c < 0$. The geometric vector $c\vec{v}$ is called a* **scalar multiple** *of $\vec{v}$.*

Example   As shown in the figure, $2\vec{v}$ and $\frac{1}{2}\vec{v}$ are geometric vectors in the same direction as $\vec{v}$ such that

$$\|2\vec{v}\| = 2\|\vec{v}\| \qquad \text{and} \qquad \|\tfrac{1}{2}\vec{v}\| = \tfrac{1}{2}\|\vec{v}\|,$$

while $-2\vec{v}$ is in the opposite direction such that

$$\|-2\vec{v}\| = |-2|\,\|\vec{v}\| = 2\|\vec{v}\|.$$

EXERCISE SET 5.1

A

In Exercises 1–40, use the geometric vectors below.

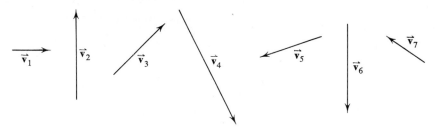

If $\|\vec{v}_1\| = 1$, estimate the magnitude and the measure of the direction angle for each of the following geometric vectors, where $\vec{v}_1$ is in the direction of the positive x-axis.

1. $\vec{v}_2$ **2.** $\vec{v}_3$ **3.** $\vec{v}_4$ **4.** $\vec{v}_5$ **5.** $\vec{v}_6$ **6.** $\vec{v}_7$

Using freehand methods, draw a single vector representing each of the following sums.

Example $(\vec{v}_1 + \vec{v}_3) + \vec{v}_7$

Solution   See the figure.

7. $\vec{\mathbf{v}}_1 + \vec{\mathbf{v}}_2$ **8.** $\vec{\mathbf{v}}_1 + \vec{\mathbf{v}}_3$ **9.** $\vec{\mathbf{v}}_1 + \vec{\mathbf{v}}_4$

10. $\vec{\mathbf{v}}_1 + \vec{\mathbf{v}}_5$ **11.** $\vec{\mathbf{v}}_1 + \vec{\mathbf{v}}_6$ **12.** $\vec{\mathbf{v}}_1 + \vec{\mathbf{v}}_7$

13. $\vec{\mathbf{v}}_2 + \vec{\mathbf{v}}_3$ **14.** $\vec{\mathbf{v}}_2 + \vec{\mathbf{v}}_4$ **15.** $(\vec{\mathbf{v}}_2 + \vec{\mathbf{v}}_5) + \vec{\mathbf{v}}_6$

16. $(\vec{\mathbf{v}}_2 + \vec{\mathbf{v}}_6) + \vec{\mathbf{v}}_7$ **17.** $(\vec{\mathbf{v}}_2 + \vec{\mathbf{v}}_3) + \vec{\mathbf{v}}_5$ **18.** $\vec{\mathbf{v}}_3 + (\vec{\mathbf{v}}_4 + \vec{\mathbf{v}}_7)$

Using freehand methods, draw each of the following.

19. $-\vec{\mathbf{v}}_1$ **20.** $-\vec{\mathbf{v}}_3$ **21.** $2\vec{\mathbf{v}}_7$

22. $4\vec{\mathbf{v}}_1$ **23.** $-2\vec{\mathbf{v}}_5$ **24.** $-\vec{\mathbf{v}}_6$

25. $\frac{1}{2}\vec{\mathbf{v}}_3$ **26.** $-\frac{1}{3}\vec{\mathbf{v}}_4$ **27.** $2\vec{\mathbf{v}}_1 + 3\vec{\mathbf{v}}_5$

28. $4\vec{\mathbf{v}}_7 + 2\vec{\mathbf{v}}_6$ **29.** $\frac{1}{2}\vec{\mathbf{v}}_4 + \vec{\mathbf{v}}_7$ **30.** $\vec{\mathbf{v}}_2 + \frac{3}{4}\vec{\mathbf{v}}_3$

31. $2\vec{\mathbf{v}}_3 + (-3\vec{\mathbf{v}}_1)$ **32.** $\vec{\mathbf{v}}_7 + (-2\vec{\mathbf{v}}_2)$

B

By appropriate freehand sketches show that each statement is true.

33. $\vec{\mathbf{v}}_2 + \vec{\mathbf{v}}_3 = \vec{\mathbf{v}}_3 + \vec{\mathbf{v}}_2$ **34.** $\vec{\mathbf{v}}_5 + \vec{\mathbf{v}}_4 = \vec{\mathbf{v}}_4 + \vec{\mathbf{v}}_5$

35. $(\vec{\mathbf{v}}_6 + \vec{\mathbf{v}}_7) + \vec{\mathbf{v}}_2 = \vec{\mathbf{v}}_6 + (\vec{\mathbf{v}}_7 + \vec{\mathbf{v}}_2)$ **36.** $\vec{\mathbf{v}}_3 + (\vec{\mathbf{v}}_7 + \vec{\mathbf{v}}_5) = (\vec{\mathbf{v}}_3 + \vec{\mathbf{v}}_7) + \vec{\mathbf{v}}_5$

37. $2(\vec{\mathbf{v}}_3 + \vec{\mathbf{v}}_5) = 2\vec{\mathbf{v}}_3 + 2\vec{\mathbf{v}}_5$ **38.** $(2 + 3)\vec{\mathbf{v}}_7 = 2\vec{\mathbf{v}}_7 + 3\vec{\mathbf{v}}_7$

39. $\frac{1}{3}\vec{\mathbf{v}}_1 + 2\vec{\mathbf{v}}_1 = \frac{7}{3}\vec{\mathbf{v}}_1$ **40.** $\vec{\mathbf{v}}_2 + (-2\vec{\mathbf{v}}_2) = -\vec{\mathbf{v}}_2$

5.2 Applications of Geometric Vectors

The solution to many kinds of practical problems involving such quantities as velocities, accelerations, forces, and directed line segments can be found by means of geometric vectors.

Projections

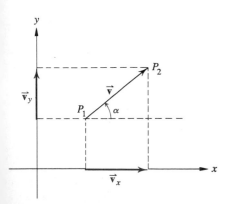

Figure 5.6

First, consider the *projections* of a geometric vector on the x- and y-axes of a Cartesian coordinate system. Figure 5.6 depicts a geometric vector $\vec{\mathbf{v}}$ and perpendicular lines drawn to the x- and y-axes from the endpoints P_1 and P_2 of $\vec{\mathbf{v}}$. The geometric vector projections $\vec{\mathbf{v}}_x$ and $\vec{\mathbf{v}}_y$ of $\vec{\mathbf{v}}$ on the x-axis and y-axis, respectively, are called the **rectangular components** of $\vec{\mathbf{v}}$; $\vec{\mathbf{v}}_x$ is called the **horizontal component** of $\vec{\mathbf{v}}$, and $\vec{\mathbf{v}}_y$ is called the **vertical component**. These components are determined by the magnitude of $\vec{\mathbf{v}}$ and its direction.

Figure 5.7

Furthermore,

$$\vec{v} = \vec{v}_x + \vec{v}_y.$$

In Figure 5.7, $\overline{P_1P_3}$ is drawn parallel to the x-axis, and $\overline{P_2P_3}$ is drawn parallel to the y-axis. Since

$$|\sin \alpha| = \frac{\|\vec{v}_y\|}{\|\vec{v}\|},$$

$$\|\vec{v}_y\| = \|\vec{v}\| \cdot |\sin \alpha|. \tag{1}$$

Also, since

$$|\cos \alpha| = \frac{\|\vec{v}_x\|}{\|\vec{v}\|},$$

$$\|\vec{v}_x\| = \|\vec{v}\| \cdot |\cos \alpha|. \tag{2}$$

In Figure 5.7, α is shown to be a positive acute angle. From Definition 5.3, we have that $-180° < \alpha \le 180°$ and hence $\cos \alpha$ or $\sin \alpha$ may be negative. Therefore because $\|\vec{v}_x\|$, $\|\vec{v}_y\|$, and $\|\vec{v}\|$ are all positive, we use $|\sin \alpha|$ and $|\cos \alpha|$ in Equations (1) and (2).

If the geometric vectors $\vec{v}_x$ and $\vec{v}_y$ are known, the magnitude $\|\vec{v}\|$ and direction angle α of $\vec{v}$ can be determined from the relationships in a right triangle, as shown in Figure 5.7:

$$\|\vec{v}\| = \sqrt{\|\vec{v}_x\|^2 + \|\vec{v}_y\|^2},$$

and

$$|\sin \alpha| = \frac{\|\vec{v}_y\|}{\|\vec{v}\|} \quad \text{or} \quad |\cos \alpha| = \frac{\|\vec{v}_x\|}{\|\vec{v}\|}.$$

Example   The horizontal component, $\vec{v}_x$, has its initial point at $(4, 0)$ with $\|\vec{v}_x\| = 6$. The vertical component, $\vec{v}_y$, has its initial point at $(0, 3)$ with $\|\vec{v}_y\| = 8$. Find α and $\|\vec{v}\|$ and sketch $\vec{v}$. Assume $\vec{v}_x$ and $\vec{v}_y$ are directed in a positive direction.

Solution   The magnitude

$$\|\vec{v}\| = \sqrt{\|\vec{v}_x\|^2 + \|\vec{v}_y\|^2}$$
$$= \sqrt{6^2 + 8^2} = \sqrt{100} = 10.$$

Because

$$\cos \alpha = \frac{6}{10} = 0.6000,$$

$$\alpha \approx 53.1°.$$

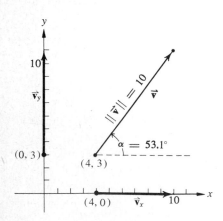

Thus, the geometric vector $\vec{v}$ has its initial point at (4, 3), a direction angle whose measure is approximately 53.1°, and a magnitude of 10.

Force

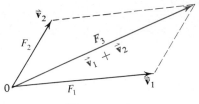

Figure 5.8

Consider two forces, F_1 and F_2, acting on a particle at the point O, as in Figure 5.8. Since a force has magnitude and direction, we can represent it with a geometric vector, with magnitude given in terms of the units of force and with direction angle given by the direction of the force. It is shown in physics that the two forces represented by $\vec{v}_1$ and $\vec{v}_2$ can be replaced by a single equivalent force, F_3, associated with the vector $\vec{v}_1 + \vec{v}_2$, the resultant of the forces represented by $\vec{v}_1$ and $\vec{v}_2$. The resultant $\vec{v}_1 + \vec{v}_2$ can be obtained, as discussed in Section 5.1, by constructing the parallelogram with adjacent sides $\vec{v}_1$ and $\vec{v}_2$ and drawing the diagonal; $\vec{v}_1 + \vec{v}_2$ is then associated with the resultant force F_3.

Example   Two forces, the first one of 5 pounds and the second of 12 pounds, act on a body at right angles to each other. Find the magnitude of the resultant force and the measure of the angle that the resultant force makes with the 5-pound force.

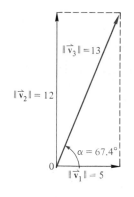

Solution   Because $\|\vec{v}_1\| = 5$ and $\|\vec{v}_2\| = 12$,

$$\|\vec{v}\| = \sqrt{5^2 + 12^2} = \sqrt{169} = 13,$$

and

$$\cos \alpha = \frac{5}{13} \approx 0.3846.$$

Hence, $\alpha = 67.4°$. Thus, the magnitude of the resultant force is 13 pounds, and the angle it makes with the 5-pound force has a measure of approximately 67.4°.

In cases where the measure of an angle between two forces is not 90°, it is sometimes convenient to use the Law of Cosines and the Law of Sines in order to find the resultant force.

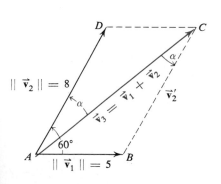

Example   A force of 5 pounds and another force of 8 pounds act on an object at an angle of 60° with respect to each other. Find the magnitude of the resultant force and the angle it forms with respect to the 8-pound force.

Solution   The figure shows the forces of 5 pounds and 8 pounds represented by the geometric vectors $\vec{v}_1$ and $\vec{v}_2$, respectively, and the resultant $\vec{v}_3 = \vec{v}_1 + \vec{v}_2$. Since $\vec{v}_2'$ is equivalent to $\vec{v}_2$, and since the

(continued)

consecutive angles of a parallelogram are supplementary,

$$\angle ABC = 180° - 60° = 120°,$$

and it follows from the Law of Cosines that

$$\|\vec{\mathbf{v}}_3\|^2 = \|\vec{\mathbf{v}}_1\|^2 + \|\vec{\mathbf{v}}_2\|^2 - 2\|\vec{\mathbf{v}}_1\| \cdot \|\vec{\mathbf{v}}_2\| \cos 120°$$
$$= 5^2 + 8^2 - 2(5)(8)(-\cos 60°)$$
$$= 25 + 64 + 80(0.5000) = 129.$$

Therefore, $\|\vec{\mathbf{v}}_3\| = \sqrt{129} \approx 11.4$. From the Law of Sines,

$$\frac{\sin \alpha}{\|\vec{\mathbf{v}}_1\|} = \frac{\sin 120°}{\|\vec{\mathbf{v}}_3\|},$$

or, equivalently,

$$\sin \alpha = \frac{\|\vec{\mathbf{v}}_1\| \sin 120°}{\|\vec{\mathbf{v}}_3\|}$$
$$\approx \frac{5(0.8660)}{11.4} \approx 0.3798.$$

Hence, $\alpha \approx 22.3°$. Thus, the forces of 5 and 8 pounds, acting at an angle of 60° with respect to each other, can be represented by the single resultant force of 11.4 pounds. The angle it makes with the 8-pound force has a measure of approximately 22.3°.

Velocity

Geometric vectors also can facilitate the solution of problems involving velocities and accelerations. As examples, we shall look at several air navigation problems. First, however, we consider some terminology that is commonly used. The **heading** of an airplane is the direction in which it is pointed, and the **course** is the direction in which it is actually moving over the ground. Its **air speed** is the speed relative to the air, and its **ground speed** is the speed relative to the ground. The two directions and two speeds may differ because of wind effect.

If we use one geometric vector, say $\vec{\mathbf{v}}_1$, to represent the heading and air speed of an airplane (see Figure 5.9) and another, say $\vec{\mathbf{v}}_2$, to represent the wind direction and speed, then $\vec{\mathbf{v}}_3 = \vec{\mathbf{v}}_1 + \vec{\mathbf{v}}_2$ represents the course and ground speed. The angle α between the geometric vectors $\vec{\mathbf{v}}_1$ and $\vec{\mathbf{v}}_3$ is called the **drift angle.** The measures of the angles for the heading, wind, and course directions in air navigation are generally taken clockwise from the ray

Figure 5.9   Figure 5.10

directed toward true north. The angle is called the **bearing** and has a positive measure. In graphical representations, we shall use the positive y-axis as the true north direction, and the positive x-axis as the true east direction. For example, in Figure 5.10, the bearing of the geometric vector $\vec{v}_1$ is 120°, that of $\vec{v}_2$ is 285°, and that of $\vec{v}_3$ is 354°.*

Example An airplane is headed northeast (bearing 45°) with an air speed of 500 miles per hour, with a wind blowing from the southeast (bearing 315°) at a speed of 75 miles per hour. Find the drift angle, the ground speed, and the course of the plane.

Solution The figure on page 196 shows three vectors: $\vec{v}_1$ representing the heading and air speed, $\vec{v}_2$ representing the wind direction and speed, and $\vec{v}_3$ representing the course and ground speed. From geometry we have that

$$\angle ABC = 315° - 45° - 180° = 90°.$$

Because $\angle ABC$ is a right angle,

$$\tan \alpha = \frac{75}{500} = 0.1500,$$

and the drift angle $\alpha \approx 8.5°$. The bearing, or direction, of the course is

$$45° - \alpha = 45° - 8.5° = 36.5°.$$

(*continued*)

* Other types of bearings, each of which has its own reference, or initial, ray from which the angle is measured, are also defined. A bearing used in surveying is designated by an acute angle measured to the east or west of a ray pointing toward true north or true south, for example, N 35° E, N 38° W, S 22° E, and S 47° W.

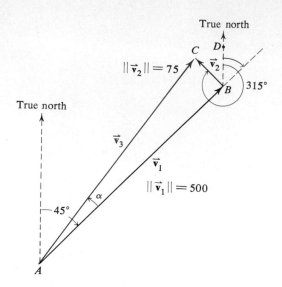

Now, because $\cos \alpha = 500/\|\vec{v}_3\|$,

$$\|\vec{v}_3\| = \frac{500}{\cos 8.5°} \approx \frac{500}{0.9890} \approx 506.$$

Thus, the ground speed (magnitude of the course vector) is approximately 506 miles per hour.

EXERCISE SET 5.2

A

The direction angle α and $\|\vec{v}\|$ are given. Find $\|\vec{v}_x\|$ and $\|\vec{v}_y\|$.

Example $\alpha = 60°$ and $\|\vec{v}\| = 8$

Solution The figure shows the geometric vector $\vec{v}$ and its rectangular components $\vec{v}_x$ and $\vec{v}_y$. From Equations (1) and (2) on page 192,

$$\|\vec{v}_x\| = \|\vec{v}\| \,|\cos 60°| = 8\left(\frac{1}{2}\right) = 4,$$

$$\|\vec{v}_y\| = \|\vec{v}\| \,|\sin 60°| = 8\left(\frac{\sqrt{3}}{2}\right) = 4\sqrt{3}.$$

1. $\alpha = 45°$ and $\|\vec{v}\| = 5\sqrt{2}$ **2.** $\alpha = 30°$ and $\|\vec{v}\| = 6$

3. $\alpha = 90°$ and $\|\vec{v}\| = 15$ **4.** $\alpha = 120°$ and $\|\vec{v}\| = 2$

5. $\alpha = -30°$ and $\|\vec{v}\| = 12$ **6.** $\alpha = -45°$ and $\|\vec{v}\| = 2\sqrt{2}$

7. $\alpha = -120°$ and $\|\vec{v}\| = 8$ **8.** $\alpha = -150°$ and $\|\vec{v}\| = 4$

In Exercises 9–14, the two given forces act on a point in a plane at an angle with the specified measure. Find the magnitude of the resultant force and the angles the resultant force makes with the given forces.

Example   Forces of 10 and 20 newtons act on a point at an angle of 45° with respect to each other.

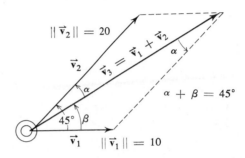

Solution   The figure shows the given forces $\vec{v}_1$ and $\vec{v}_2$ and the resultant $\vec{v}_3 = \vec{v}_1 + \vec{v}_2$. Since $\alpha + \beta = 45°$, the measure of the angle opposite $\vec{v}_3$ is 135°. From the fact that $\cos 135° = -\cos 45°$ and from the Law of Cosines,

$$\|\vec{v}_3\|^2 = 10^2 + 20^2 - 2(10)(20)(-\cos 45°)$$

$$\approx 100 + 400 + 283 \approx 783;$$

$$\|\vec{v}_3\| \approx 28.0.$$

From the Law of Sines,

$$\frac{\sin \alpha}{\|\vec{v}_1\|} = \frac{\sin 135°}{\|\vec{v}_3\|},$$

so

$$\sin \alpha = \frac{\|\vec{v}_1\| \sin 135°}{\|\vec{v}_3\|} \approx \frac{10(0.7071)}{28.0} \approx 0.2525.$$

Hence, $\alpha \approx 14.6°$. If $\alpha + \beta = 45°$, then

$$\beta \approx 45° - 14.6° = 30.4°.$$

Thus, the resultant force is approximately 28.0 newtons. The angles this force makes with the force of 10 newtons and with the force of 20 newtons have measures of approximately 30.4° and 14.6°, respectively.

9. Forces of 3 and 4 pounds, acting at an angle of 90° with respect to each other.

10. Forces of 5 and 7 pounds, acting at an angle of 30° with respect to each other.

11. Forces of 15 and 20 pounds, acting at an angle of 75° with respect to each other.

12. Forces of 8 and 11 pounds, acting at an angle of 132° 48′ with respect to each other.

13. Forces of 6 and 10 newtons acting at an angle of 120° with respect to each other.

14. Forces of 20 and 30 newtons, acting at an angle of 100° with respect to each other.

15. The resultant of two forces acting at an angle of 90° with respect to each other is 50 newtons. If one of the forces is 30 newtons, find the other force and the angle it makes with the resultant.

16. Two equal forces acting at an angle of 60° with respect to each other have a resultant of 80 newtons. Find the magnitude of each force.

17. Two forces acting at an angle of 60° with respect to each other have a resultant of 50 pounds. If one force acts at an angle of 20° with respect to the resultant, find the magnitude of each force.

18. Two forces, one of 100 pounds and the other 200 pounds, have a resultant of 250 pounds. Find the angle formed by the two forces.

Example   An auto weighing 3000 pounds is on a slope that makes an angle of 9° 30′ with the horizontal. Find the force that pulls the auto down the hill.

Solution   The force due to gravity can be resolved into two forces, one force in the direction of the incline and one force perpendicular to the plane. See the figure. In this case, $\overrightarrow{AB}$, the force due to gravity, is resolved into $\overrightarrow{AC}$ along the incline and $\overrightarrow{CB}$, perpendicular to $\overrightarrow{AC}$. From geometry, $\angle B = \angle D$ because the sides of $\angle B$ are perpendicular to the sides of $\angle D$. Thus,

$$\sin 9° \, 30' = \frac{\|\overrightarrow{AC}\|}{3000},$$

from which

$$\|\overrightarrow{AC}\| = 3000 \sin 9° \, 30' = 3000 \sin 9.5° \approx 495.$$

Hence, a force of 495 pounds pulls the auto down the hill.

19. An auto, weighing 3500 pounds, is on a slope that makes an angle of 8° 24′ with the horizontal. Find the force that pulls the auto down the hill.

20. A steel ball, weighing 50 pounds, is on an inclined surface. What angle does the plane make with the horizontal if a force of 10 pounds is pulling the ball down the surface?

21. A force of 110 pounds is needed to keep a weight of 200 pounds from sliding down an inclined plane. What angle does the plane make with the horizontal?   *Note:* The force that keeps it from sliding is opposite to the force pulling it down the plane, and has the same magnitude.

22. A force of 200 pounds is needed to keep a weight of 320 pounds from sliding down an inclined ramp. What angle does the plane make with the horizontal?

23. A bullet has an acceleration of 2 meters per second^2 in a direction making an angle of 35° with the horizontal. Find the horizontal and vertical components ($\|\vec{v}_x\|$ and $\|\vec{v}_y\|$) of its acceleration.

24. A space rocket has an acceleration of 4.2 meters per second^2 in a direction making an angle of 73° with the horizontal. Find the horizontal and vertical components of the rocket's acceleration.

25. A crew can row a boat at a speed of 9.2 kilometers per hour in still water. If the crew heads across a river at right angles to the current and finds its "drift angle" to be 6°, find the speed of the current.

26. A person can swim at a speed of 2 kilometers per hour in still water. If he heads across a river at right angles to a current of 5 kilometers per hour, find his speed in relation to the land and the direction in which he actually moves.

27. An airplane is headed southeast (bearing 135°) with an air speed of 600 miles per hour, with the wind blowing from the northeast (bearing 225°) at a speed of 120 miles per hour. Find the drift angle, the ground speed, and the course of the airplane.

28. Solve Exercise 27 if the wind is from the southwest (bearing 45°) at a speed of 60 miles per hour.

29. An airplane is headed on a bearing of 250° with an air speed of 425 miles per hour. The course has a bearing of 262°. The ground speed is 475 miles per hour. Find the drift angle, the wind direction, and the wind speed.

30. Solve Exercise 29 if the course has a bearing of 241°.

31. A pilot wishes to fly on a course bearing 90° and with a ground speed of 600 kilometers per hour. If a wind is blowing from the north (bearing

180°) with a speed of 50 kilometers an hour, what must be the heading and air speed of the aircraft?

32. Solve Exercise 31 if the wind is from the northwest (bearing 135°).

B

33. Two ships leave a harbor at the same time, the first traveling at 20 knots (nautical miles per hour) on a course bearing 40°, and the second at 25 knots on a course bearing 300°. After one hour, how far apart are they and what is the bearing of a course from the first ship to the second?

34. Three forces of 25, 35, and 45 pounds act in directions having bearings of 240°, 290°, and 340°, respectively. Find a force, acting due east, and a force, acting due south, that would counterbalance these three forces.

35. A 100-pound weight hangs from a rope that makes an angle of 90° where it is attached to the weight (see figure). What is the tension (force) on each half of the rope?

36. A 50-pound weight hangs from a rope whose two ends are tied at 60° angles to a support (see figure). What is the tension on each half of the rope?

Ex. 35

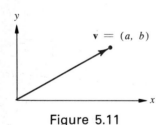

Ex. 36

5.3 Vectors as Ordered Pairs

In Sections 5.1 and 5.2 we considered geometric vectors. In this section we will consider the more general notion of a two-dimensional vector as an ordered pair of real numbers, some of its properties, and the close relationship that exists between such a vector and a geometric vector in the plane.

Geometric vectors whose initial points are located at a fixed position in the plane are called **bound geometric vectors,** or **position vectors.** The terminal point of such a geometric vector with initial point at the origin has coordinates that are the components of exactly one ordered pair (a, b) of real numbers. This relationship, shown in Figure 5.11, suggests the following definition.

Figure 5.11

$\mathbf{v} = (a, b)$

Definition 5.8 *A* **two-dimensional vector** $\mathbf{v}$ *is an ordered pair of real numbers*

$$\mathbf{v} = (a, b), \quad \text{where } a, b \in R.$$

The set of all such vectors is designated by **V**.

For a vector $\mathbf{v} = (a, b)$, the numbers a and b are called the **scalar components** of $\mathbf{v}$.

Notice that boldface type *without* an arrow above the symbol is being used to differentiate a vector **v**, an ordered pair of real numbers, from a geometric vector $\vec{v}$.

Example The figure shows the geometric vectors associated with the vectors $\mathbf{v}_1 = (2, 3)$, $\mathbf{v}_2 = (-3, 5)$, $\mathbf{v}_3 = (-5, -1)$, and $\mathbf{v}_4 = (0, -4)$, respectively.

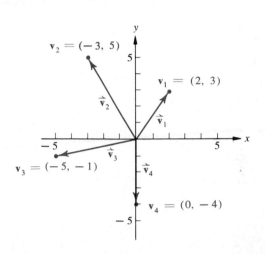

The following definitions are suggested by the close association of a vector with a geometric vector.

Definition 5.9 *For all $\mathbf{v}_1, \mathbf{v}_2 \in \mathbf{V}$, where $\mathbf{v}_1 = (a_1, b_1)$ and $\mathbf{v}_2 = (a_2, b_2)$,*

$$\mathbf{v}_1 = \mathbf{v}_2 \quad \textit{if and only if} \quad a_1 = a_2 \quad \textit{and} \quad b_1 = b_2.$$

Definition 5.10 *For all $\mathbf{v} \in \mathbf{V}$, where $\mathbf{v} = (a, b)$, the* **norm** *or* **magnitude** *of* **v** *is*

$$\|\mathbf{v}\| = \|(a, b)\| = \sqrt{a^2 + b^2}.$$

Definition 5.11 *For all $\mathbf{v} \in \mathbf{V}$, where $\mathbf{v} = (a, b)$ and $\|\mathbf{v}\| \neq 0$, the* **direction angle** *of* **v** *is the angle α such that*

$$\cos \alpha = \frac{\alpha}{\|\mathbf{v}\|} \quad \textit{and} \quad \sin \alpha = \frac{b}{\|\mathbf{v}\|},$$

where $-180° < \alpha \leq 180°$.

Although Definition 5.11 defines the direction angle of $\mathbf{v} = (a, b)$ in terms of both cos α and sin α, in practice we generally determine only one function value in order to find α. This is possible because the quadrant in which **v** lies can readily be determined from the ordered pair (a, b).

Example   By Definition 5.10, the norm of the vector **v** $= (2, -5)$ is

$$\|\mathbf{v}\| = \sqrt{2^2 + (-5)^2} = \sqrt{29}.$$

By Definition 5.11,

$$\sin \alpha = \frac{-5}{\sqrt{29}} \approx -0.9285.$$

and because $(2, -5)$ is in Quadrant IV, then $-90° < \alpha < 0°$, and

$$\alpha \approx -68.2° = -68° \ 12'.$$

Notice that the norm of a vector **v** is the length of the corresponding geometric vector $\vec{v}$, and its direction angle is the angle α such that $-180° < \alpha \leq 180°$, associated with the geometric vector.

EXERCISE SET 5.3

A

Find real numbers x and y for which each statement is true.

Example $(x, -y) = (-2, -8)$

Solution   By Definition 5.9, $x = -2$ and $y = 8$.

1. $(x, y) = (-3, 6)$ **2.** $(-x, y) = (5, 1)$

3. $(2x, 7) = (-7, -y)$ **4.** $(6, 3y) = (x, -6)$

5. $(x + y, x - y) = (-4, 0)$ **6.** $(2x - y, x + 2y) = (0, -5)$

For each ordered pair (**a**) *sketch each corresponding geometric vector with initial point at the origin and* (**b**) *find the norm and the direction angle (to the nearest six minutes) of each vector.*

Example $(-2, 3)$

Solution
   a. See the figure.

   b. By Definition 5.10.

$$\|(-2, 3)\| = \sqrt{(-2)^2 + (3)^2} = \sqrt{13}.$$

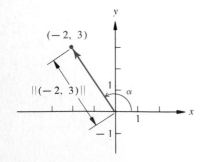

By Definition 5.11,

$$\sin \alpha = \frac{3}{\sqrt{13}} \approx 0.8321,$$

and because $(-2, 3)$ is in Quadrant II, $90° < \alpha < 180°$, and

$$\alpha \approx 123.7° = 123° \; 42'.$$

7. $(3, 4)$ **8.** $(-3, 4)$ **9.** $(1, 1)$ **10.** $(-1, 1)$

11. $(-3, -3)$ **12.** $(0, -5)$ **13.** $(4, 0)$ **14.** $(\sqrt{3}, 1)$

Given the following norms and direction angles, find the vector $\mathbf{v}$, *and sketch the corresponding geometric vector.*

Example $\|\mathbf{v}\| = 5$ and $\alpha = 30°$

Solution   By Definition 5.11, if $\mathbf{v} = (a, b)$, then

$$a = \|\mathbf{v}\| \cos \alpha,$$
$$b = \|\mathbf{v}\| \sin \alpha.$$

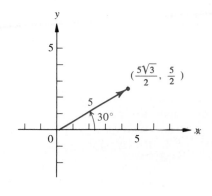

Thus,

$$\mathbf{v} = (5 \cos 30°, 5 \sin 30°) = \left(\frac{5\sqrt{3}}{2}, \frac{5}{2}\right).$$

15. $\|\mathbf{v}\| = 2\sqrt{2}$ and $\alpha = 45°$ **16.** $\|\mathbf{v}\| = 2$ and $\alpha = 60°$

17. $\|\mathbf{v}\| = 1$ and $\alpha = 90°$ **18.** $\|\mathbf{v}\| = 4$ and $\alpha = 150°$

19. $\|\mathbf{v}\| = 5$ and $\alpha = -150°$ **20.** $\|\mathbf{v}\| = 3\sqrt{2}$ and $\alpha = -30°$

B

21. Use Definition 5.11 to show that the direction angles of (a, b) and $(-b, a)$ differ by $90°$, given $a, b > 0$.

22. Show that if $a, b, c, d > 0$ and $ac - bd = 0$, then the direction angles of (a, b) and $(-c, d)$ differ by $90°$.

5.4 Fundamental Operations

Two basic operations on vectors are closely related to the operations on geometric vectors that have been defined in Section 5.1.

Vector Addition

As was the case for geometric vectors, the operation of forming the sum of two vectors is called **vector addition.**

Definition 5.12 *For all* $v_1, v_2 \in V$, *where* $v_1 = (a_1, b_1)$ *and* $v_2 = (a_2, b_2)$, *the* **sum** $v_1 + v_2$ *is a vector such that*

$$v_1 + v_2 = (a_1, b_1) + (a_2, b_2) = (a_1 + a_2, b_1 + b_2).$$

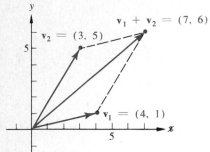

Example   Given that $v_1 = (4, 1)$ and $v_2 = (3, 5)$,

$$
\begin{aligned}
v_1 + v_2 &= (4, 1) + (3, 5) \\
&= (4 + 3, 1 + 5) \\
&= (7, 6).
\end{aligned}
$$

The corresponding geometric vectors for v_1, v_2, and their sum $v_1 + v_2$ are shown in the figure.

Note that, in the above example, the resultant $v_1 + v_2$ is associated with the geometric vector that is one of the diagonals of a parallelogram.

Multiplication of a Vector by a Scalar

The second basic operation on vectors concerns multiplication of a scalar (real number) and a vector and is also related to the similar operation on geometric vectors.

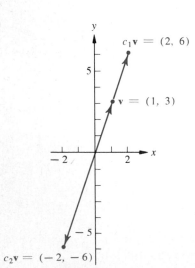

Definition 5.13   *For all* $v \in V$, *where* $v = (a, b)$ *and* $c \in R$,

$$cv = c(a, b) = (ca, cb).$$

Example   If $v = (1, 3)$, $c_1 = 2$, and $c_2 = -2$, then

$$c_1 v = 2(1, 3) = (2, 6)$$

and

$$c_2 v = -2(1, 3) = (-2, -6).$$

Notice from the above example that cv has a magnitude $|c| \cdot \|v\|$ if c is a positive or negative real number. The direction angles of v and cv are the same if c is positive, and they differ by $180°$ if c is negative.

Properties of Vectors

The following two definitions establish the identity element and the additive inverse element in our system of vectors.

Definition 5.14   *The* **zero vector** *is*

$$\mathbf{0} = (0, 0).$$

Any direction may be assigned to the zero vector.
  Since

$$\mathbf{v} + \mathbf{0} = (a, b) + (0, 0)$$
$$= (a + 0), (b + 0)$$
$$= (a, b) = \mathbf{v},$$

the zero vector is an *additive identity element*.

Examples

a. $(-3, 4) + \mathbf{0} = (-3, 4)$ b. $(6, -1) + \mathbf{0} = (6, -1)$

Definition 5.15   *For all* $\mathbf{v} \in \mathbf{V}$, *where* $\mathbf{v} = (a, b)$,

$$-\mathbf{v} = -(a, b) = (-a, -b).$$

Since

$$\mathbf{v} + (-\mathbf{v}) = (a, b) + (-a, -b)$$
$$= [a + (-a), b + (-b)]$$
$$= (0, 0) = \mathbf{0}.$$

$-\mathbf{v}$ is an *additive inverse element*.

Example   Given $\mathbf{v} = (-5, 7)$, then $-\mathbf{v} = (5, -7)$. Thus,

$$\mathbf{v} + (-\mathbf{v}) = (-5, 7) + (5, -7)$$
$$= (0, 0) = \mathbf{0}.$$

Difference of Two Vectors

The difference of two vectors, $\mathbf{v}_2 - \mathbf{v}_1$, is defined similarly to the difference of two real numbers.

Definition 5.16   *For all vectors* v_1 *and* v_2, *the* **difference** $v_2 - v_1$ *is a vector such that*

$$v_2 - v_1 = v_2 + (-v_1).$$

Example   Given that $v_1 = (2, 5)$ and $v_2 = (7, 1)$,

$$v_2 - v_1 = (7, 1) - (2, 5)$$
$$= (7, 1) + (-2, -5)$$
$$= (7 - 2, 1 - 5)$$
$$= (5, -4).$$

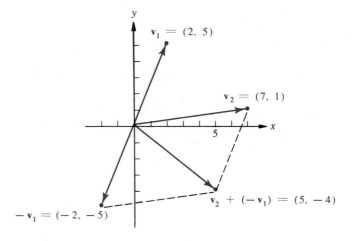

Quotient of a Vector and a Scalar

The quotient of a vector divided by a scalar is defined in terms of a product in much the same way as is the quotient of two real numbers.

Definition 5.17   *For all* $v \in V$ *and* $c \in R$ $(c \neq 0)$,

$$\frac{v}{c} = \frac{1}{c} v.$$

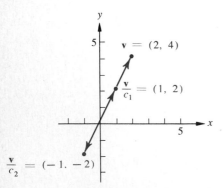

Example   If $v = (2, 4)$, $c_1 = 2$, and $c_2 = -2$, then

$$\frac{v}{c_1} = \frac{(2, 4)}{2} = \frac{1}{2}(2, 4) = (1, 2),$$

$$\frac{v}{c_2} = \frac{(2, 4)}{-2} = -\frac{1}{2}(2, 4) = (-1, -2).$$

ai + bj Form

If vectors $\mathbf{v}_1 = (1, 0)$ and $\mathbf{v}_2 = (0, 1)$, called **unit vectors,** are designated $\mathbf{i}$ and $\mathbf{j}$, respectively, then each vector

$$\begin{aligned}
\mathbf{v} &= (a, b) \\
&= (a, 0) + (0, b) \\
&= a(1, 0) + b(0, 1) \\
&= a\mathbf{i} + b\mathbf{j}.
\end{aligned}$$

Here again the real numbers a and b are called the **scalar components** of the vector.

Example Express each vector as the sum of scalar multiples of $\mathbf{i}$ and $\mathbf{j}$ and sketch the corresponding geometric vectors.

a. $(-4, 0)$ b. $(-4, 3)$

Solutions

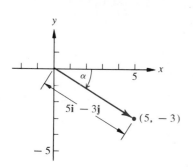

a. $(-4, 0) = -4(1, 0)$ b. $(-4, 3) = (-4, 0) + (0, 3)$

$\qquad\qquad\quad = -4\mathbf{i}$ $\qquad\qquad\quad = -4(1, 0) + 3(0, 1)$

$\qquad\qquad\qquad\qquad\qquad\qquad = -4\mathbf{i} + 3\mathbf{j}$

Example Find the norm and the direction angle of $5\mathbf{i} - 3\mathbf{j}$.

Solution $5\mathbf{i} - 3\mathbf{j} = 5(1, 0) - 3(0, 1) = (5, -3)$. Thus,

$$\|5\mathbf{i} - 3\mathbf{j}\| = \|(5, -3)\|,$$

from which, by Definition 5.10, we have

$$\|5\mathbf{i} - 3\mathbf{j}\| = \sqrt{5^2 + (-3)^2} = \sqrt{34}.$$

From Definition 5.11,

$$\sin \alpha = \frac{-3}{\sqrt{34}} \approx -0.5145.$$

Furthermore, $(5, -3)$ corresponds to a point in Quadrant IV. Hence

$$\alpha \approx -31.0°.$$

EXERCISE SET 5.4

A

Write the sum of each pair of vectors as an ordered pair and show the corresponding geometric vectors.

Example $v_1 = (2, -5)$ and $v_2 = (-4, 2)$

Solution $v_1 + v_2 = (2, -5) + (-4, 2)$

$$= (2 + (-4), -5 + 2)$$

$$= (-2, -3)$$

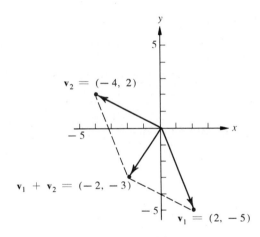

1. $v_1 = (-4, 0)$ and $v_2 = (0, 6)$ 2. $v_1 = (-6, 0)$ and $v_2 = (0, 1)$

3. $v_1 = (1, -4)$ and $v_2 = (2, 5)$ 4. $v_1 = (7, -8)$ and $v_2 = (-3, -7)$

5. $v_1 = (-5, 3)$ and $v_2 = (1, -2)$ 6. $v_1 = (-6, 6)$ and $v_2 = (3, -3)$

Let $v_1 = (3, 6)$, $v_2 = (5, -1)$, $v_3 = (-4, 2)$, $c_1 = 2$, *and* $c_2 = -3$. *Write each expression as an ordered pair and show the corresponding geometric vectors.*

Example $c_1 v_2 + c_1 v_3$

Solution $c_1 v_2 + c_1 v_3 = 2(5, -1) + 2(-4, 2)$

$$= (10, -2) + (-8, 4)$$

$$= (10 + (-8), -2 + 4)$$

$$= (2, 2)$$

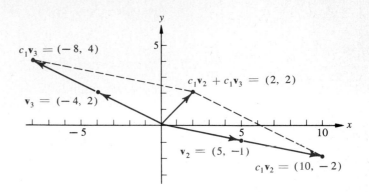

7. $-\mathbf{v}_1$

8. $-\mathbf{v}_2$

9. $\mathbf{v}_3 + \mathbf{0}$

10. $\mathbf{0} - \mathbf{v}_1$

11. $c_1\mathbf{v}_1$

12. $c_2\mathbf{v}_2$

13. $c_2\mathbf{v}_1 + \mathbf{v}_3$

14. $c_1\mathbf{v}_2 + \mathbf{v}_1$

15. $c_1\mathbf{v}_1 + c_2\mathbf{v}_2$

16. $c_1\mathbf{v}_2 + c_2\mathbf{v}_3$

17. $c_1(\mathbf{v}_1 + \mathbf{v}_2)$

18. $c_2(\mathbf{v}_2 + \mathbf{v}_3)$

19. $c_1\mathbf{v}_2 - \mathbf{v}_1$

20. $c_2\mathbf{v}_3 - \mathbf{v}_2$

21. $c_2\mathbf{v}_2 - c_1\mathbf{v}_3$

22. $c_2\mathbf{v}_3 - c_1\mathbf{v}_1$

23. $c_2(\mathbf{v}_2 - \mathbf{v}_3)$

24. $c_1(\mathbf{v}_3 - \mathbf{v}_1)$

25. $\dfrac{\mathbf{v}_2}{c_1}$

26. $\dfrac{\mathbf{v}_3}{c_2}$

27. $\dfrac{\mathbf{v}_1}{c_1} - \dfrac{\mathbf{v}_2}{c_2}$

28. $\dfrac{\mathbf{v}_2}{c_2} - \dfrac{\mathbf{v}_3}{c_1}$

29. $\dfrac{1}{4}c_1\mathbf{v}_1 - \dfrac{1}{3}c_2\mathbf{v}_3$

30. $\dfrac{1}{8}c_1\mathbf{v}_2 + \dfrac{1}{6}c_2\mathbf{v}_1$

Express each vector as a sum of scalar multiples of $\mathbf{i}$ *and* $\mathbf{j}$.

Example $(2, 9)$

Solution $(2, 9) = (2, 0) + (0, 9)$

$$= 2(1, 0) + 9(0, 1)$$

$$= 2\mathbf{i} + 9\mathbf{j}$$

31. $(3, 5)$

32. $(-2, 1)$

33. $(1, -4)$

34. $(-2, -3)$

35. $(-3, 0)$

36. $(0, 4)$

Find the norm and the direction angle, to the nearest six minutes, of each vector. Sketch the corresponding geometric vector.

Example $2\mathbf{i} - 3\mathbf{j}$

(continued)

Solution $\quad 2\mathbf{i} - 3\mathbf{j} = 2(1, 0) - 3(0, 1)$

$$= (2, 0) + (0, -3)$$

$$= (2, -3)$$

Hence, by Definition 5.10,

$$\|2\mathbf{i} - 3\mathbf{j}\| = \|(2, -3)\|$$

$$= \sqrt{2^2 + (-3)^2}$$

$$= \sqrt{13}.$$

From Definition 5.11,

$$\sin \alpha = \frac{-3}{\sqrt{13}} \approx -0.8321.$$

Since $(2, -3)$ is in Quadrant IV, $-90° < \alpha < 0°$. Thus,

$$\alpha \approx -56.3° = -56° \; 18'.$$

37. $4\mathbf{i} - 4\mathbf{j}$ **38.** $\sqrt{3}\mathbf{i} + \mathbf{j}$ **39.** $2\mathbf{i} + \mathbf{j}$ **40.** $-\mathbf{i} - 5\mathbf{j}$

41. $-\mathbf{i} - 3\mathbf{j}$ **42.** $-6\mathbf{i} + \mathbf{j}$ **43.** $-\mathbf{i} + \sqrt{3}\mathbf{j}$ **44.** $3\mathbf{i} - 2\mathbf{j}$

5.5 Inner Products

In Section 5.4 we considered two basic operations, vector addition and the multiplication of a vector by a scalar. The result of either operation is a vector and the set of vectors **V** is said to be closed for each of these operations. Another operation on two vectors is defined in such a way as to be useful in a variety of problems related to physical and geometric concepts.

Definition 5.18   *For all* $\mathbf{v}_1, \mathbf{v}_2 \in \mathbf{V}, \; \mathbf{v}_1 = (a_1, b_1) \; and \; \mathbf{v}_2 = (a_2, b_2),$ *the* **inner product**

$$\mathbf{v}_1 \cdot \mathbf{v}_2 = (a_1, b_1) \cdot (a_2, b_2) = a_1 a_2 + b_1 b_2. \tag{1}$$

Since the symbol for the operation is a dot, the inner product is also called a **dot product**. Notice that the dot as used here has a different meaning than when it is used for the operation of multiplication in the set of real numbers.

Examples
a. If $v_1 = (2, 3)$ and $v_2 = (-5, 4)$, then

$$v_1 \cdot v_2 = (2, 3) \cdot (-5, 4)$$
$$= (2)(-5) + (3)(4) = 2.$$

b. If $v_1 = 3i - 4j$ and $v_2 = i + 2j$, then

$$v_1 \cdot v_2 = (3i - 4j) \cdot (i + 2j)$$
$$= (3)(1) + (-4)2 = -5.$$

Notice that the set of vectors **V** is not closed for this operation. The inner product of two vectors is a scalar, not a vector. For this reason the inner product is also called the **scalar product.**

Another useful form of the inner product can be obtained from the Law of Cosines:

If $v_1 = (a_1, b_1)$ *and* $v_2 = (a_2, b_2)$, *where* v_1 *and* v_2 *are nonzero vectors, and γ is the angle formed by the geometric vectors corresponding to* v_1 *and* v_2, *then*

$$v_1 \cdot v_2 = \|v_1\| \, \|v_2\| \, \cos \gamma. \tag{2}$$

The above property follows from the following argument. Consider the line segment with length c drawn between the points corresponding to (a_1, b_1) and (a_2, b_2), as shown in Figure 5.12. From the Law of Cosines,

$$c^2 = \|v_1\|^2 + \|v_2\|^2 - 2\|v_1\| \, \|v_2\| \, \cos \gamma,$$

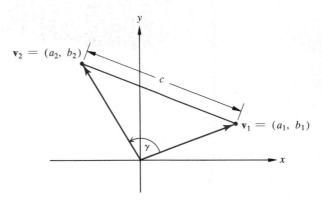

Figure 5.12

from which, for nonzero vectors v_1 and v_2,

$$\cos \gamma = \frac{\|v_1\|^2 + \|v_2\|^2 - c^2}{2\|v_1\| \, \|v_2\|}.$$

Substituting $\sqrt{a_1^2 + b_1^2}$ for $\|v_1\|$, $\sqrt{a_2^2 + b_2^2}$ for $\|v_2\|$, and

$$\sqrt{(a_2 - a_1)^2 + (b_2 - b_1)^2}$$

for c and simplifying, we obtain

$$\cos \gamma = \frac{a_1 a_2 + b_1 b_2}{\|v_1\| \, \|v_2\|}.$$

Substituting $v_1 \cdot v_2$ for $a_1 a_2 + b_1 b_2$ from Definition 5.18 yields

$$v_1 \cdot v_2 = \|v_1\| \, \|v_2\| \cos \gamma.$$

We can now express the inner product of two vectors $v_1 = (a_1, b_1)$ and $v_2 = (a_2, b_2)$ either, as shown in Equation (1), in terms of their scalar components or, as in Equation (2), in terms of the norms of the two vectors and the cosine of the angle γ formed by their corresponding geometric vectors.

Example   Given that $\|v_1\| = 3$, and v_1 has a direction angle, $\alpha_1 = 20°$, $\|v_2\| = 8$, and v_2 has a direction angle, $\alpha_2 = 170°$, find $v_1 \cdot v_2$.

Solution   From Equation (2) above

$$v_1 \cdot v_2 = (3)(8) \cos(170° - 20°) = (3)(8) \cos 150°$$

$$= 24 \left(-\frac{\sqrt{3}}{2} \right) = -12\sqrt{3}.$$

Scalar Projections

The property of inner products exhibited for two nonzero vectors in Equation (2) is particularly useful in several equivalent forms:

$$\cos \gamma = \frac{v_1 \cdot v_2}{\|v_1\| \, \|v_2\|}, \tag{3}$$

$$\|v_1\| \cos \gamma = \frac{v_1 \cdot v_2}{\|v_2\|}, \tag{4}$$

$$\|v_2\| \cos \gamma = \frac{v_1 \cdot v_2}{\|v_1\|}. \tag{5}$$

Equation (3) enables us to find the angle γ between any two nonzero vectors.

Example The angle γ between the geometric vectors corresponding to $(2, 3)$ and $(0, 2)$ is given by

$$\cos \gamma = \frac{(2, 3) \cdot (0, 2)}{\|(2, 3)\| \, \|(0, 2)\|} = \frac{2 \cdot 0 + 3 \cdot 2}{\sqrt{4 + 9} \sqrt{0 + 4}}$$

$$= \frac{3}{\sqrt{13}} \approx 0.8321.$$

Thus, $\gamma \approx 33.7° = 33° \, 42'$.

In Section 5.2 we obtained projections of a vector on the x- and y-axes of a Cartesian coordinate system. Equations (4) and (5) above now give us a way to find the projection of a vector on any other vector.

The product $\|\mathbf{v}_1\| \cos \gamma$ is called the **scalar projection** of $\mathbf{v}_1$ on $\mathbf{v}_2$ and the product $\|\mathbf{v}_2\| \cos \gamma$ is called the scalar projection of $\mathbf{v}_2$ on $\mathbf{v}_1$. In a geometric sense, these can be interpreted as the lengths of the line segments shown in Figure 5.13 (a and b).

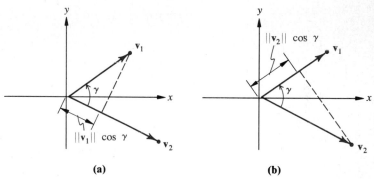

(a)   (b)

Figure 5.13

Example Given that $\mathbf{v}_1 = (-3, -2)$ and $\mathbf{v}_2 = (-5, 5)$, the scalar projection of $\mathbf{v}_1$ on $\mathbf{v}_2$ [Equation (4) on page 212] is

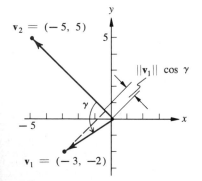

$$\|\mathbf{v}_1\| \cos \gamma = \frac{\mathbf{v}_1 \cdot \mathbf{v}_2}{\|\mathbf{v}_2\|} = \frac{(-3, -2) \cdot (-5, 5)}{\|(-5, 5)\|}$$

$$= \frac{15 - 10}{\sqrt{(-5)^2 + 5^2}} = \frac{5}{\sqrt{50}}$$

$$= \frac{5}{5\sqrt{2}} = \frac{1}{\sqrt{2}}.$$

(continued)

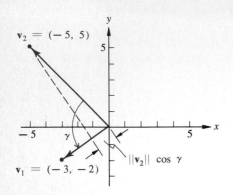

The scalar projection of $\mathbf{v}_2$ on $\mathbf{v}_1$ [Equation (5) on page 212] is

$$\|\mathbf{v}_2\| \cos \gamma = \frac{\mathbf{v}_1 \cdot \mathbf{v}_2}{\|\mathbf{v}_1\|} = \frac{(-3, -2) \cdot (-5, 5)}{\|(-3, -2)\|}$$

$$= \frac{15 - 10}{\sqrt{(-3)^2 + (-2)^2}} = \frac{5}{\sqrt{13}}.$$

Orthogonal Vectors

If $\mathbf{v}_1$ and $\mathbf{v}_2$ are nonzero vectors and orthogonal (perpendicular), then $\|\mathbf{v}_1\| \neq 0$, $\|\mathbf{v}_2\| \neq 0$, $\cos \gamma = 0$, and

$$\mathbf{v}_1 \cdot \mathbf{v}_2 = \|\mathbf{v}_1\| \|\mathbf{v}_2\| \cos \gamma = 0.$$

We also note that for $0° \leq \gamma \leq 180°$, $\cos \gamma = 0$ only if $\gamma = 90°$. From Definition 5.18,

$$\mathbf{v}_1 \cdot \mathbf{v}_2 = a_1 a_2 + b_1 b_2.$$

Hence, we can test for the orthogonality (perpendicularity) of two vectors directly from their scalar components using the following property:

The vectors $\mathbf{v}_1 = (a_1, b_1)$ *and* $\mathbf{v}_2 = (a_2, b_2)$ *are orthogonal if and only if*

$$\mathbf{v}_1 \cdot \mathbf{v}_2 = a_1 a_2 + b_1 b_2 = 0. \tag{6}$$

Examples

a. $(2, -4)$ and $(6, 3)$ are orthogonal because

$$(2, -4) \cdot (6, 3) = (2)(6) + (-4)(3) = 0.$$

b. $-\mathbf{i} - 2\mathbf{j}$ and $4\mathbf{i} - 2\mathbf{j}$ are orthogonal because

$$(-\mathbf{i} - 2\mathbf{j}) \cdot (4\mathbf{i} - 2\mathbf{j}) = (-1)(4) + (-2)(-2) = 0.$$

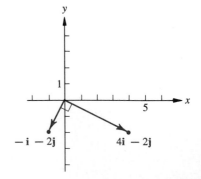

a.   b.

Applications

If a force is applied to an object so as to cause motion in a particular direction, and if the force is applied throughout the motion and in the same direction as the motion, then the product of the magnitude of the force (F) and the distance (d) the object moves is called **work.** In many situations, the direction of an applied force is not the same as the direction in which the object moves. In this case, the work is the product of the magnitude of the *component of the force in the direction of motion* and the distance the object moves. Since the inner product of two vectors can be used to find such a product whether or not the two vectors have the same direction, we provide a more general definition of work in Definition 5.19.

Definition 5.19 *If* **F** *is a force vector and* **d** *a displacement vector, then the work W is given by*

$$W = \mathbf{F} \cdot \mathbf{d}.$$

Example A force of 40 pounds at an angle of 30° to the horizontal is used to move a box 8 feet on a level floor. Find the work done.

Solution The figure shows the vectors in a Cartesian system. Since $\|\mathbf{F}\| = 40$ pounds, $\|\mathbf{d}\| = 8$ feet, and $\gamma = 30°$,

$$W = \mathbf{F} \cdot \mathbf{d} = \|\mathbf{F}\|\,\|\mathbf{d}\| \cos \gamma$$

$$= (40)(8) \cos 30°$$

$$= (40)(8) \left(\frac{\sqrt{3}}{2}\right)$$

$$= 160\sqrt{3} \approx 277.$$

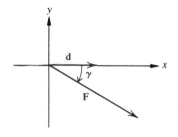

If the force magnitude is measured in pounds and the displacement magnitude in feet, as in this case, the work is measured in foot-pounds. Hence the work done is approximately 277 foot-pounds.

Example How much work is done in lifting an 80-pound weight to a height of 3 feet?

(continued)

Solution Since the force and displacement vectors are in the same direction, the angle $\gamma = 0°$ and $\cos \gamma = 1$. Hence,

$$W = \mathbf{F} \cdot \mathbf{d} = \|\mathbf{F}\| \, \|\mathbf{d}\| \cos \gamma = (80)(3)(1) = 240.$$

Hence, the work done is 240 foot-pounds. In this case the work is simply the product of the force times the distance.

EXERCISE SET 5.5

A

Find the inner product, $\mathbf{v}_1 \cdot \mathbf{v}_2$, *of the given vectors.*

Examples
 a. $\mathbf{v}_1 = (7, -9)$ and $\mathbf{v}_2 = (-6, -5)$
 b. $\mathbf{v}_1 = -9\mathbf{i} + 8\mathbf{j}$ and $\mathbf{v}_2 = 3\mathbf{i} - 2\mathbf{j}$

Solutions From Definition 5.18,

a. $\mathbf{v}_1 \cdot \mathbf{v}_2 = (7, -9) \cdot (-6, -5)$ b. $\mathbf{v}_1 \cdot \mathbf{v}_2 = (-9\mathbf{i} + 8\mathbf{j}) \cdot (3\mathbf{i} - 2\mathbf{j})$

$\qquad = (7)(-6) + (-9)(-5)$ $\qquad = (-9)(3) + (8)(-2)$

$\qquad = 3$ $\qquad = -43$

1. $\mathbf{v}_1 = (6, 0)$ and $\mathbf{v}_2 = (8, 10)$ 2. $\mathbf{v}_1 = (9, 3)$ and $\mathbf{v}_2 = (8, -5)$
3. $\mathbf{v}_1 = (7, -8)$ and $\mathbf{v}_2 = (2, 2)$ 4. $\mathbf{v}_1 = (3, 0)$ and $\mathbf{v}_2 = (0, -5)$
5. $\mathbf{v}_1 = 9\mathbf{i} - 4\mathbf{j}$ and $\mathbf{v}_2 = 5\mathbf{i}$ 6. $\mathbf{v}_1 = 8\mathbf{j}$ and $\mathbf{v}_2 = 7\mathbf{i} + 7\mathbf{j}$
7. $\mathbf{v}_1 = \mathbf{i} - 2\mathbf{j}$ and $\mathbf{v}_2 = 3\mathbf{i} - 5\mathbf{j}$ 8. $\mathbf{v}_1 = -4\mathbf{i} + 3\mathbf{j}$ and $\mathbf{v}_2 = 10\mathbf{j}$

Find the exact value of $\mathbf{v}_1 \cdot \mathbf{v}_2$ *given the norms and direction angles for* $\mathbf{v}_1$ *and* $\mathbf{v}_2$.

Example $\|\mathbf{v}_1\| = 10$, $\alpha_1 = 75°$ and $\|\mathbf{v}_2\| = 3$, $\alpha_2 = 45°$

Solution From Equation (2),

$$\mathbf{v}_1 \cdot \mathbf{v}_2 = (10)(3) \cos(75° - 45°) = 30 \cos 30°$$

$$= 30 \left(\frac{\sqrt{3}}{2} \right) = 15\sqrt{3}.$$

9. $\|\mathbf{v}_1\| = 8$, $\alpha_1 = 60°$ and $\|\mathbf{v}_2\| = 9$, $\alpha_2 = 15°$
10. $\|\mathbf{v}_1\| = 7$, $\alpha_1 = 160°$ and $\|\mathbf{v}_2\| = 6$, $\alpha_2 = -170°$

11. $\|\mathbf{v}_1\| = 5$, $\alpha_1 = -155°$ and $\|\mathbf{v}_2\| = \dfrac{3}{10}$, $\alpha_2 = 145°$

12. $\|\mathbf{v}_1\| = \dfrac{1}{2}$, $\alpha_1 = -60°$ and $\|\mathbf{v}_2\| = 10$, $\alpha_2 = -150°$

Determine the angle γ, where $0° \leq \gamma \leq 180°$, to the nearest six minutes, formed by each pair of vectors.

Example $\mathbf{v}_1 = (1, -7)$ and $\mathbf{v}_2 = (9, -1)$

Solution   From Equation (3),

$$\cos \gamma = \frac{(1, -7) \cdot (9, -1)}{\sqrt{1^2 + (-7)^2}\sqrt{9^2 + (-1)^2}}$$

$$= \frac{16}{\sqrt{50}\sqrt{82}} = \frac{16}{5\sqrt{2} \cdot \sqrt{2} \cdot \sqrt{41}} = \frac{8}{5\sqrt{41}} \approx 0.2499,$$

from which $\gamma \approx 75.5° = 75° \ 30'$.

13. $\mathbf{v}_1 = (3, 0)$ and $\mathbf{v}_2 = (-7, 3)$ **14.** $\mathbf{v}_1 = (1, -3)$ and $\mathbf{v}_2 = (8, -2)$

15. $\mathbf{v}_1 = (4, 0)$ and $\mathbf{v}_2 = (0, -2)$ **16.** $\mathbf{v}_1 = (-2, 0)$ and $\mathbf{v}_2 = (0, -3)$

17. $\mathbf{v}_1 = 9\mathbf{i} + 4\mathbf{j}$ and $\mathbf{v}_2 = 2\mathbf{i}$ **18.** $\mathbf{v}_1 = -4\mathbf{i} - 8\mathbf{j}$ and $\mathbf{v}_2 = \mathbf{i} - 5\mathbf{j}$

Find the scalar projection of $\mathbf{v}_1$ on $\mathbf{v}_2$ and of $\mathbf{v}_2$ on $\mathbf{v}_1$ for the given vectors. Sketch the corresponding geometric vectors and show the scalar projections.

Example $\mathbf{v}_1 = (1, 2)$ and $\mathbf{v}_2 = (-3, 4)$

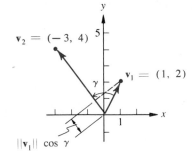

Solution   From Equation (4), the scalar projection of $\mathbf{v}_1$ on $\mathbf{v}_2$ is

$$\|\mathbf{v}_1\| \cos \gamma = \frac{\mathbf{v}_1 \cdot \mathbf{v}_2}{\|\mathbf{v}_2\|} = \frac{(1)(-3) + (2)(4)}{\sqrt{(-3)^2 + (4)^2}}$$

$$= \frac{-3 + 8}{\sqrt{25}} = \frac{5}{5} = 1.$$

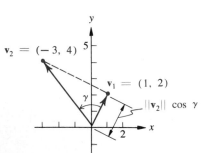

From Equation (5), the scalar projection of $\mathbf{v}_2$ on $\mathbf{v}_1$ is

$$\|\mathbf{v}_2\| \cos \gamma = \frac{\mathbf{v}_1 \cdot \mathbf{v}_2}{\|\mathbf{v}_1\|} = \frac{5}{\sqrt{1^2 + 2^2}}$$

$$= \frac{5}{\sqrt{5}} = \sqrt{5}.$$

19. $v_1 = -4i$ and $v_2 = -5i - 2j$ **20.** $v_1 = i - 6j$ and $v_2 = -5j$

21. $v_1 = (2, 3)$ and $v_2 = (-4, 5)$ **22.** $v_1 = (-9, 10)$ and $v_2 = (2, 4)$

23. $v_1 = (-3, 5)$ and $v_2 = (3, -1)$ **24.** $v_1 = (-7, 4)$ and $v_2 = (3, -6)$

25. $v_1 = 3i$ and $v_2 = 2j$ **26.** $v_1 = 2i$ and $v_2 = -2j$

Verify that the following pairs of vectors are orthogonal.

Example $i + 4j$ and $8i - 2j$

Solution $(i + 4j) \cdot (8i - 2j) = (1)(8) + (4)(-2)$

$$= 8 - 8 = 0.$$

Hence, by Equation (6), the vectors are orthogonal.

27. $(0, 5)$ and $(-3, 0)$ **28.** $(-2, 2)$ and $(5, 5)$

29. $(-4, 1)$ and $(1, 4)$ **30.** $(9, 1)$ and $(-1, 9)$

31. $6i - 9j$ and $3i + 2j$ **32.** $-3i + \sqrt{3}\,j$ and $i + \sqrt{3}\,j$

Find the work done, given each force vector and displacement vector.

Example $F = (1, 5)$ and $d = (-2, 3)$

Solution   By Definition 5.19,

$$W = F \cdot d = (1, 5) \cdot (-2, 3)$$
$$= 1 \cdot (-2) + 5 \cdot 3 = 13$$

33. $F = (3, 6)$ and $d = (4, 2)$

34. $F = (5, -2)$ and $d = (-1, -7)$

35. $F = 7i - 2j$ and $d = 4i - 4j$

36. $F = -12i + 3j$ and $d = -3i + 12j$

37. A small wagon is pulled horizontally a distance of 100 meters by a short rope. If the rope is at an angle of 30° to the horizontal and a constant force of 20 newtons is used, find the work done.

38. How much work is done in moving a string of cars 500 meters along a track by a truck exerting a force of 2000 newtons at an angle of 20° to the direction of the track?

39. In Exercise 38 how much work is done if the force is applied in the direction of motion?

40. How much work is done by a 175-pound man climbing 12 feet up a vertical ladder?

B

41. Show that $\mathbf{v}_1 \cdot \mathbf{v}_1 = \|\mathbf{v}_1\|^2$.   **42.** Show that $\mathbf{v}_1 \cdot \mathbf{v}_2 = \mathbf{v}_2 \cdot \mathbf{v}_1$.

43. Show that $c(\mathbf{v}_1 \cdot \mathbf{v}_2) = (c\mathbf{v}_1) \cdot \mathbf{v}_2$.

44. Show that $\mathbf{v}_1 \cdot (\mathbf{v}_2 + \mathbf{v}_3) = \mathbf{v}_1 \cdot \mathbf{v}_2 + \mathbf{v}_1 \cdot \mathbf{v}_3$.

Chapter Summary

5.1 A **geometric vector** $\vec{\mathbf{v}}$ is a line segment with a specified direction. Its **norm** or **magnitude** is designated by $\|\vec{\mathbf{v}}\|$. The **direction angle** is the angle α, where $-180° < \alpha \le 180°$, such that:

> The initial side of α is the ray from the initial point of $\vec{\mathbf{v}}$ parallel to the x-axis and directed in the positive x direction; the terminal side of α is the ray from the initial point of $\vec{\mathbf{v}}$ and containing $\vec{\mathbf{v}}$.

The **sum** $\vec{\mathbf{v}}_1 + \vec{\mathbf{v}}_2$ is defined as the geometric vector having as its initial point the initial point of $\vec{\mathbf{v}}_1$ and as its terminal point the terminal point of $\vec{\mathbf{v}}_2$, and where the terminal point of $\vec{\mathbf{v}}_1$ is the initial point of $\vec{\mathbf{v}}_2$. When viewed in this way, geometric vectors are said to be added according to "the triangle law." When viewed as the diagonal of a parallelogram with adjacent sides $\vec{\mathbf{v}}_1$ and $\vec{\mathbf{v}}_2$, geometric vectors are said to be added according to "the parallelogram law."

If $\vec{\mathbf{v}}$ is any geometric vector and c is any real number (scalar), then $c\vec{\mathbf{v}}$ is a geometric vector collinear with $\vec{\mathbf{v}}$ with magnitude $|c| \cdot \|\vec{\mathbf{v}}\|$.

5.2 The magnitudes of the **geometric vector projections**, $\vec{\mathbf{v}}_x$ and $\vec{\mathbf{v}}_y$, of $\vec{\mathbf{v}}$ on the x-axis and y-axis, respectively, are given by

$$\|\vec{\mathbf{v}}_x\| = \|\vec{\mathbf{v}}\| \cdot |\cos \alpha| \qquad \text{and} \qquad \|\vec{\mathbf{v}}_y\| = \|\vec{\mathbf{v}}\| \cdot |\sin \alpha|.$$

Geometric vectors can help to visualize and thus facilitate the solution of problems involving such quantities as force, velocity, and acceleration, since these quantities have both magnitude and direction.

5.3 A **two-dimensional vector** $\mathbf{v}$ is an ordered pair of real numbers. That is, if $a, b \in R$, then $\mathbf{v} = (a, b)$. The set of all such vectors is designated by $\mathbf{V}$. Each vector in the set corresponds to a unique bound geometric vector with initial point at the origin.

For all $v_1, v_2 \in V$, where $v_1 = (a_1, b_1)$ and $v_2 = (a_2, b_2)$,

$$v_1 = v_2 \quad \text{if and only if} \quad a_1 = a_2 \text{ and } b_1 = b_2;$$

the **norm,** or **magnitude,** of $v = (a, b)$ is the real number (scalar)

$$\|v\| = \sqrt{a^2 + b^2};$$

for $\|v\| \neq 0$, the **direction angle** of $v = (a, b)$ is the angle α such that

$$\cos \alpha = \frac{a}{\|v\|} \quad \text{and} \quad \sin \alpha = \frac{b}{\|v\|} \quad (-180° < \alpha \leq 180°).$$

The **norm** (or **magnitude**) of a vector, $\|v\|$, is the length of the associated geometric vector $\vec{v}$ and the **direction angle** of the vector is the angle α, where $-180° < \alpha \leq 180°$, measured from the positive x-axis to the geometric vector corresponding to v.

5.4 For all $v_1, v_2 \in V$, where $v_1 = (a_1, b_1)$, $v_2 = (a_2, b_2)$, and $c \in R$,

$$v_1 + v_2 = (a_1, b_1) + (a_2, b_2) = (a_1 + a_2, b_1 + b_2);$$
$$cv = c(a_1, b_1) = (ca_1, cb_1).$$

The **zero vector** $0 = (0, 0)$ is the additive identity for addition and the vector $-v = (-a, -b)$ is the additive inverse of $v = (a, b)$.

For all $v_1, v_2 \in V$, where $v_1 = (a_1, b_1)$, $v_2 = (a_2, b_2)$, and $c \in R$,

$$v_1 - v_2 = v_1 + (-v_2);$$
$$\frac{v_1}{c} = \frac{1}{c} v_1 \ (c \neq 0).$$

If $i = (1, 0)$ and $j = (0, 1)$, then each vector v can be expressed as

$$v = (a, b) = ai + bj.$$

5.5 The **inner (dot, scalar) product** of vectors v_1 and v_2 is

$$v_1 \cdot v_2 = (a_1, b_1) \cdot (a_2, b_2) = a_1 a_2 + b_1 b_2.$$

If γ is the angle between the geometric vectors corresponding to v_1 and v_2, then

$$v_1 \cdot v_2 = \|v_1\| \, \|v_2\| \cos \gamma.$$

Equivalently,

$$\cos \gamma = \frac{\mathbf{v}_1 \cdot \mathbf{v}_2}{\|\mathbf{v}_1\| \, \|\mathbf{v}_2\|};$$

$\|\mathbf{v}_1\| \cos \gamma = \dfrac{\mathbf{v}_1 \cdot \mathbf{v}_2}{\|\mathbf{v}_2\|}$, the scalar projection of $\mathbf{v}_1$ on $\mathbf{v}_2$;

$\|\mathbf{v}_2\| \cos \gamma = \dfrac{\mathbf{v}_1 \cdot \mathbf{v}_2}{\|\mathbf{v}_1\|}$, the scalar projection of $\mathbf{v}_2$ on $\mathbf{v}_1$.

The vectors $\mathbf{v}_1 = (a_1, b_1)$ and $\mathbf{v}_2 = (a_2, b_2)$ are **orthogonal** if and only if the inner product.

$$\mathbf{v}_1 \cdot \mathbf{v}_2 = a_1 a_2 + b_1 b_2 = 0.$$

A list of symbols introduced in this chapter is shown inside the front cover, and important properties are shown inside the back cover.

Review Exercises

A

5.1   *Use free-hand methods and the geometric vectors shown at the left to draw a single vector to represent each of the following.*

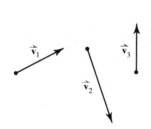

1. $(\vec{\mathbf{v}}_1 + \vec{\mathbf{v}}_2) + \vec{\mathbf{v}}_3$ **2.** $-3\vec{\mathbf{v}}_1$ **3.** $2\vec{\mathbf{v}}_2 + \vec{\mathbf{v}}_3$ **4.** $2\vec{\mathbf{v}}_1 - 3\vec{\mathbf{v}}_3$

5.2   **5.** If the direction angle $\alpha = 60°$, and $\|\vec{\mathbf{v}}\| = 12$, find $\|\vec{\mathbf{v}}_x\|$ and $\|\vec{\mathbf{v}}_y\|$.

6. If $\vec{\mathbf{v}}_x$ and $\vec{\mathbf{v}}_y$ are directed in a positive direction; $\|\vec{\mathbf{v}}_x\| = 4$, with initial point at $(1, 0)$; and $\|\vec{\mathbf{v}}_y\| = 3$, with initial point at $(0, 2)$, find $\|\vec{\mathbf{v}}\|$ and the direction angle α.

7. The resultant of two forces acting at an angle of $90°$ with respect to each other is 130 newtons. If one of the forces is 50 newtons, find the other force and the angle it makes with the resultant.

8. A person can swim at a speed of 3 miles per hour in still water. If he heads across a river at right angles to a current of 6 miles per hour, find his speed in relation to the land and the direction in which he actually moves.

9. An airplane is headed on a bearing of $280°$ with an air speed of 800 kilometers per hour. The course has a bearing of $292°$. The ground speed is 860 kilometers per hour. Find the wind direction and wind speed.

10. A force of 150 pounds is needed to keep a steel ball weighing 250 pounds from rolling down an inclined plane. What angle does the plane make with the horizontal?

5.3 *Find real numbers x and y for which each statement is true.*

11. $(x, y) = (-1, 3)$ **12.** $(x + y, x - y) = (5, 3)$

(a) *Sketch the geometric vector corresponding to each vector, and*
(b) *find the norm and the direction angle (to the nearest six minutes) of each vector.*

13. $(-4, 3)$ **14.** $(5, -5)$

Given the following norms and direction angles, find the vector **v**.

15. $\|\mathbf{v}\| = 10$ and $\alpha = -135°$ **16.** $\|\mathbf{v}\| = 2$ and $\alpha = 150°$

5.4 *Write the sum of each pair of vectors as an ordered pair.*

17. $\mathbf{v}_1 = (4, -5)$ and $\mathbf{v}_2 = (8, 1)$ **18.** $\mathbf{v}_1 = (-3, 7)$ and $\mathbf{v}_2 = (-7, 3)$

Let $\mathbf{v}_1 = (3, -2)$, $\mathbf{v}_2 = (5, 4)$, $\mathbf{v}_3 = (1, -3)$, $c_1 = 2$, *and* $c_2 = -1$.
Express each of the following as an ordered pair.

19. $\mathbf{v}_2 - c_2\mathbf{v}_1$ **20.** $c_1(\mathbf{v}_3 + \mathbf{v}_2)$ **21.** $\dfrac{\mathbf{v}_3}{c_2}$ **22.** $\dfrac{\mathbf{v}_2}{c_1\|\mathbf{v}_1\|}$

Express each vector as the sum of scalar multiples of **i** *and* **j**.

23. $(7, -9)$ **24.** $(-8, -1)$

Find the norm and direction angle (to the nearest six minutes) of each vector.

25. $-10\mathbf{i} + 4\mathbf{j}$ **26.** $7\mathbf{i} - 9\mathbf{j}$

5.5 *Find the inner product* $\mathbf{v}_1 \cdot \mathbf{v}_2$ *for given vectors.*

27. $\mathbf{v}_1 = (6, 6)$ and $\mathbf{v}_2 = (-5, -4)$

28. $\mathbf{v}_1 = 8\mathbf{i} - \mathbf{j}$ and $\mathbf{v}_2 = 9\mathbf{i} - 3\mathbf{j}$

29. $\|\mathbf{v}_1\| = 8$ and $\alpha_1 = 60°$; $\|\mathbf{v}_2\| = 4$ and $\alpha_2 = 30°$

30. $\|\mathbf{v}_1\| = 10$ and $\alpha_1 = -45°$; $\|\mathbf{v}_2\| = 5$ and $\alpha_2 = -15°$

Determine the angle γ to the nearest 6', where $0° \le \gamma \le 180°$, formed by each pair of vectors.

31. $\mathbf{v}_1 = (-7, 5)$ and $\mathbf{v}_2 = (1, -5)$

32. $\mathbf{v}_1 = -7\mathbf{i} - \mathbf{j}$ and $\mathbf{v}_2 = -8\mathbf{i} + 9\mathbf{j}$

Find the scalar projections of $\mathbf{v}_1$ on $\mathbf{v}_2$ and of $\mathbf{v}_2$ on $\mathbf{v}_1$.

33. $\mathbf{v}_1 = (1, 7)$ and $\mathbf{v}_2 = (10, -1)$

34. $\mathbf{v}_1 = 9\mathbf{i} + 2\mathbf{j}$ and $\mathbf{v}_2 = -8\mathbf{i} - 5\mathbf{j}$

Verify that each pair of vectors is orthogonal.

35. $(-4, -5)$ and $\left(4, -\dfrac{16}{5}\right)$ **36.** $7\mathbf{i} + 6\mathbf{j}$ and $5\mathbf{i} - \dfrac{35}{6}\mathbf{j}$

6 COMPLEX NUMBERS; POLAR COORDINATES

The set of real numbers does not contain an element whose square is a negative real number. Thus, any symbol of the form $\sqrt{-b}$, $b \in R$, $b > 0$, does not represent a real number, and equations such as

$$x^2 + 4 = 0 \qquad \text{equivalent to} \qquad x = \pm\sqrt{-4} \qquad (1)$$

and

$$x^2 - 2x + 5 = 0 \qquad \text{equivalent to} \qquad x = \frac{2 \pm \sqrt{-16}}{2} \qquad (2)$$

have no solutions in this set. In this chapter we consider a set of numbers, called the **set of complex numbers** and designated by C, that does provide solutions to such equations.

6.1 Sums and Products

The elements in the set C that we now consider are represented by various symbols. In this section we consider one form that is frequently used. In the following sections we consider other symbolic forms for these numbers.

Definition 6.1   *For all $a, b \in R$, a* **complex number** *is*

$$z = a + bi, \qquad where \quad i^2 = -1.$$

In this chapter we shall restrict the use of the variable z to represent elements in the set C. It is sometimes convenient to refer to a as the **real part** of the complex number and b as the **imaginary part,** with i the **imaginary unit.**

Properties of Equality

Equality in the set of complex numbers is established as follows.

Definition 6.2   *For every pair of complex numbers* $z_1 = a_1 + b_1 i$ *and* $z_2 = a_2 + b_2 i$,

$$z_1 = z_2 \quad \textit{if and only if} \quad a_1 = a_2 \quad \textit{and} \quad b_1 = b_2.$$

Example $2x + y^2 i = 8 + 9i$ if and only if $2x = 8$ and $y^2 = 9$ or, equivalently, $x = 4$ and $y = \pm 3$.

Properties similar to the equality axioms in the set of real numbers R (see Appendix A.2) and the substitution law are valid for the set of complex numbers. These laws follow from the corresponding properties for real numbers. Thus, if $z_1, z_2, z_3 \in C$, then

a. $z_1 = z_1$ Reflexive law

b. If $z_1 = z_2$, then $z_2 = z_1$ Symmetric law

c. If $z_1 = z_2$ and $z_2 = z_3$, then $z_1 = z_3$ Transitive law

Computing Sums and Products

If complex numbers are treated as if i were a variable and $a + bi$ were a binomial in the variable i, then the sum or product of two complex numbers can be found by applying the ordinary rules of the algebra of real numbers. Thus,

$$(2 + 3i) + (5 - 2i) = (2 + 5) + (3 - 2)i$$
$$= 7 + i;$$
$$(2 + 3i)(5 - 2i) = 10 + 11i - 6i^2.$$

However, replacing i^2 with -1, we have

$$(2 + 3i)(5 - 2i) = 10 + 11i - 6(-1)$$
$$= 16 + 11i.$$

More formally we have the following definition.

Definition 6.3 *If* $z_1 = a_1 + b_1 i$ *and* $z_2 = a_2 + b_2 i$, *then*

$$z_1 + z_2 = (a_1 + a_2) + (b_1 + b_2)i,$$

and

$$z_1 z_2 = (a_1 a_2 - b_1 b_2) + (a_1 b_2 + a_2 b_1)i.$$

Using such an approach for sums and products, we find that

1. C will contain elements $a + 0i$, or a, that can be identified with the real numbers for these operations, and
2. C will contain elements $0 + bi$, or bi, whose squares are negative real numbers.

For example,

$$(2) + (3) = (2 + 0i) + (3 + 0i)$$
$$= 5 + 0i = 5,$$
$$(2)(3) = (2 + 0i)(3 + 0i)$$
$$= 6 + 0i + 0i^2 = 6,$$
$$(3i)^2 = 9i^2 = -9.$$

Because the complex number $a + bi$, where $b = 0$, behaves like the real number a, we shall simply write a for $a + 0i$ as convenient and refer to a or $a + 0i$ as a real number.

If $b \neq 0$, the complex number $a + bi$ is called an **imaginary number** and for the special case where $a = 0$ and $b \neq 0$, the complex number is called a **pure imaginary number.** The relationship of the set C and other sets of numbers is shown in Figure 6.1. For example, 3 and -5 are elements of C that are real; $2 + 5i$ and $5i$ are elements of C that are imaginary numbers. Furthermore, $5i$ is a pure imaginary number.

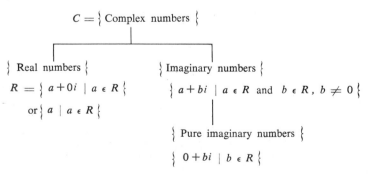

Figure 6.1

Periodic Property of i

Powers of the imaginary unit i have an interesting periodic property. Observe that

$$i^1 = i,$$
$$i^2 = -1$$
$$i^3 = i^2 \cdot i = -1 \cdot i = -i,$$
$$i^4 = i^2 \cdot i^2 = (-1)(-1) = 1,$$
$$i^5 = i^4 \cdot i = 1 \cdot i = i,$$
$$i^6 = i^4 \cdot i^2 = 1 \cdot (-1) = -1,$$
$$i^7 = i^4 \cdot i^3 = 1 \cdot (-i) = -i,$$

etc.

Thus, any integral power of i can be expressed as one of the numbers i, -1, $-i$, or 1.

Radical Notation

Another notation used to represent a complex number involves the square root symbol. In the set C, we make the following definition.

Definition 6.4   *For all* $b \in R$, $b > 0$,

$$\sqrt{-b} = \sqrt{-1}\sqrt{b} = i\sqrt{b}.$$

Observe that for the special case where $b = 1$,

$$\sqrt{-1} = \sqrt{-1}\sqrt{1} = i \cdot 1 = i.$$

Note that for a and b, positive real numbers,

$$\sqrt{-a}\sqrt{-b} = (i\sqrt{a})(i\sqrt{b}) = i^2\sqrt{a}\sqrt{b} = -\sqrt{ab},$$

a negative real number. Hence, the property of real numbers

$$\sqrt{a}\sqrt{b} = \sqrt{ab}, \quad a, b \geq 0,$$

is not applicable for imaginary numbers. For example,

$$\sqrt{-4}\sqrt{-9} \neq \sqrt{(-4)(-9)} = 6.$$

In this case,

$$\sqrt{-4}\sqrt{-9} = 2i \cdot 3i = 6i^2 = -6.$$

To avoid difficulty in rewriting products, expressions of the form $\sqrt{-b}$ ($b > 0$) should first be expressed in the form $\sqrt{b}\, i$ or $i\, \sqrt{b}$.

Examples

a. $\sqrt{-3}(5 + \sqrt{-3})$

$= i\sqrt{3}(5 + i\sqrt{3})$

$= 5i\sqrt{3} + i^2(3)$

$= 5i\sqrt{3} + (-1)(3)$

$= -3 + 5i\sqrt{3}$

b. $(3 - \sqrt{-2})(3 + \sqrt{-2})$

$= (3 - i\sqrt{2})(3 + i\sqrt{2})$

$= 9 - i^2(2)$

$= 9 - (-1)(2)$

$= 11$

EXERCISE SET 6.1

A

Find real numbers x and y for which each statement is true.

Example   By Definition 6.2, $x + 2yi = -5 + 8i$ if and only if

$$x = -5 \quad \text{and} \quad 2y = 8,$$

or, equivalently, if

$$x = -5 \quad \text{and} \quad y = 4.$$

1. $-5x + yi = -5 - 3i$

2. $x + 4i = 2yi$

3. $y + 4xi = 8 - 2i$

4. $x - 4i = 2y - x^2i$

Write each expression in the form $a + bi$.

Example $(3 + 2i) + (1 - 3i) = (3 + 1) + [2 + (-3)]i$
$$= 4 - i.$$

5. $(4 + 8i) + (-3 + 2i)$

6. $(-6 + 6i) + (4 + 2i)$

7. $(-9 - 5i) + (-1 - 6i)$

8. $(-4 + 3i) + (5 - i)$

9. $(3 + 3i) + (-4 + 3i)$

10. $(-1 + 6i) + (2 - 2i)$

11. $(a + bi) + (-a - bi)$

12. $(a + bi) + (a - bi)$

Examples

a. $2(1 - 5i) = 2(1) + 2(-5)i$ b. $-3(4 + i) = -3(4) - 3(1)i$
$\quad\quad = 2 - 10i$ $\quad\quad = -12 - 3i$

13. $4(3 - 3i)$ **14.** $5(2 + 4i)$ **15.** $-6(6 + 2i)$

16. $-4(8 - i)$ **17.** $\dfrac{1}{2}(4 + 6i)$ **18.** $\dfrac{2}{3}(9 - 6i)$

Example $(-3 + 4i)(1 + 6i) = -3 + 4i - 18i + 24i^2$
$\quad\quad\quad\quad\quad\quad\quad = -3 - 14i + 24(-1) = -27 - 14i.$

19. $(-3 + 8i)(1 + 2i)$ **20.** $(4 + 2i)(-8 - 2i)$
21. $(-6 - i)(-9 + 8i)$ **22.** $(-3 - 9i)(3 + 4i)$
23. $(5 - 6i)(4 + i)$ **24.** $(5 + i)(-1 + i)$
25. $(a + bi)(a - bi)$ **26.** $(a + bi)(a + bi)$

Write each expression as one of the complex numbers, i, -1, $-i$, or 1.

27. i^5 **28.** i^8 **29.** i^{10} **30.** i^{11}
31. i^{14} **32.** i^{16} **33.** i^{19} **34.** i^{23}

Write each expression in the form bi or $a + bi$.

Examples

a. $\sqrt{-16} = \sqrt{16}\sqrt{-1}$ b. $4 + \sqrt{-8} = 4 + \sqrt{4}\sqrt{2}\sqrt{-1}$
$\quad\quad = 4i$ $\quad\quad\quad = 4 + 2\sqrt{2}i$

35. $\sqrt{-25}$ **36.** $\sqrt{-49}$ **37.** $\sqrt{-12}$ **38.** $\sqrt{-27}$
39. $1 + \sqrt{-9}$ **40.** $2 - \sqrt{-64}$ **41.** $3 - \sqrt{-20}$ **42.** $5 + \sqrt{-48}$

Examples

a. $\sqrt{-9}(2 - 3\sqrt{-9}) = 3i(2 - 3 \cdot 3i)$
$\quad\quad\quad\quad\quad\quad = 3i(2 - 9i)$
$\quad\quad\quad\quad\quad\quad = 6i - 27i^2$
$\quad\quad\quad\quad\quad\quad = 6i - 27(-1)$
$\quad\quad\quad\quad\quad\quad = 27 + 6i$

(continued)

b. $(1 + \sqrt{-5})(1 - 2\sqrt{-5}) = (1 + \sqrt{5}i)(1 - 2\sqrt{5}i)$

$$= 1 + \sqrt{5}i - 2\sqrt{5}i - 2(\sqrt{5})^2 i^2$$

$$= 1 - \sqrt{5}i - 2 \cdot 5(-1) = 11 - \sqrt{5}i$$

43. $\sqrt{-4}(1 - 2\sqrt{-4})$ **44.** $\sqrt{-8}(3 + 4\sqrt{-2})$

45. $\sqrt{-16}(2 + \sqrt{-4})$ **46.** $\sqrt{-9}(3 - \sqrt{-25})$

47. $(\sqrt{-50} + 2)\sqrt{-10}$ **48.** $(\sqrt{-48} - 1)\sqrt{-21}$

49. $(5 + 2\sqrt{-4})(3 - \sqrt{-4})$ **50.** $(1 - 4\sqrt{-9})(7 + 2\sqrt{-9})$

51. $(2 + \sqrt{-3})(3 - \sqrt{-3})$ **52.** $(3 - \sqrt{-7})(3 + \sqrt{-7})$

53. Show that $2i$ and $-2i$ satisfy $x^2 + 4 = 0$.

54. Show that $1 + 2i$ and $1 - 2i$ satisfy $x^2 - 2x + 5 = 0$.

B

55. Show that $\left(-\dfrac{1}{2} + i\dfrac{\sqrt{3}}{2}\right)^3 = 1$ and $\left(-\dfrac{1}{2} - i\dfrac{\sqrt{3}}{2}\right)^3 = 1$.

56. Show that $\left(\dfrac{1}{2} + i\dfrac{\sqrt{3}}{2}\right)^3 = -1$ and $\left(\dfrac{1}{2} - i\dfrac{\sqrt{3}}{2}\right)^3 = -1$.

Find a and b for which each statement is true.

57. $(a + bi)^2 = -1$.   *Hint:* First express the given equation as $a^2 - b^2 + 2abi = -1$ and then use Definition 6.2.

58. $(a + bi)^2 = 1$.

The formula $e^{ix} = \cos x + i \sin x$, *where* $e = 2.718\ldots$, *is called* **Euler's formula** (*it is usually derived in a course in calculus*). *Use this formula to show that the following equations are identities.*

59. $e^{\pi i} = -1$ **60.** $e^{-ix} = \cos x - i \sin x$

61. $\sin x = \dfrac{e^{ix} - e^{-ix}}{2i}$ **62.** $\cos x = \dfrac{e^{ix} + e^{-ix}}{2}$

63. $\sin^2 x + \cos^2 x = 1$.   *Hint:* Use Exercises 61 and 62.

6.2 Differences and Quotients

The operations of addition and multiplication of complex numbers have been defined so that many of the properties of real numbers are valid in the set of complex numbers.

Two other useful operations, subtraction and division, are defined in terms of addition and multiplication. These definitions are similar to the definitions for a difference and a quotient in the set of real numbers.

Definition 6.5   *For all* $z_1, z_2 \in C$, the **difference**

$$z_1 - z_2 = z_1 + (-z_2).$$

Example $(3 + 4i) - (5 + 2i) = (3 + 4i) + (-5 - 2i)$

$$= (3 - 5) + (4 - 2)i = -2 + 2i.$$

Definition 6.6   *For all* $z_1, z_2 \in C$ *where* $z_2 \neq 0$, *the* **quotient**

$$\frac{z_1}{z_2} = z_1\left(\frac{1}{z_2}\right).$$

Since the axioms used for the real numbers together with equality axioms are also valid in C, for each property that follows from these axioms in the set R (see Appendix A.2), there is a corresponding property in the set C. One such property,

If $z_1, z_2, z_3 \in C$, *such that* $z_2, z_3 \neq 0$, *then*

$$\frac{z_1}{z_2} = \frac{z_1 z_3}{z_2 z_3}, \tag{1}$$

is particularly useful in rewriting quotients in the set C. We first make the following definition.

Definition 6.7   *For all* $z = a + bi$, *the* **conjugate** *of* z *is*

$$\bar{z} = a - bi.$$

Examples
   a. The conjugate of $5 + 4i$ is $5 - 4i$.

   b. The conjugate of $2 - 3i$ is $2 + 3i$.

A quotient of complex numbers, z_1/z_2, can be expressed in the form $a + bi$ by first multiplying both the numerator and the denominator by the conjugate of the denominator. If the denominator is a pure imaginary number, then it is necessary only to multiply the numerator and denominator by i. If the denominator is a real number and the numerator is of the form $a + bi$, then it is necessary only to write the quotient as two terms.

Examples

a. $\dfrac{3}{1 + 2i} = \dfrac{3(1 - 2i)}{(1 + 2i)(1 - 2i)}$

$= \dfrac{3 - 6i}{1 - 4i^2}$

$= \dfrac{3 - 6i}{5}$

$= \dfrac{3}{5} - \dfrac{6}{5} i$

b. $\dfrac{4 - i}{2 - 3i} = \dfrac{(4 - i)(2 + 3i)}{(2 - 3i)(2 + 3i)}$

$= \dfrac{8 + 10i - 3i^2}{4 - 9i^2}$

$= \dfrac{11 + 10i}{13}$

$= \dfrac{11}{13} + \dfrac{10}{13} i$

c. $\dfrac{3 - i}{3i} = \dfrac{(3 - i)i}{3i \cdot i} = \dfrac{3i - i^2}{3i^2}$

$= \dfrac{3i + 1}{-3} = -\dfrac{1}{3} - i$

d. $\dfrac{4 + 5i}{7} = \dfrac{4}{7} + \dfrac{5}{7} i$

In Example a, note that the product of the complex number $1 + 2i$ and its conjugate $1 - 2i$ is the real number 5. In Example b, note that the product of the complex number $2 - 3i$ and its conjugate $2 + 3i$ is the real number 13. In fact, observe that for all $a + bi \in C$,

$$(a + bi)(a - bi) = a^2 - b^2 i^2 = a^2 + b^2,$$

a real number.

As we noted in Section 6.1, to avoid difficulty in rewriting products, expressions of the form $\sqrt{-b}\ (b > 0)$ should first be expressed in the form $\sqrt{b}\,i$ or $i\sqrt{b}$. Hence, quotients that involve this form should also be rewritten before using Equation (1).

Examples

a. $\dfrac{5}{\sqrt{-4}} = \dfrac{5}{2i}$

$= \dfrac{5 \cdot i}{2i \cdot i}$

$= \dfrac{5i}{2(-1)}$

$= -\dfrac{5}{2} i$

b. $\dfrac{4}{3 - \sqrt{-2}}$

$= \dfrac{4(3 + i\sqrt{2})}{(3 - i\sqrt{2})(3 + i\sqrt{2})}$

$= \dfrac{12 + 4i\sqrt{2}}{9 - i^2(2)}$

$= \dfrac{12 + 4i\sqrt{2}}{9 - (-1)(2)}$

$= \dfrac{12}{11} + \dfrac{4\sqrt{2}}{11} i$

EXERCISE SET 6.2

A

Express each difference in the form a + bi.

Example $(1 - 4i) - (-1 + 10i) = (1 - 4i) + (1 - 10i)$

$$= (1 + 1) + (-4 - 10)i = 2 - 14i$$

1. $(6 + i) - (-7 + 5i)$ **2.** $(-7 - 3i) - (-5 + 5i)$

3. $(3 - i) - (1 - 3i)$ **4.** $(3 + 2i) - (-5 + 8i)$

5. $(5 + 2i) - 3i$ **6.** $4 - (2 - 3i)$

Write the conjugate of each complex number in the form a + bi.

Examples a. $z = -3 + 4i$ b. $z = 5 - 2i$

Solutions a. $\bar{z} = -3 - 4i$ b. $\bar{z} = 5 + 2i$

7. $6 + 3i$ **8.** $5 - 5i$ **9.** $-3 - 3i$

10. $-4 + i$ **11.** $3i$ **12.** 7

Express each quotient in the form a + bi.

Examples

a. $\dfrac{8 - 2i}{8i} = \dfrac{(8 - 2i)i}{8i \cdot i}$ b. $\dfrac{-4}{-3 + i} = \dfrac{(-4)(-3 - i)}{(-3 + i)(-3 - i)}$

$\qquad = \dfrac{8i - 2i^2}{8i^2}$ $= \dfrac{12 + 4i}{9 - i^2}$

$\qquad = \dfrac{8i + 2}{-8}$ $= \dfrac{6}{5} + \dfrac{2}{5} i$

$\qquad = -\dfrac{1}{4} - i$

13. $\dfrac{-9 - 6i}{-4i}$ **14.** $\dfrac{-2 + 3i}{3i}$ **15.** $\dfrac{4}{2 - 9i}$ **16.** $\dfrac{2}{-8 + 5i}$

17. $\dfrac{2 + i}{4 + i}$ **18.** $\dfrac{2 - 5i}{-1 + 6i}$ **19.** $\dfrac{-5 + 9i}{6 - 2i}$ **20.** $\dfrac{3 - 2i}{3 + 2i}$

Examples

a. $\dfrac{3}{\sqrt{-2}} = \dfrac{3(i)}{i\sqrt{2}(i)}$

$= \dfrac{3i}{i^2\sqrt{2}}$

$= \dfrac{3i}{(-1)\sqrt{2}}$

$= -\dfrac{3}{\sqrt{2}}\, i$

b. $\dfrac{5}{2 + \sqrt{-3}}$

$= \dfrac{5(2 - i\sqrt{3})}{(2 + i\sqrt{3})(2 - i\sqrt{3})}$

$= \dfrac{10 - 5i\sqrt{3}}{4 - i^2(3)}$

$= \dfrac{10 - 5i\sqrt{3}}{4 - (-1)(3)}$

$= \dfrac{10}{7} - \dfrac{5\sqrt{3}}{7}\, i$

21. $\dfrac{1}{\sqrt{-5}}$

22. $\dfrac{2}{\sqrt{-7}}$

23. $\dfrac{4}{\sqrt{-8}}$

24. $\dfrac{7}{\sqrt{-12}}$

25. $\dfrac{3}{4 + \sqrt{-5}}$

26. $\dfrac{7}{5 - \sqrt{-3}}$

27. $\dfrac{6}{\sqrt{-2} + 3}$

28. $\dfrac{4}{\sqrt{-7} - 6}$

29. $\dfrac{6}{\sqrt{2} + \sqrt{-5}}$

30. $\dfrac{10}{\sqrt{3} - \sqrt{-6}}$

31. $\dfrac{2}{\sqrt{5} + \sqrt{-8}}$

32. $\dfrac{3}{\sqrt{6} - \sqrt{-12}}$

6.3 Graphical Representation

The elements in $\{a + bi \mid a, b \in R\}$ can be placed in a one-to-one correspondence with the set of points in a plane by associating a with the abscissa of a point and b with the ordinate. In this way each point in a plane can be viewed as the graph of a complex number. A plane on which complex numbers are thus represented is called a **complex plane**. The elements in $\{a + 0i \mid a \in R\}$, or simply $\{a \mid a \in R\}$, are associated with the points on the x-axis, called the **real axis** and the elements in $\{0 + bi \mid b \in R\}$, or simply $\{bi \mid b \in R\}$, are associated with points on the y-axis, called the **imaginary axis**.

Examples   Graph each number z, its negative, and its conjugate.

a. $z = 3 - 2i$

b. $z = -2 - 4i$

Solutions

a. The negative of $3 - 2i$ is $-3 + 2i$; the conjugate is $3 + 2i$.

b. The negative of $-2 - 4i$ is $2 + 4i$; the conjugate is $-2 + 4i$.

a.   b.

Sometimes the graph of a complex number is shown as a geometric vector with its initial point located at the origin. With this representation, complex numbers can be added graphically by the parallellogram law applicable to geometric vectors.

Examples   Represent each sum graphically using geometric vectors and check the results by analytic methods.

a. $(2 + 3i) + (4 - 5i)$ b. $(2 + 2i) + (-4 + 5i)$

Solutions

a.   b.

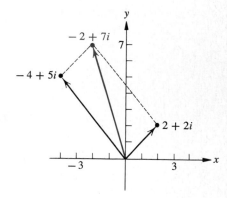

a. Check: $(2 + 3i) + (4 - 5i) = 6 - 2i$

b. Check: $(2 + 2i) + (-4 + 5i) = -2 + 7i$

Complex Numbers as Ordered Pairs

The elements in both the set $C = \{a + bi \mid a, b \in R\}$ and the elements in the set $\{(a, b) \mid a, b \in R\}$ have now been associated with the points in a plane. Hence, elements in C can also be associated with, and represented by, ordered pairs of real numbers (a, b).

Examples

a. $2 + 3i = (2, 3)$

b. $-5 = -5 + 0i = (-5, 0)$

c. $6i = 0 + 6i = (0, 6)$

Absolute Value and Argument

Corresponding to the norm $\|\vec{v}\|$ and the direction angle α for each geometric vector $\vec{v}$, there are a real number and a *set* of angles associated with each complex number.

Figure 6.2

Definition 6.8   *For all* $z \in C$, *the* **absolute value** *or* **modulus** *of* z, *where* $z = a + bi = (a, b)$, *is*

$$\rho = |z| = |a + bi|$$
$$= |(a, b)|$$
$$= \sqrt{a^2 + b^2}.$$

The modulus ρ (the Greek letter "rho") is the distance from the origin to the graph of (a, b), as shown in Figure 6.2.

Definition 6.9   *For all* $z \in C$, *an* **argument** *or* **amplitude** *of* z, *where* $z = a + bi = (a, b)$, *denoted by* $\arg(a + bi)$ *or* $\arg(a, b)$, *is an angle* θ *such that*

$$\cos \theta = \frac{a}{\sqrt{a^2 + b^2}} = \frac{a}{\rho} \quad and \quad \sin \theta = \frac{b}{\sqrt{a^2 + b^2}} = \frac{b}{\rho} \quad (a^2 + b^2 \neq 0).$$

Thus, $\arg(a + bi)$ is an angle θ with initial side the positive x-axis and terminal side the ray from the origin through the graph of (a, b), as shown in Figure 6.2.

If θ is an argument of $a + bi$, then for $k \in J$, so is $(\theta + k \cdot 360)°$. The angle θ with the least nonnegative measure is sometimes called the **principal argument.** For $a + bi = 0 + 0i$, any angle θ can be used as an argument.

Example   Find the absolute value and principal argument of $\sqrt{3} + i$.

Solution $\rho = |\sqrt{3} + i| = \sqrt{(\sqrt{3})^2 + 1^2} = \sqrt{3 + 1} = 2;$

$$\cos \theta = \frac{a}{\rho} = \frac{\sqrt{3}}{2} \quad \text{and} \quad \sin \theta = \frac{b}{\rho} = \frac{1}{2}.$$

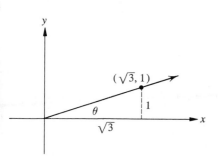

Because both $\cos \theta$ and $\sin \theta$ are positive, θ is in Quadrant I, and hence $\theta = 30°$.

There is another method that is often used to determine θ. Consider the number $a + bi = \sqrt{3} + i$ from the example above. First plot the point $(a, b) = (\sqrt{3}, 1)$ and form angle θ. From the sketch, θ is in Quadrant I, and $\tan \theta = 1/\sqrt{3}$. Hence $\theta = 30°$.

EXERCISE SET 6.3

A

Express each complex number as an ordered pair.

1. $-1 + 6i$	**2.** $5 - 9i$	**3.** $8 + i$
4. $-2 + 7i$	**5.** $6i$	**6.** -5

Express each complex number in the form $a + bi$.

7. $(-8, 7)$	**8.** $(0, 5)$	**9.** $(-4, 0)$
10. $(6, -3)$	**11.** $(7, -\sqrt{2})$	**12.** $(-\sqrt{3}, 1)$

Graph each complex number z, its negative $-z$, and its conjugate $\bar{z}$.

Examples

a. $-5 - 4i$ b. $(3, 2)$

Solutions

a. The negative is $5 + 4i$; the conjugate is $-5 + 4i$.

b. The negative is $(-3, -2)$; the conjugate is $(3, -2)$.

(continued)

a.

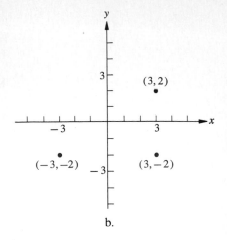

b.

13. $2 - 3i$ **14.** $-6 + 6i$ **15.** 5 **16.** $2i$

17. $(4, 6)$ **18.** $(-5, -3)$ **19.** $(-1, 0)$ **20.** $(0, 3)$

Represent each sum graphically. Check results by algebraic methods.

Example $(8 + 3i) + (1 - 5i)$

Solution

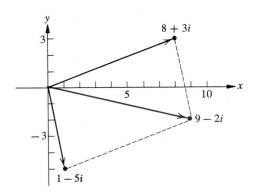

Check: $(8 + 3i) + (1 - 5i) = (8 + 1) + (3 - 5)i = 9 - 2i.$

21. $2 + (-3 + 6i)$ **22.** $(7 - i) + 2i$

23. $(4 - 7i) + (-8 + 2i)$ **24.** $(2 + 4i) + (1 - 5i)$

25. $(-6 + 7i) + (5 - 3i)$ **26.** $(-3 - 5i) + (3 - 2i)$

Find the absolute value and principal argument of each complex number.

Examples

a. $-7 + 5i$ b. $(3, 9)$

Solutions From Definitions 6.8 and 6.9,

$\quad$ a. $\rho = |-7 + 5i| = \sqrt{(-7)^2 + 5^2} = \sqrt{74};$

$$\cos \theta = -\frac{7}{\sqrt{74}} \approx -0.8137 \quad \text{and} \quad \sin \theta = \frac{5}{\sqrt{74}} \approx 0.5812.$$

Because $\cos \theta$ is negative and $\sin \theta$ is positive, θ is in Quadrant II; hence, $\theta \approx 144.5°$.

$\quad$ b. $\rho = |(3, 9)| = \sqrt{3^2 + 9^2} = \sqrt{90} = 3\sqrt{10};$

$$\cos \theta = \frac{3}{3\sqrt{10}} \approx 0.3162 \quad \text{and} \quad \sin \theta = \frac{9}{3\sqrt{10}} \approx 0.9487.$$

Because $\cos \theta$ and $\sin \theta$ are both positive, θ is in Quadrant I; hence, $\theta \approx 71.6°$.

27. $5 + 2i$ $\qquad$ **28.** $6 + 8i$ $\qquad$ **29.** $-2 - 2i$ $\qquad$ **30.** $8 - 3i$

31. 7 $\qquad$ **32.** -4 $\qquad$ **33.** $-5i$ $\qquad$ **34.** $6i$

35. $(3, -9)$ $\qquad$ **36.** $(-6, -2)$ $\qquad$ **37.** $(4, 0)$ $\qquad$ **38.** $(0, -7)$

39. Determine the conditions that a and b must satisfy if the graph of the complex number $a + bi$ is
$\quad$ **a.** on the real axis. $\qquad\qquad$ **b.** on the imaginary axis.
$\quad$ **c.** above the real axis. $\qquad\quad$ **d.** below the real axis.

40. What condition does $a^2 + b^2$ satisfy if the graph of $a + bi$ is
$\quad$ **a.** on a circle with radius of length 5?
$\quad$ **b.** inside a circle with radius of length 5?
$\quad$ **c.** outside a circle with radius of length 5?

B

Graphically, multiplication or division of a complex number $a + bi$ by the complex number i produces a 90° rotation of the geometric vector corresponding to $a + bi$, counterclockwise for multiplication and clockwise for division. Graph
(a) *the given complex number $a + bi$,* **(b)** $i \cdot (a + bi)$, *and* **(c)** $\dfrac{a + bi}{i}$.

Example $2 + 3i$

Solutions

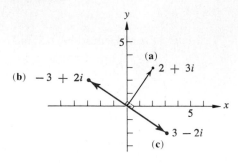

b. $i \cdot (2 + 3i) = 2i + 3i^2$

$$= -3 + 2i$$

c. $\dfrac{2 + 3i}{i} = \dfrac{(2 + 3i) \cdot i}{i \cdot i}$

$$= \dfrac{2i + 3i^2}{i^2}$$

$$= \dfrac{2i - 3}{-1} = 3 - 2i$$

41. $5 + 4i$ **42.** $1 - 2i$ **43.** $-6 + i$

44. $-3 - 4i$ **45.** 2 **46.** $-5i$

6.4 Trigonometric Form

In previous sections we have represented complex numbers in the form $a + bi$, $a + \sqrt{-b^2}$, and (a, b). It is sometimes convenient to represent a complex number in terms of its absolute value $\rho = \sqrt{a^2 + b^2}$ and its argument θ. From Definition 6.9,

$$\cos \theta = \frac{a}{\rho} \quad \text{and} \quad \sin \theta = \frac{b}{\rho}.$$

Therefore, as shown in Figure 6.3,

$$a = \rho \cos \theta \quad \text{and} \quad b = \rho \sin \theta,$$

from which we have

$$a + bi = \rho \cos \theta + (\rho \sin \theta)i,$$

$$a + bi = \rho(\cos \theta + i \sin \theta).$$

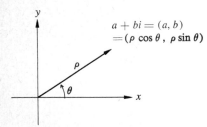

Figure 6.3

(The expression $\cos \theta + i \sin \theta$ is sometimes abbreviated as cis θ.)

The sine and cosine functions are periodic with period of $360°$ or $2\pi^R$. Hence, it is also true that

$$a + bi = \rho[\cos(\theta + k \cdot 360°) + i \sin(\theta + k \cdot 360°)], \quad k \in J,$$

or

$$a + bi = \rho[\cos(\theta + k \cdot 2\pi^R) + i\sin(\theta + k \cdot 2\pi^R)], \quad k \in J$$

where θ is the angle of smallest nonnegative measure. Either equation is called the **trigonometric form** of a complex number.

Example   Represent $-2 + 2i$ in trigonometric form.

Solution   From Definition 6.8,

$$\rho = |-2 + 2i| = \sqrt{(-2)^2 + 2^2} = \sqrt{8} = 2\sqrt{2}.$$

From Definition 6.9,

$$\cos\theta = \frac{-2}{2\sqrt{2}} = -\frac{1}{\sqrt{2}} \quad \text{and} \quad \sin\theta = \frac{2}{2\sqrt{2}} = \frac{1}{\sqrt{2}}.$$

Thus, θ is in Quadrant II and $\theta = 135°$. For $k \in J$,

$$-2 + 2i = 2\sqrt{2}[\cos(135° + k \cdot 360°) + i\sin(135° + k \cdot 360°)].$$

For $k = 0$,

$$-2 + 2i = 2\sqrt{2}(\cos 135° + i\sin 135°).$$

Example   Represent $3(\cos 60° + i\sin 60°)$ graphically and express the number in the form $a + bi$.

Solution   Since

$$a = \rho\cos\theta = 3\cos 60° = 3\left(\frac{1}{2}\right) = \frac{3}{2},$$

and

$$b = \rho\sin\theta = 3\sin 60° = 3\left(\frac{\sqrt{3}}{2}\right) = \frac{3\sqrt{3}}{2},$$

we have

$$3(\cos 60° + i\sin 60°) = \frac{3}{2} + \frac{3\sqrt{3}}{2} \cdot i.$$

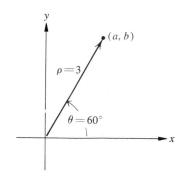

Products and Quotients

Products and quotients of complex numbers expressed in trigonometric form can be found quite readily. If

$$z_1 = \rho_1(\cos\theta_1 + i\sin\theta_1) \quad \text{and} \quad z_2 = \rho_2(\cos\theta_2 + i\sin\theta_2),$$

then

$$z_1 \cdot z_2 = \rho_1(\cos \theta_1 + i \sin \theta_1) \cdot \rho_2(\cos \theta_2 + i \sin \theta_2)$$
$$= \rho_1 \rho_2 [(\cos \theta_1 \cos \theta_2 - \sin \theta_1 \sin \theta_2)$$
$$+ i(\cos \theta_1 \sin \theta_2 + \sin \theta_1 \cos \theta_2)].$$

Substituting

$$\cos(\theta_1 + \theta_2) \text{ for } \cos \theta_1 \cos \theta_2 - \sin \theta_1 \sin \theta_2$$

and then

$$\sin(\theta_1 + \theta_2) \text{ for } \cos \theta_1 \sin \theta_2 + \sin \theta_1 \cos \theta_2,$$

we have

$$z_1 \cdot z_2 = \rho_1 \rho_2 [\cos(\theta_1 + \theta_2) + i \sin(\theta_1 + \theta_2)]. \tag{1}$$

It can also be shown in a similar way that

$$\frac{z_1}{z_2} = \frac{\rho_1}{\rho_2} [\cos(\theta_1 - \theta_2) + i \sin(\theta_1 - \theta_2)], \quad z_2 \neq (0, 0). \tag{2}$$

Example Express $6(\cos 75° + i \sin 75°) \cdot 3(\cos 15° + i \sin 15°)$ in the form $a + bi$.

Solution From Equation (1),

$$6(\cos 75° + i \sin 75°) \cdot 3(\cos 15° + i \sin 15°)$$
$$= 6 \cdot 3[\cos(75° + 15°) + i \sin(75° + 15°)]$$
$$= 18(\cos 90° + i \sin 90°)$$
$$= 18(0 + i) = 18i.$$

Example Express $\dfrac{6(\cos 75° + i \sin 75°)}{3(\cos 15° + i \sin 15°)}$ in the form $a + bi$.

Solution From Equation (2),

$$\frac{6(\cos 75° + i \sin 75°)}{3(\cos 15° + i \sin 15°)} = \frac{6}{3} [\cos(75° - 15°) + i \sin(75° - 15°)]$$
$$= 2(\cos 60° + i \sin 60°)$$
$$= 2\left(\frac{1}{2} + i \frac{\sqrt{3}}{2}\right)$$
$$= 1 + \sqrt{3}i.$$

EXERCISE SET 6.4

A

Represent each complex number in trigonometric form, using the angle with least nonnegative measure.

Example $\sqrt{3} - i$

Solution From Definition 6.8,

$$\rho = |\sqrt{3} - i| = \sqrt{3 + 1} = 2.$$

From Definition 6.9, $\cos \theta = \sqrt{3}/2$ and $\sin \theta = -1/2$. Hence θ is in Quadrant IV and $\theta = 330°$. Therefore,

$$\sqrt{3} - i = 2[\cos(330° + k \cdot 360°) + i \sin(330° + k \cdot 360°)], \quad k \in J.$$

For $k = 0$, we obtain

$$\sqrt{3} - i = 2(\cos 330° + i \sin 330°).$$

1. $-2 + 2\sqrt{3}i$ **2.** $6 + 6i$ **3.** $-5\sqrt{3} - 5i$ **4.** $3\sqrt{3} + 3i$

5. $2 - 2i$ **6.** $-4 - 4i$ **7.** -8 **8.** $-9i$

Represent each complex number graphically and write as a + bi.

Example $6(\cos 315° + i \sin 315°)$

Solution Since

$$a = \rho \cos \theta = 6\left(\frac{\sqrt{2}}{2}\right) = 3\sqrt{2},$$

and

$$b = \rho \sin \theta = 6\left(\frac{-\sqrt{2}}{2}\right) = -3\sqrt{2},$$

we have

$$a + bi = 3\sqrt{2} - 3\sqrt{2}i.$$

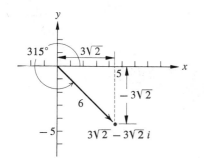

9. $2(\cos 240° + i \sin 240°)$ **10.** $2(\cos 90° + i \sin 90°)$

11. $2(\cos 135° + i \sin 135°)$ **12.** $3(\cos 30° + i \sin 30°)$

13. $\cos(-45°) + i \sin(-45°)$ **14.** $\cos(-210°) + i \sin(-210°)$

15. $\frac{1}{2}(\cos 150° + i \sin 150°)$ **16.** $\frac{3}{4}(\cos 330° + i \sin 330°)$

Express each product in the form a + bi.

Example $2(\cos 140° + i \sin 140°) \cdot 5(\cos 70° + i \sin 70°)$

Solution From Equation (1) in this section,

$$2(\cos 140° + i \sin 140°) \cdot 5(\cos 70° + i \sin 70°)$$
$$= 2 \cdot 5[\cos(140° + 70°) + i \sin(140° + 70°)]$$
$$= 10(\cos 210° + i \sin 210°)$$
$$= 10\left[\frac{-\sqrt{3}}{2} + i\left(-\frac{1}{2}\right)\right]$$
$$= -5\sqrt{3} - 5i.$$

17. $5(\cos 100° + i \sin 100°) \cdot 2(\cos 35° + i \sin 35°)$

18. $3(\cos 130° + i \sin 130°) \cdot \frac{2}{3}(\cos 140° + i \sin 140°)$

19. $4(\cos 85° + i \sin 85°) \cdot \frac{3}{4}(\cos 245° + i \sin 245°)$

20. $4(\cos 20° + i \sin 20°) \cdot \frac{1}{2}(\cos 40° + i \sin 40°)$

21. $2(\cos 40° + i \sin 40°) \cdot (\cos 95° + i \sin 95°)$

22. $2(\cos 5° + i \sin 5°) \cdot 3(\cos 40° + i \sin 40°)$

Express each quotient in the form a + bi.

Example $\dfrac{6(\cos 85° + i \sin 85°)}{2(\cos 40° + i \sin 40°)}$

Solution From Equation (2) in this section,

$$\frac{6(\cos 85° + i \sin 85°)}{2(\cos 40° + i \sin 40°)} = \frac{6}{2}[\cos(85° - 40°) + i \sin(85° - 40°)]$$
$$= 3(\cos 45° + i \sin 45°)$$
$$= \frac{3}{\sqrt{2}} + \frac{3}{\sqrt{2}} i.$$

23. $\dfrac{8(\cos 260° + i \sin 260°)}{2(\cos 80° + i \sin 80°)}$ **24.** $\dfrac{6(\cos 215° + i \sin 215°)}{3(\cos 80° + i \sin 80°)}$

25. $\dfrac{2(\cos 255° + i \sin 255°)}{5(\cos 15° + i \sin 15°)}$ **26.** $\dfrac{2(\cos 280° + i \sin 280°)}{3(\cos 70° + i \sin 70°)}$

27. $\dfrac{3(\cos 70° + i \sin 70°)}{\cos 130° + i \sin 130°}$

28. $\dfrac{\cos 19° + i \sin 19°}{4(\cos 259° + i \sin 259°)}$

29. Write $[2(\cos 20° + i \sin 20°)]^3$ in the form $a + bi$.

30. Write $\left(\dfrac{1}{\sqrt{2}} + \dfrac{1}{\sqrt{2}} i\right)^3$ in the form $a + bi$.

31. Show that if
$$a + bi = \rho(\cos \theta + i \sin \theta),$$
then $\qquad (a + bi)^2 = \rho^2(\cos 2\theta + i \sin 2\theta).$

32. Use the result of Exercise 31 to show that if
$$a + bi = \rho(\cos \theta + i \sin \theta),$$
then $\qquad (a + bi)^3 = \rho^3(\cos 3\theta + i \sin 3\theta).$

33. Use the result of Exercise 32 to show that if
$$a + bi = \rho(\cos \theta + i \sin \theta),$$
then $\qquad (a + bi)^4 = \rho^4(\cos 4\theta + i \sin 4\theta).$

34. Using the results of Exercises 31–33, what can you conjecture about $(a + bi)^n$, $n \in N$, if $a + bi = \rho(\cos \theta + i \sin \theta)$?

35. Show that $\cos \theta + i \sin \theta$ and $\cos(-\theta) + i \sin(-\theta)$ are conjugates of each other.

36. Show that $\cos \theta + i \sin \theta$ and $\cos(-\theta) + i \sin(-\theta)$ are reciprocals of each other.

6.5 DeMoivre's Theorem; Powers and Roots

Consider the complex number $z = a + bi = \rho(\cos \theta + i \sin \theta)$. From Equation (1) in Section 6.4.

$$[\rho(\cos \theta + i \sin \theta)]^2 = \rho(\cos \theta + i \sin \theta) \cdot \rho(\cos \theta + i \sin \theta)$$
$$= \rho^2[\cos(\theta + \theta) + i \sin(\theta + \theta)]$$
$$= \rho^2(\cos 2\theta + i \sin 2\theta);$$
$$[\rho(\cos \theta + i \sin \theta)]^3 = \rho^2(\cos 2\theta + i \sin 2\theta) \cdot \rho(\cos \theta + i \sin \theta)$$
$$= \rho^3[\cos(2\theta + \theta) + i \sin(2\theta + \theta)]$$
$$= \rho^3(\cos 3\theta + i \sin 3\theta).$$

It can be shown that similar results are valid for each such power for $n \in N$. That is, if $z = \rho(\cos \theta + i \sin \theta)$ and $n \in N$, then

$$z^n = \rho^n (\cos n\theta + i \sin n\theta). \qquad (1)$$

This statement is known as **DeMoivre's theorem.** The proof of the statement and extensions of it which are presented in this section involve the process of mathematical induction and will not be shown. However, the informal argument above should make the validity of Equation (1) plausible.

Example   Express $(\sqrt{3} + i)^5$ in the form $a + bi$.

Solution   From Definition 6.8,

$$\rho = |\sqrt{3} + i| = \sqrt{3 + 1} = 2.$$

By Definition 6.9, $\cos \theta = \sqrt{3}/2$ and $\sin \theta = 1/2$. Therefore θ is in Quadrant I and $\theta = 30°$. Thus,

$$
\begin{aligned}
(\sqrt{3} + i)^5 &= [2(\cos 30° + i \sin 30°)]^5 \\
&= 2^5(\cos 5 \cdot 30° + i \sin 5 \cdot 30°) \\
&= 32(\cos 150° + i \sin 150°) \\
&= 32\left(-\frac{\sqrt{3}}{2} + \frac{1}{2}i\right) \\
&= -16\sqrt{3} + 16i.
\end{aligned}
$$

Equation (1) was stated for powers with natural number exponents. If this statement is to be valid for integral exponents, consistent meanings must be assigned to z^0 and z^{-n} for $n \in N$.

Substituting 0 for n in Equation (1), we have

$$
\begin{aligned}
z^0 = \rho^0(\cos 0 \cdot \theta + i \sin 0 \cdot \theta) &= 1 \cdot (\cos 0 + i \sin 0) \\
&= 1 \cdot (1 + 0) = 1.
\end{aligned}
$$

Substituting $-n$ for n in Equation (1), we have

$$z^{-n} = \rho^{-n}[\cos(-n\theta) + i \sin(-n\theta)].$$

Since $\cos(-n\theta) = \cos n\theta$ and $\sin(-n\theta) = -\sin n\theta$, we have

$$z^{-n} = \frac{1}{\rho^n} \cdot (\cos n\theta - i \sin n\theta).$$

Multiplying the numerator and the denominator of the right-hand member by the conjugate of the numerator yields

$$z^{-n} = \frac{1}{\rho^n} \cdot \frac{(\cos n\theta - i \sin n\theta) \cdot (\cos n\theta + i \sin n\theta)}{\cos n\theta + i \sin n\theta}$$

$$= \frac{1}{\rho^n} \cdot \frac{\cos^2 n\theta + \sin^2 n\theta}{\cos n\theta + i \sin n\theta}$$

$$= \frac{1}{\rho^n} \cdot \frac{1}{\cos n\theta + i \sin n\theta}$$

$$= \frac{1}{\rho^n(\cos n\theta + i \sin n\theta)}.$$

Substituting z^n for $\rho^n(\cos n\theta + i \sin n\theta)$ yields

$$z^{-n} = \frac{1}{z^n}.$$

Thus, if we want Equation (1) to be valid for every exponent $n \in J$, z^0 and z^{-n} must be given the following interpretations:

Definition 6.10 *For all $z \neq 0 + 0i$,*

$$\text{I.} \qquad z^0 = 1 + 0i = 1.$$

$$\text{II.} \qquad z^{-n} = \frac{1}{z^n}, \quad n \in N.$$

DeMoivre's theorem [Equation (1)] can now be applied to powers with $n \in J$.

Example   Express $(-\sqrt{3} + i)^{-3}$ in the form $a + bi$.

Solution   From Definitions 6.8 and 6.9, if $z = -\sqrt{3} + i$,

$$\rho = |-\sqrt{3} + i| = \sqrt{3 + 1} = 2;$$

since

$$\cos \theta = -\frac{\sqrt{3}}{2} \qquad \text{and} \qquad \sin \theta = \frac{1}{2},$$

θ is in Quadrant II and $\theta = 150°$. Thus,

$$(-\sqrt{3} + i)^{-3} = [2(\cos 150° + i \sin 150°)]^{-3},$$

(continued)

and from DeMoivre's theorem we have

$$(-\sqrt{3} + i)^{-3} = 2^{-3}[\cos(-3 \cdot 150°) + i\sin(-3 \cdot 150°)]$$

$$= \frac{1}{8}[\cos(-450°) + i\sin(-450°)]$$

$$= \frac{1}{8}[0 - 1 \cdot i] = -\frac{1}{8}i.$$

Another extension of DeMoivre's theorem is possible for rational number exponents $1/n$, where $n \in N$. First we define an nth root of z in the same way that an nth root of a real number a is defined, when such a number exists.

Definition 6.11 *For $z \in C$, $n \in N$, the number ω is an nth root of z if $\omega^n = z$.*

We can now state another extension of DeMoivre's theorem applicable to roots of complex numbers:

If $n \in N$ and $z = \rho(\cos\theta + i\sin\theta)$, $z \neq 0$, then

$$\rho^{1/n}\left[\cos\left(\frac{\theta + k \cdot 360°}{n}\right) + i\sin\left(\frac{\theta + k \cdot 360°}{n}\right)\right], \quad k \in J \qquad (2)$$

specifies all nth roots of z.

Using this extension of DeMoivre's theorem, we can now find n distinct complex nth roots for each $z \in C$, where $z \neq 0$. For $k = 0$, one nth root is

$$\rho^{1/n}\left(\cos\frac{\theta}{n} + i\sin\frac{\theta}{n}\right),$$

where θ is the angle with the least positive measure. This root is called the **principal nth root** of z.

Example Express each of the five fifth roots of $z = \sqrt{2} + \sqrt{2}i$ in trigonometric form.

Solution From Definition 6.8,

$$\rho = |\sqrt{2} + \sqrt{2}i| = \sqrt{2 + 2} = 2;$$

since, from Definition 6.9,

$$\cos\theta = \frac{\sqrt{2}}{2} \quad \text{and} \quad \sin\theta = \frac{\sqrt{2}}{2},$$

the argument θ is in Quadrant I and $\theta = 45°$. Thus, in trigonometric form,

$$z = 2(\cos 45° + i \sin 45°),$$

and by Equation (2),

$$2^{1/5}\left[\cos\left(\frac{45° + k \cdot 360°}{5}\right) + i \sin\left(\frac{45° + k \cdot 360°}{5}\right)\right], \quad k \in J,$$

is a fifth root of z. Taking $k = 0, 1, 2, 3,$ and 4 in turn yields the roots

$$2^{1/5}(\cos 9° + i \sin 9°),$$
$$2^{1/5}(\cos 81° + i \sin 81°),$$
$$2^{1/5}(\cos 153° + i \sin 153°),$$
$$2^{1/5}(\cos 225° + i \sin 225°),$$
$$2^{1/5}(\cos 297° + i \sin 297°).$$

Notice that the substitution of any other integer for k will simply produce one of these five complex numbers. For example, if $k = 5$,

$$2^{1/5}[\cos(9 + 360)° + i \sin(9 + 360)°] = 2^{1/5}(\cos 9° + i \sin 9°).$$

EXERCISE SET 6.5

A

Express each power in the form $a + bi$.

Example $(\sqrt{3} + i)^{-2}$

Solution   From Definition 6.8,

$$\rho = |\sqrt{3} + i| = \sqrt{(\sqrt{3})^2 + (1)^2} = \sqrt{4} = 2;$$

since, from Definition 6.9,

$$\cos \theta = \frac{\sqrt{3}}{2} \quad \text{and} \quad \sin \theta = \frac{1}{2},$$

the argument θ is in Quadrant I and $\theta = 30°$. Thus,

$$(\sqrt{3} + i)^{-2} = [2(\cos 30° + i \sin 30°)]^{-2}.$$

(*continued*)

From DeMoivre's theorem,

$$(\sqrt{3} + i)^{-2} = 2^{-2} \left[\cos(-2 \cdot 30°) + i \sin(-2 \cdot 30°) \right]$$

$$= \frac{1}{4} \left[\cos(-60°) + i \sin(-60°) \right]$$

$$= \frac{1}{4} \left(\frac{1}{2} - \frac{\sqrt{3}}{2} i \right) = \frac{1}{8} - \frac{\sqrt{3}}{8} i.$$

1. $[3(\cos 20° + i \sin 20°)]^3$ **2.** $[2(\cos 42° + i \sin 42°)]^5$

3. $(-1 + i)^6$ **4.** $(2\sqrt{3} - 2i)^5$

5. $\left(\dfrac{1}{4} + \dfrac{\sqrt{3}}{4} i \right)^3$ **6.** $\left(-\dfrac{\sqrt{3}}{2} + \dfrac{1}{2 \cdot} i \right)^6$

7. $[9(\cos 315° + i \sin 315°)]^{-2}$ **8.** $[8(\cos 120° + i \sin 120°)]^{-3}$

9. $(2 - 2i)^{-3}$ **10.** $(\sqrt{3} - i)^{-3}$

11. $(-1 - \sqrt{3}i)^{-5}$ **12.** $(-2 - 2i)^{-4}$

In Exercises 13–18, find the roots indicated and express each of them in trigonometric form.

Example   The two square roots of $\dfrac{\sqrt{3}}{2} - \dfrac{1}{2} i$

Solution

$$\rho = \left| \frac{\sqrt{3}}{2} - \frac{1}{2} i \right| = \sqrt{\frac{3}{4} + \frac{1}{4}} = 1.$$

Thus, $\cos \theta = \sqrt{3}/2$ and $\sin \theta = -1/2$, and θ is in Quadrant IV. Hence $\theta = 330°$ and

$$\frac{\sqrt{3}}{2} - \frac{1}{2} i = \cos(330° + k \cdot 360°) + i \sin(330° + k \cdot 360°).$$

The square roots of $\dfrac{\sqrt{3}}{2} - \dfrac{1}{2} i$, using DeMoivre's theorem, are

$$1^{1/2} \left[\cos \left(\frac{330° + k \cdot 360°}{2} \right) + i \sin \left(\frac{330° + k \cdot 360°}{2} \right) \right],$$

where $k = 0$ and $k = 1$. For $k = 0$, we have

$$\cos 165° + i \sin 165°,$$

and for $k = 1$, we have

$$\cos 345° + i \sin 345°.$$

13. The two square roots of

   a. $\cos 80° + i \sin 80°$ **b.** $\dfrac{1}{2} + \dfrac{\sqrt{3}}{2} i$

14. The two square roots of

   a. $9(\cos 140° + i \sin 140°)$ **b.** $-4i$

15. The three cube roots of

   a. $8(\cos 120° + i \sin 120°)$ **b.** $\dfrac{1}{2} + \dfrac{\sqrt{3}}{2} i$

16. The three cube roots of

   a. $2(\cos 75° + i \sin 75°)$ **b.** $-27i$

17. The four fourth roots of

   a. $\cos 240° + i \sin 240°$ **b.** $\dfrac{1}{2} - \dfrac{\sqrt{3}}{2} i$

18. The four fourth roots of

   a. $81(\cos 60° + i \sin 60°)$ **b.** $-16i$

Solve each equation for all $x \in C$.

19. $x^3 + 1 = 0$.   *Hint:* $x^3 = -1 = -1 + 0i$. Thus, x is a cube root of $-1 + 0i$.

20. $x^5 + 1 = 0$ **21.** $x^3 - 1 = 0$

22. $x^5 - 1 = 0$ **23.** $x^3 = \dfrac{\sqrt{3}}{4} - \dfrac{1}{4} i$

24. $x^5 = 12 - 5i$

B

25. Graph each of the following on a separate Cartesian coordinate system; make a conjecture about the graphs of all nth roots of 1.
   a. The two square roots of 1   **b.** The three cube roots of 1
   c. The four fourth roots of 1   **d.** The five fifth roots of 1
   e. The six sixth roots of 1   **f.** The eight eighth roots of 1

26. Use the results of Exercise 21 to show that the reciprocal of any cube root of unity is itself a cube root of unity.

27. Use DeMoivre's theorem and Definition 6.2 to find an identity for $\cos 3\theta$ in terms of $\cos \theta$.

Hint: $(\cos \theta + i \sin \theta)^3 = \cos 3\theta + i \sin 3\theta$.

28. Use the hint in Exercise 27 to find an identity for $\sin 3\theta$ in terms of $\sin \theta$.

6.6 Polar Coordinates

The basis that we used to graph functions in Chapter 3 was the fact that we associated a point in the plane with an ordered pair of real numbers (x, y) called the Cartesian coordinates of the point, the first component of the ordered pair being the directed distance of the point from the y-axis and the second component being the directed distance from the x-axis.

Graphing Ordered Pairs (ρ, θ)

The location of a point A in the plane can also be specified by ρ (rho), the distance (a positive number) from the pole O to A and the measure of an angle θ* that the ray through point A makes with a ray called the **polar axis,** as shown in Figure 6.4. If we extend the replacement set of ρ to include the negative real numbers and the replacement set of θ to include the set of all angles, we can specify point A by many ordered pairs of the form (ρ, θ). The two components of each of these ordered pairs are called **polar coordinates** of A. For example, in Figure 6.4a, if (ρ, θ) are polar coordinates of A, then so are $[\rho, (\theta + k \cdot 360)^\circ], \; k \in J$. Also, if $-\rho \; (\rho > 0)$ denotes the directed distance from O to A along the negative extension of the ray $0A'$, as shown in Figure 6.4b, then $[-\rho, (\theta + 180 + k \cdot 360)^\circ]$ are also acceptable coordinates of A. For example, in Figure 6.5, the point having polar coordinates

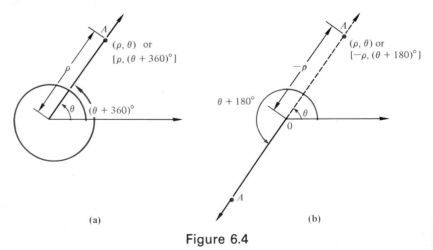

(a)   (b)

Figure 6.4

* The measure of θ can be expressed in degrees or radians.

Figure 6.5

(4, 30°) also has polar coordinates (4, 390°) and (4, −330°) for positive values of ρ, and (−4, 210°) for a negative value of ρ. Of course, there are infinitely many other possible polar coordinates for this point.

If point A is at the pole, then $\rho = 0$, and the coordinates of A are (0, θ), where θ is an arbitrary angle, not necessarily zero. For example, (0, 90°) and (0, 60°) are both coordinates of the pole.

Cartesian and Polar Coordinates

If the pole coincides with the origin, and the polar axis coincides with the positive x-axis, as in Figure 6.6, then for any point on the terminal ray of angle θ with Cartesian coordinates x and y, $\cos \theta = x/\rho$ and $\sin \theta = y/\rho$. Hence, the Cartesian and polar coordinates can be related as follows:

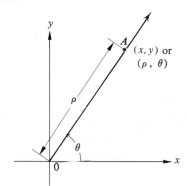

Figure 6.6

$$\left. \begin{array}{l} x = \rho \cos \theta \\ y = \rho \sin \theta \end{array} \right\} \quad (1)$$

$$\left. \begin{array}{l} \rho = \pm \sqrt{x^2 + y^2} \\[2mm] \cos \theta = \dfrac{x}{\rho} \\[2mm] \sin \theta = \dfrac{y}{\rho}, \quad \rho \neq 0 \end{array} \right\} \quad (2)$$

Equations (1) can be used to find the rectangular coordinates for a point if the polar coordinates are known, and Equations (2) can be used to find the polar coordinates for a point if the Cartesian coordinates are known. The positive or negative radical expression is chosen depending on the particular problem.

Example   Find the Cartesian coordinates of the point with polar coordinates (3, 60°).

Solution   From Equations (1), it follows that

$$x = 3 \cos 60° = 3 \cdot \frac{1}{2} = \frac{3}{2},$$

$$y = 3 \sin 60° = 3 \cdot \frac{\sqrt{3}}{2} = \frac{3\sqrt{3}}{2}.$$

The point and the corresponding ordered pairs are shown on page 254.

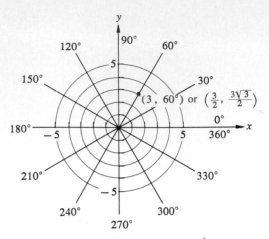

Example Find an approximation for the polar coordinates of the point with Cartesian coordinates $(5, -3)$, where $0° < \theta < 360°$ and $\rho > 0$. Graph the point.

Solution From Equations (2) and since $\rho > 0$,

$$\rho = \sqrt{x^2 + y^2} = \sqrt{25 + 9} = \sqrt{34} \approx 5.831;$$

$$\cos \theta = \frac{5}{5.831} \approx 0.8575.$$

Since the point is in Quadrant IV, then $\theta \approx 329.0°$. Therefore, the pair of polar coordinates is $(\sqrt{34}, 329°)$. The point and the corresponding ordered pairs are shown in the figure.

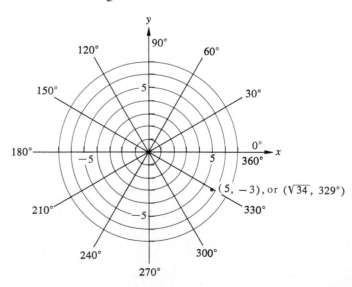

Rewriting Equations

Equations (1) and (2) on page 253 can be used to transform an equation in Cartesian form to polar form, and vice versa.

Example   Rewrite $x^2 + y^2 - 4x = 0$ in polar form.

Solution   Substituting ρ^2 for $x^2 + y^2$ and $\rho \cos \theta$ for x in

$$x^2 + y^2 - 4x = 0$$

yields

$$\rho^2 - 4\rho \cos \theta = 0,$$

from which

$$\rho(\rho - 4 \cos \theta) = 0.$$

Hence,

$$\rho = 0 \quad \text{or} \quad \rho - 4 \cos \theta = 0.$$

Note that the graph of $\rho = 0$ is the origin and also that the graph of $\rho = 4 \cos \theta$ includes the origin as one of its points (for example, for $\theta = 90°$, we have $\rho = 0$). Hence the equation $\rho = 4 \cos \theta$ corresponds to $x^2 + y^2 - 4x = 0$, and it is not necessary to write $\rho = 0$ as part of the answer.

Example   Rewrite $\rho(1 - 3 \cos \theta) = 2$ in Cartesian form.

Solution   Substituting $\pm\sqrt{x^2 + y^2}$ for ρ and x/ρ, or $x/\pm\sqrt{x^2 + y^2}$, for $\cos \theta$, we have

$$\pm\sqrt{x^2 + y^2}\left[1 - 3\left(\frac{x}{\pm\sqrt{x^2 + y^2}}\right)\right] = 2$$

from which

$$\pm\sqrt{x^2 + y^2} - 3x = 2,$$
$$\pm\sqrt{x^2 + y^2} = 3x + 2.$$

Squaring each member of this equation, we have

$$x^2 + y^2 = 9x^2 + 12x + 4,$$

from which

$$8x^2 - y^2 + 12x + 4 = 0.$$

EXERCISE SET 6.6

A

Find four additional sets of polar coordinates $(-360° < \theta \le 360°)$ *for each point whose coordinates are given.*

1. $(3, 380°)$ **2.** $(6, 840°)$ **3.** $(5, -450°)$

4. $(4, -540°)$ **5.** $(-4, 420°)$ **6.** $(-5, -600°)$

Find the Cartesian coordinates of each point with polar coordinates as given.

7. $\left(4, \dfrac{\pi}{6}^{R}\right)$ **8.** $\left(6, \dfrac{2\pi}{3}^{R}\right)$ **9.** $\left(3, -\dfrac{3\pi}{4}^{R}\right)$

10. $\left(0, \dfrac{13\pi}{3}^{R}\right)$ **11.** $\left(-3, \dfrac{5\pi}{6}^{R}\right)$ **12.** $\left(-4, \dfrac{\pi}{4}^{R}\right)$

Find the pair of polar coordinates (θ in radians) using an angle θ such that $0^{R} < \theta < 2\pi^{R}$ *and* $\rho > 0$ *for each point with Cartesian coordinates as given.*

13. $(4, 0)$ **14.** $(0, -3)$ **15.** $(2, 2)$

16. $(-1, -\sqrt{3})$ **17.** $\left(-\dfrac{\sqrt{3}}{2}, \dfrac{1}{2}\right)$ **18.** $\left(\dfrac{1}{2}, -\dfrac{\sqrt{3}}{2}\right)$

Transform each Cartesian equation to an equation in polar form.

Example $y = 3x$

Solution   Substituting $\rho \sin \theta$ for y and $\rho \cos \theta$ for x yields

$$\rho \sin \theta = 3 \rho \cos \theta$$

Multiplying each member by $1/(\rho \cos \theta)$, $\rho \ne 0$, $\theta \ne 90° + k \cdot 180°$, yields

$$\frac{\sin \theta}{\cos \theta} = 3,$$

$$\tan \theta = 3.$$

Notice that for values of θ for which the equation is satisfied, $\tan \theta$ is a constant, namely 3.

19. $x = 4$ **20.** $y = -3$ **21.** $x^2 = 8y$

22. $y^2 = 4x$ **23.** $x^2 + y^2 = 16$ **24.** $4x^2 + y^2 = 4$

25. $x^2 + y^2 - 3y = 0$ **26.** $x^2 + y^2 + 4x = 0$

Transform each polar equation to an equation in Cartesian form.

Example $\rho = 2 \sec \theta$

Solution   Substituting $1/\cos \theta$ for $\sec \theta$ yields

$$\rho = \frac{2}{\cos \theta},$$

$$\rho \cos \theta = 2.$$

Because $\rho \cos \theta = x$, the Cartesian form of the equation is $x = 2$.

27. $\rho = 6$ **28.** $\rho = 9$ **29.** $\rho = 4 \sin \theta$

30. $\rho = 3 \cos \theta$ **31.** $\rho = 2 \csc \theta$ **32.** $\rho = -4 \sec \theta$

33. $\rho = \dfrac{4}{1 - \sin \theta}$ **34.** $\rho = \dfrac{4}{1 + 2 \cos \theta}$

6.7 Graphing Equations in Polar Form

In our earlier work we have seen that an equation in two variables, say x and y, serves to pair elements from the replacement sets of the variables and has a graph in the geometric plane in a Cartesian coordinate system. An equation in ρ and θ also has a graph in a polar coordinate system.

Example   Graph $\rho = 4 \sin \theta$ in the interval $0° \leq \theta < 360°$.

Solution   Some ordered pairs with commonly used angles $30°, 45°$, etc., are first obtained. The tabulation of the data in the arrangement shown is sometimes helpful. The approximations shown for some values of $\sin \theta$ are sufficient for our purpose. These ordered pairs (ρ, θ) are shown in figure (a).

θ	$\sin \theta$	ρ or $4 \sin \theta$
0°	0	0
30°	0.50	2.0
45°	0.71	2.8
60°	0.87	3.5
90°	1.00	4.0
120°	0.87	3.5
135°	0.71	2.8
150°	0.50	2.0
180°	0	0

(a)

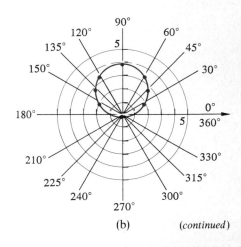

(b)   (*continued*)

The complete curve is obtained for $0° \le \theta < 180°$. For values of θ where $180° \le \theta < 360°$, the components of ordered pairs (ρ, θ) are also coordinates of points on the curve shown in figure (b).

Example Graph $\rho = \dfrac{3}{\sin \theta + \cos \theta}$.

Solution Ordered pairs are first obtained as shown in the table and in figure (a). The graph, a straight line, is shown in figure (b).

θ	$\sin \theta$	$\cos \theta$	$\sin \theta + \cos \theta$	$\rho = \dfrac{3}{\sin \theta + \cos \theta}$
0°	0	1.00	1.00	3.0
30°	0.50	0.87	1.37	2.2
45°	0.71	0.71	1.42	2.1
60°	0.87	0.50	1.37	2.2
90°	1.00	0	1.00	3.0
120°	0.87	−0.50	0.37	8.1
135°	0.71	−0.71	0	undef.

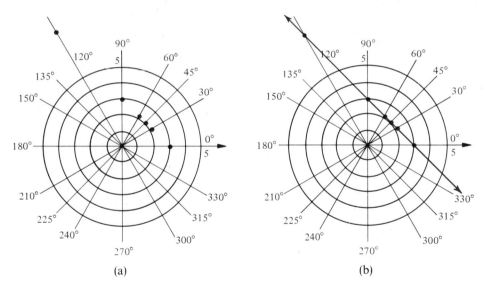

(a)   (b)

Frequently, the polar form of an equation is more useful than the rectangular form; furthermore, its graph may be simpler to sketch from the polar representation.

Excluded Values of θ

An analysis of the polar equation for particular features can facilitate the procedure.

For example, if

$$\rho^2 = 9 \cos 2\theta \qquad (1)$$

and we select $\theta = 60°$, then

$$\rho^2 = 9 \cos 120° = \frac{-9}{2},$$

and $\rho = \pm\sqrt{-9/2}$, an imaginary number. Therefore, $60°$ must be excluded as a value of θ. In fact, for all θ such that

$$(90° + k \cdot 360°) < 2\theta < (270° + k \cdot 360°),$$

or

$$(45° + k \cdot 180°) < \theta < (135° + k \cdot 180°), \quad k \in J,$$

it follows that $\cos 2\theta < 0$, and hence these values of θ must be excluded (see Figure 6.7).

Hence, in order to graph $\rho^2 = 9 \cos 2\theta$, it is necessary to select values of θ that are not excluded. Some ordered pairs (ρ, θ) that are solutions of the equation are shown in Table 6.1, and its graph is shown in Figure 6.8. The graph is called a **lemniscate**.

Figure 6.7

Figure 6.8

TABLE 6.1

θ	2θ	$\cos 2\theta$	ρ^2 or $9 \cos 2\theta$	ρ
$0°$	$0°$	1	9	± 3
$15°$	$30°$	$\dfrac{\sqrt{3}}{2}$	$\dfrac{9\sqrt{3}}{2}$	± 2.8
$30°$	$60°$	$\dfrac{1}{2}$	$\dfrac{9}{2}$	± 2.1
$45°$	$90°$	0	0	0
$135°$	$270°$	0	$0 \quad 0$	0
$150°$	$300°$	$\dfrac{1}{2}$	$\dfrac{9}{2}$	± 2.1
$165°$	$330°$	$\dfrac{\sqrt{3}}{2}$	$\dfrac{9\sqrt{3}}{2}$	± 2.8
$180°$	$360°$	1	9	± 3

Tangent Lines at the Pole

If, for some particular value of θ, say α, the value of ρ becomes 0, then the line $\theta = \alpha$ is tangent to the graph at the pole. For example, for the line to be tangent to the graph of $\rho^2 = 9 \cos 2\theta$ at the pole, ρ must be 0. Hence,

$$\cos 2\theta = 0,$$

from which

$$2\theta = \frac{\pi}{2} \quad \text{or} \quad 2\theta = \frac{3\pi}{2},$$

$$\theta = \frac{\pi}{4} \quad \text{or} \quad \theta = \frac{3\pi}{4}.$$

Thus, there are two lines tangent to the graph at the pole as shown in Figure 6.8.

Symmetry

The graph of an equation in polar form is often symmetric with respect to a line or a point. In particular, considerations of symmetry with respect to the polar axis ($\theta = 0°$), the vertical line through the pole, sometimes called the vertical axis ($\theta = 90°$), and the pole ($\rho = 0$) are helpful when graphing equations in polar form.

Testing an equation for each of these symmetries consists of making certain replacements of ρ and θ. Then, if the new equation is equivalent to the original, the graph has that particular symmetry. Table 6.2 lists various replacements that are helpful. Note that for each possible symmetry, there are two tests. If either test is successful, the graph has that particular symmetry.

Note that tests (a), (d), and (f) from Table 6.2 are successful when applied to the equation $\rho^2 = 9 \cos 2\theta$. Thus, for test (a), if θ is replaced by $-\theta$,

$$\rho^2 = 9 \cos 2(-\theta) = 9 \cos(-2\theta)$$

$$= 9 \cos 2\theta.$$

Then for test (d), replacing ρ with $-\rho$ and θ with $-\theta$, we have

$$(-\rho)^2 = 9 \cos 2(-\theta)$$

or

$$\rho^2 = 9 \cos 2\theta.$$

Finally, for test (f), replacing ρ with $-\rho$,

$$(-\rho)^2 = 9 \cos 2\theta$$

TABLE 6.2

Symmetry with respect to		Replacement for	
		ρ	θ
Polar axis	(a)		$-\theta$
($\theta = 0°$)	(b)	$-\rho$	$\pi - \theta$
Vertical axis	(c)		$\pi - \theta$
($\theta - 90°$)	(d)	$-\rho$	$-\theta$
Pole	(e)		$\pi + \theta$
($\rho = 0$)	(f)	$-\rho$	

or

$$\rho^2 = 9 \cos 2\theta.$$

Hence, as indicated in Figure 6.8, the graph is symmetric with respect to the polar axis, the vertical axis, and the pole.

Example Test the equation $\rho = 1 + \cos \theta$ for the symmetries in Table 6.2, and then graph the equation.

Test (a): Replacing θ with $-\theta$:

$$\rho = 1 + \cos(-\theta) = 1 + \cos \theta.$$

The new equation is equivalent to the original, hence the graph is symmetric with respect to the polar axis.

Test (c): Replacing θ with $\pi - \theta$, we have

$$\rho = 1 + \cos(\pi - \theta) = 1 - \cos \theta.$$

This is not equivalent to the original equation, so we try the second test.

Test (d): Replacing ρ with $-\rho$, and θ with $-\theta$, we have

$$-\rho = 1 + \cos(-\theta) = 1 + \cos \theta.$$

Since this equation also is not equivalent to the original, we cannot conclude that the graph is symmetric with respect to the vertical axis.

Test (e): Replacing θ with $\pi + \theta$, we have

$$\rho = 1 + \cos(\pi + \theta) = 1 - \cos \theta,$$

which is not equivalent to the original equation. Hence, we try the second test.

Test (f): Replacing ρ with $-\rho$, we have

$$-\rho = 1 + \cos \theta.$$

This again is not equivalent to the original equation, so we cannot conclude that the graph is symmetric with respect to the pole.

We have shown only that the graph is symmetric with respect to the polar axis. Now, using selected values of θ between $0°$ and $180°$, we first graph the

θ	$\cos \theta$	ρ or $1 + \cos \theta$
$0°$	1	2
$30°$	$\dfrac{\sqrt{3}}{2}$	1.87
$60°$	$\dfrac{1}{2}$	1.5
$90°$	0	1
$120°$	$-\dfrac{1}{2}$	$\dfrac{1}{2}$
$150°$	$-\dfrac{\sqrt{3}}{2}$	0.13
$180°$	-1	0

(*continued*)

equation in this interval. Then, using the fact that the graph is symmetric with respect to the polar axis from test (a), we sketch the entire graph. The graph is known as a **cardioid.**

Several additional interesting types of curves that are graphs of polar equations are shown in Figures 6.9 and 6.10.

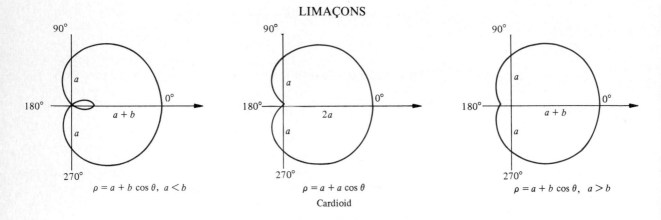

LIMAÇONS

$\rho = a + b \cos \theta, \ a < b$

$\rho = a + a \cos \theta$
Cardioid

$\rho = a + b \cos \theta, \ a > b$

LEMNISCATES

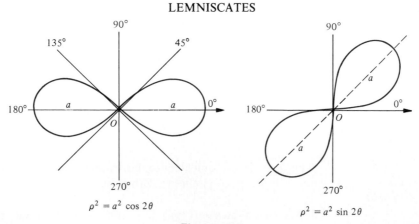

$\rho^2 = a^2 \cos 2\theta$

$\rho^2 = a^2 \sin 2\theta$

Figure 6.9

THREE-LEAVED ROSE

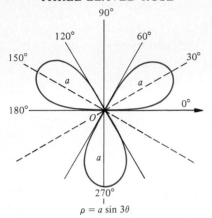

$\rho = a \sin 3\theta$

FOUR-LEAVED ROSE

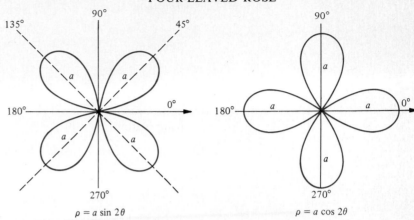

$\rho = a \sin 2\theta$

$\rho = a \cos 2\theta$

SPIRAL OF ARCHIMEDES

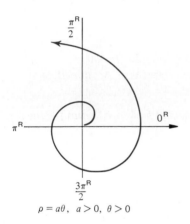

$\rho = a\theta, \quad a > 0, \ \theta > 0$

HYPERBOLIC SPIRAL

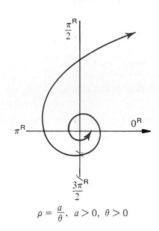

$\rho = \dfrac{a}{\theta}, \quad a > 0, \ \theta > 0$

LOGARITHMIC SPIRAL

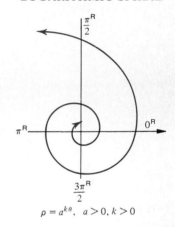

$\rho = a^{k\theta}, \quad a > 0, k > 0$

Figure 6.10

EXERCISE SET 6.7

A

Graph each equation.

1. $\rho = 4 \cos \theta$
(circle)

2. $\rho = 3 \sin \theta$
(circle)

3. $\rho = \cos 3\theta$
(three-leaved rose)

4. $\rho = 2 \sin 3\theta$
(three-leaved rose)

5. $\rho = \sin 2\theta$
(four-leaved rose)

6. $\rho = 2 \cos 2\theta$
(four-leaved rose)

7. $\rho = 1 + 2 \cos \theta$
(limaçon)

8. $\rho = 3 + 3 \cos \theta$
(cardioid)

9. $\rho = 2 + 2 \sin \theta$
(cardioid)

10. $\rho = 2 + \cos \theta$
(limaçon)

11. $\rho^2 = 4 \cos 2\theta$
(lemniscate)

12. $\rho^2 = 4 \sin 2\theta$
(lemniscate)

13. $\rho = 4$ *Hint:* Consider the equation $\rho = 0 \cdot \theta + 4$.

14. $\rho = 5$

15. $\theta = 30°$ *Hint:* Consider the equation $\theta = 0 \cdot \rho + 30°$.

16. $\theta = 45°$

B

Graph each equation.

17. $\rho = \dfrac{1}{\theta}, \ \theta > 0$
(hyperbolic spiral)

18. $\rho = 2\theta, \ \theta > 0$
(spiral of Archimedes)

Graph each of the given pairs of equations and find coordinates for the points of intersection.

19. $\rho = \sin \theta$
$\rho = \cos \theta$

20. $\rho = 1 + \sin \theta$
$\rho = 1$

21. $\rho = 1 + \sin \theta$
$\rho = 1 + \cos \theta$

22. $\rho = 2 \sin \theta$
$\rho = 2 \sin 2\theta$

23. Solve the system in Exercise 19 analytically. Explain why the coordinates of the origin are not in the solution set of the system.

24. Solve the system in Exercise 20 analytically.

25. Solve the system in Exercise 21 analytically.

26. Solve the system in Exercise 22 analytically.

27. Show that the distance between the points with coordinates (ρ_1, θ_1) and (ρ_2, θ_2) is given by $d = \sqrt{\rho_1{}^2 + \rho_2{}^2 - 2\rho_1\rho_2 \cos(\theta_1 - \theta_2)}$.

28. Find an equation in polar form for the circle with center at $(a, 0^R)$ and radius of length a.   *Hint:* Recall that an angle inscribed in a semicircle is a right angle.

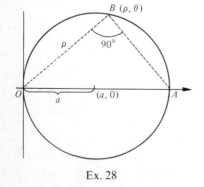

Ex. 28

Chapter Summary

6.1 A **complex number** is defined by

$$z = a + bi, \quad a, b \in R \text{ and } i^2 = -1.$$

If $z_1 = a_1 + b_1 i$ and $z_2 = a_2 + b_2 i$, then

$$z_1 = z_2 \quad \text{if and only if } a_1 = a_2 \text{ and } b_1 = b_2.$$

Sums and products of complex numbers are found by treating $a + bi$ as a binomial in the variable i and replacing i^2 with -1.

A complex number $a + bi$ can be identified with a **real number** if $b = 0$; if $b \neq 0$, the number is an **imaginary number,** and for the special case where $a = 0$, the number is a **pure imaginary number.**

For all $b \in R$, $b > 0$,

$$\sqrt{-b} = \sqrt{b}\, i = i\sqrt{b}.$$

Hence, for a and b positive real numbers,

$$\sqrt{-a}\sqrt{-b} = i^2\sqrt{ab} = -\sqrt{ab}.$$

6.2 The properties used for the set of complex numbers are similar to the corresponding properties of real numbers. Furthermore,

$$z_1 - z_2 = z_1 + (-z_2);$$

$$\frac{z_1}{z_2} = z_1 \cdot \frac{1}{z_2} \quad (z_2 \neq 0).$$

If $z_2, z_3 \neq 0$, then

$$\frac{z_1}{z_2} = \frac{z_1 z_3}{z_2 z_3}.$$

If $z = a + bi$, then the **conjugate** of z is $\bar{z} = a - bi$. The product of $a + bi$ and its conjugate equals $a^2 + b^2$, a real number.

6.3 A plane on which complex numbers are graphed is called a **complex plane.** The x-axis is called the **real axis** and the y-axis is called the **imaginary axis.**

Complex numbers can be added graphically by the parallelogram law applicable to geometric vectors.

The **absolute value** or **modulus** of the complex number z is

$$\rho = |z| = |a + bi| = |(a, b)| = \sqrt{a^2 + b^2}.$$

The modulus is the distance from the origin to the graph of (a, b). An **argument** or **amplitude** of the complex number $z = a + bi = (a, b)$, denoted by **arg($a + bi$)** or **arg(a, b)**, is an angle θ such that

$$\cos \theta = \frac{a}{\rho} \quad \text{and} \quad \sin \theta = \frac{b}{\rho} \quad (a^2 + b^2 \neq 0).$$

If θ is an argument of $a + bi$, then $(\theta + k \cdot 360)°$, $k \in J$, is also an argument of $a + bi$. The angle θ with the least positive measure is called the **principal argument**.

6.4 A nonzero complex number $a + bi$ can be expressed in **trigonometric form** as

$$a + bi = \rho[\cos(\theta + k \cdot 360°) + i \sin(\theta + k \cdot 360°)], \quad k \in J.$$

If $z_1 = \rho_1(\cos \theta_1 + i \sin \theta_1)$ and $z_2 = \rho_2(\cos \theta_2 + i \sin \theta_2)$, then

$$z_1 \cdot z_2 = \rho_1 \cdot \rho_2[\cos(\theta_1 + \theta_2) + i \sin(\theta_1 + \theta_2)],$$

$$\frac{z_1}{z_2} = \frac{\rho_1}{\rho_2}[\cos(\theta_1 - \theta_2) + i \sin(\theta_1 - \theta_2)].$$

6.5 If $z = \rho(\cos \theta + i \sin \theta)$ and $n \in N$, then

$$z^n = \rho^n(\cos n\theta + i \sin n\theta).$$

This is known as **DeMoivre's theorem.**
If $z \neq 0 + 0i$, then

$$z^0 = 1 + 0i = 1 \quad \text{and} \quad z^{-n} = \frac{1}{z^n}, \quad n \in N.$$

If $n \in N$ and $z = \rho(\cos \theta + i \sin \theta)$, then

$$\rho^{1/n}\left[\cos\left(\frac{\theta + k \cdot 360°}{n}\right) + i \sin\left(\frac{\theta + k \cdot 360°}{n}\right)\right], \quad k \in J$$

is an **nth root** of z. For $k = 0$, the nth root

$$\rho^{1/n}\left(\cos \frac{\theta}{n} + i \sin \frac{\theta}{n}\right)$$

is called the **principal root.**

6.6 The components of an ordered pair $(\rho, \theta°)$ or (ρ, θ^R) that locate a point A in the plane are called the **polar coordinates** of A. Each such ordered pair specifies a unique point; however, each point in the plane corresponds to infinitely many ordered pairs of polar coordinates.

Cartesian and polar coordinates are related as follows:

$$x = \rho \cos \theta \qquad\qquad \rho = \pm\sqrt{x^2 + y^2},$$

$$y = \rho \sin \theta \qquad\qquad \cos \theta = \frac{x}{\rho} = \frac{x}{\pm\sqrt{x^2 + y^2}},$$

$$\sin \theta = \frac{y}{\rho} = \frac{y}{\pm\sqrt{x^2 + y^2}} \qquad [(x, y) \neq (0, 0)].$$

6.7 Polar equations can be graphed by obtaining a number of ordered pairs (ρ, θ). The notions of excluded values, tangent lines, and symmetry can facilitate the procedures of graphing such equations.

A list of symbols introduced in this chapter is shown inside the front cover, and important properties are shown inside the back cover.

Review Exercises

A

6.1 *Find real numbers x and y for which each statement is true.*

    1. $9x - 6yi = -3 + 4i$ **2.** $-8x - 5i = -2y - 5x^2i$

Write each expression in the form a + bi.

    3. $(-4 - i) + (-4 + 9i)$ **4.** $(-3 + i) + (3 - 2i)$

    5. $(8 + 6i)(-2 + 5i)$ **6.** $(5 + 3i)(2 - 4i)$

Write each expression as one of the complex numbers i, -1, $-i$, or 1.

    7. i^{13} **8.** i^{23}

Write each expression in the form a + bi.

    9. $\sqrt{-5}(3 - \sqrt{-2})$ **10.** $(3 - \sqrt{-5})(3 + \sqrt{-5})$

11. Show that $3i$ and $-3i$ satisfy the equation $x^2 + 9 = 0$.

12. Show that $1 + 3i$ and $1 - 3i$ satisfy $x^2 - 2x + 10 = 0$.

6.2   *Express each difference in the form a + bi.*

13. $(-7 + 2i) - (8 + 3i)$ **14.** $(6 - 9i) - (3 - 4i)$

Express each quotient in the form a + bi.

15. $\dfrac{8 + 3i}{2 + 4i}$ **16.** $\dfrac{-7 + i}{6 - 10i}$

Write each expression in the form bi or a + bi.

17. $\dfrac{-2}{\sqrt{-6}}$ **18.** $\dfrac{4}{2 + \sqrt{-5}}$

6.3   *Express each complex number as an ordered pair.*

19. $-4 + i$ **20.** $2\sqrt{3} - 2i$

Express each complex number in the form a + bi.

21. $(9, -6)$ **22.** $(3, 4)$

Graph each complex number z, its negative $-z$, and its conjugate $\bar{z}$.

23. $-3 + 5i$ **24.** $(-2, -6)$

Represent each sum graphically and check the results by algebraic methods.

25. $(-4 + 3i) + (8 - 2i)$ **26.** $(-1, -8) + (6, 7)$

Find the absolute value and the principal argument of each complex number.

27. $4 + 2i$ **28.** $(-2, -5)$

6.4   *Represent each complex number in trigonometric form, using the angle with least nonnegative measure.*

29. $1 + i$ **30.** $-3 + 3\sqrt{3}i$

Represent each complex number graphically and write it in the form a + bi.

31. $7(\cos 120° + i \sin 120°)$ **32.** $2[\cos(-315°) + i \sin(-315°)]$

Express each product in the form a + bi.

33. $3(\cos 200° + i \sin 200°) \cdot 2(\cos 25° + i \sin 25°)$

34. $\sqrt{3}(\cos 50° + i \sin 50°) \cdot \sqrt{12}[\cos(-20°) + i \sin(-20°)]$

Express each quotient in the form a + bi.

35. $\dfrac{8(\cos 309° + i \sin 309°)}{9(\cos 129° + i \sin 129°)}$

36. $\dfrac{12[\cos(-315°) + i \sin(-315°)]}{3(\cos 270° + i \sin 270°)}$

6.5 *Express each power in the form a + bi.*

37. $[3(\cos 15° + i \sin 15°)]^4$ **38.** $(\sqrt{3} - i)^3$

39. $[4(\cos 240° + i \sin 240°)]^{-2}$ **40.** $(4 - 4i)^{-3}$

41. Find the two square roots of $\dfrac{\sqrt{3}}{2} + \dfrac{1}{2}i$.

42. Find the three cube roots of $125(\cos 330° + i \sin 330°)$.

Solve each equation, $x \in C$.

43. $x^4 + 1 = 0$ **44.** $x^5 = -5 - 5\sqrt{3}i$

6.6 *Find four additional sets of polar coordinates $(-360° < \theta \le 360°)$ for each point.*

45. $(2, 610°)$ **46.** $(-3, 390°)$

Find the Cartesian coordinates for each point.

47. $\left(3, \dfrac{\pi}{3}^R\right)$ **48.** $\left(-2, \dfrac{3\pi}{4}^R\right)$

Find the pair of polar coordinates (θ in radians) using an angle θ such that $0^R < \theta < 2\pi^R$ and $\rho > 0$ for each point.

49. $(2, -2)$ **50.** $(-\sqrt{3}, -1)$

Transform each equation to an equation in polar form.

51. $x = 3$ **52.** $x^2 + 3y^2 = 5$

Transform each equation to an equation in Cartesian form.

53. $\rho = 5 \sin \theta$ **54.** $\rho = \dfrac{2}{1 - \cos \theta}$

6.7   *Graph each equation.*

55. $\rho = 1 + \sin \theta$ **56.** $\rho = 1 - \sin \theta$

57. $\rho = \cos 2\theta$ **58.** $\rho = 2 \cos 3\theta$

ALGEBRA AND GEOMETRY REVIEW

In Appendix A we review the basic ideas of algebra and geometry, using terminology and notation that are needed in the development of various topics in the text.

A.1 Set Notation

Recall that a set is simply a collection of "things." Each "thing" in the set is called a **member** or **element** of the set. The membership can be described by listing the names of the members in braces, { }, or by stating a rule that identifies the members in the set.

Definition A.1 *Two sets are* **equal** *if and only if they have the same members.*

Customarily, the slant bar, /, drawn through symbols for relations, indicates negation. Thus,

$$\{3, 4, 5\} = \{5, 3, 4\} \qquad \text{but} \qquad \{3, 4, 5\} \neq \{4, 5, 6\}.$$

The symbol $\in$ is used to denote membership in a set. For example, we can write

$$3 \in \{3, 4, 5\}$$

to represent the statement "3 is an element of $\{3, 4, 5\}$."

Sets are also designated by means of capital letters, A, B, C, R, etc.

Example   If $A = \{2, 3, 4\}$ and $B = \{4, 5, 6\}$, then

$$3 \in A \qquad \text{but} \qquad 3 \notin B.$$

An unspecified element of a set is usually denoted by a lowercase italic letter such as a, b, c, x, and y or by a lowercase letter from the Greek alphabet such as α, β, γ, and θ. Such a symbol is called a **variable**; a symbol used to denote a specific element is called a **constant.** Variables are used in conjunction with braces in another symbolism that is often useful in discussing sets. This symbolism, illustrated by

$$\{x \mid x \in A \quad \text{and} \quad x \in B\}$$

and read "the set of all x such that x is an element of A and x is an element of B," is called **set-builder notation.** The vertical line, |, is read "such that." Set-builder notation names a variable and states conditions on the variable.

Example   If $A = \{2, 3, 4\}$ and $B = \{4, 5, 6\}$, then

$$\{x \mid x \in A \quad \textbf{or} \quad x \in B\} = \{2, 3, 4, 5, 6\} \tag{1}$$

because 2, 3, 4, 5, 6 are elements of *one or the other* of the two sets, whereas

$$\{x \mid x \in A \quad \textbf{and} \quad x \in B\} = \{4\} \tag{2}$$

because 4 is an element of *both* sets.

The preceding example is associated with the following two operations on sets.

Definition A.2   *The **union** of two sets A and B is the set of all elements that belong to A or to B or to both.*

The symbol $\cup$ is used to denote this operation; $A \cup B$ is read "the union of A and B."

Example   If $A = \{2, 3, 4\}$ and $B = \{4, 5, 6\}$, then

$$A \cup B = \{2, 3, 4, 5, 6\}. \tag{3}$$

Notice that each element in $A \cup B$ is listed only once. Since the symbol 4 denotes only one number, it would be redundant to list the name of the element twice.

A second useful operation on sets is the following.

Definition A.3   *The **intersection** of two sets A and B is the set of all elements that belong to both A and B.*

The symbol $\cap$ is used to denote this operation; $A \cap B$ is read "the intersection of A and B."

Example   If $A = \{2, 3, 4\}$ and $B = \{4, 5, 6\}$, then

$$A \cap B = \{4\}. \tag{4}$$

Notice from Equations (1) and (3) that the operation of union is associated with the word *or*, and from Equations (2) and (4) that the operation of intersection is associated with the word *and*.

The set that contains no elements is called the **null set** or the **empty set** and is denoted by the symbol $\varnothing$.

Example   The intersection of two sets is a set. Therefore, if $C = \{2, 3, 4\}$ and $D = \{5, 6, 7\}$, we can write

$$C \cap D = \{2, 3, 4\} \cap \{5, 6, 7\} = \varnothing.$$

Two sets such as C and D, which do not have any elements in common, are said to be **disjoint.**

A.2 The Set of Real Numbers

Recall from your study of algebra that each real number can be associated with one and only one point on a number line and each point on the line can be associated with one and only one real number. Several examples are shown in Figure A.1. The real number corresponding to a point on the line is called the **coordinate** of the point, and the point is called the **graph** of the number. A number whose graph is to the right of the graph of zero is **positive,** and a number whose graph is to the left is **negative.**

Figure A.1

The following sets of numbers are contained in the set of **real numbers, R.** The capital letters shown are the ones that are sometimes associated with the respective sets. These are the symbols that we use in this text.

1. The set N of **natural numbers,** among whose elements are such numbers as 1, 2, 6, 25, and 624.

2. The set **J** of **integers,** whose elements consist of the natural numbers, their negatives, and zero. Included are such elements as $-24, -9, 0, 7,$ and 314.

3. The set **Q** of **rational numbers,** whose elements are all those numbers that can be represented in the form $\dfrac{a}{b}$, or a/b, where a and b are integers and b is not zero. Included are such elements as $-\frac{1}{3}, 0, \frac{1}{4}$, and 6.

4. The set **H** of **irrational numbers,** whose elements are those numbers that are the coordinates of points on the number line that are not the graphs of rational numbers. An irrational number cannot be represented in the form a/b, where a and b are integers.

Figure A.2 summarizes the relationships among these sets of numbers.

Figure A.2

Radical notation is useful to represent some irrational numbers.

Definition A.4 *For all $a \in R$ where a is positive or zero, $\sqrt{a}$ is the nonnegative number such that*

$$\sqrt{a} \cdot \sqrt{a} = a.$$

Example

a. $\sqrt{9}$ is the positive number such that $\sqrt{9}\sqrt{9} = 9$.

b. For $x > 0$, $\sqrt{x}$ is the positive number such that $\sqrt{x}\sqrt{x} = x$.

Frequently, rational numbers are used as approximations for irrational numbers. For example, by using the symbol $\approx$ for "approximately equal to," we write $\sqrt{2} \approx 1.414$, $\sqrt{3} \approx 1.732$, $\sqrt{5} \approx 2.236$, etc. Table I in Appendix C gives rational number approximations for some square roots that are irrational numbers.

If *every* element of a set A is also an element of a set B, then A is said to be a **subset** of B. Thus the sets H and Q can be viewed as subsets of R, the set J as a subset of Q or R, and the set N as a subset of J, Q, or R.

If the elements of a set can be paired with the numerals $1, 2, 3, \ldots, n$ for some natural number n so that each element is paired with exactly one numeral and each numeral is paired with exactly one element, then the set is said to be a **finite set.** A set that is not the empty set and is not finite is called an **infinite set.** Some infinite sets can be designated by using three dots and some of the members listed to establish a pattern as shown in Figure A.2.

Often we wish to consider the nonnegative member of the pair of real numbers a and $-a$, where $a \neq 0$. The special notation $|a|$ is used for this purpose. To extend the notation for all $a \in R$, we define $|0| = 0$.

Definition A.5 *For all $a \in R$,*

$$\textit{if a is positive or zero, then } |a| = a;$$

$$\textit{if a is negative, then } |a| = -a.$$

$|a|$ *is called the* **absolute value** *of a.*

Examples

a. $|4| = 4$ 　　　　　　b. $|-4| = -(-4) = 4$ 　　　　　　c. $|0| = 0$

The order of the elements of the set of real numbers can be established by the following.

Definition A.6 *For all $a, b \in R$, b is* **less than** *a if for some positive real number c,*

$$b + c = a.$$

For such conditions, a is said to be **greater than** *b.*

Examples

a. 4 is less than 7 because there is a *positive number c* (equal to 3) such that $4 + c = 7$.

b. -9 is less than -5 because there is a *positive number c* (equal to 4) such that $-9 + c = -5$.

The number line shown in Figure A.3 is very helpful in visualizing whether one real number is less than or greater than a second real number. If the graph of b lies to the *left* of the graph of a, then b is less than a.

Figure A.3

The following symbols are used in connection with the property of order:

$<$ read "is less than";

$\leq$ read "is less than or equal to";

$>$ read "is greater than";

$\geq$ read "is greater than or equal to."

Observe that $b < a$ and $a > b$ are equivalent statements.

Examples
  a. $4 < 7$ and $7 > 4$

  b. $-9 < -5$ and $-5 > -9$

  c. $x < 7$ is equivalent to $7 > x$

In your study of arithmetic and algebra, you applied the operations of addition and multiplication to elements of the set of real numbers. You also made assumptions, called **axioms,** about these operations and studied the logical consequences of these axioms. For your reference, we include a list of some of these axioms and properties on page 277.

A.3 Equations in Two Variables; Relations and Functions

Recall from your study of algebra that the ordered pair (2, 9) is called a **solution** of the equation $y = x + 7$ because the replacement of x by the first component 2 and the replacement of y by the second component 9 results in a true statement, $9 = 2 + 7$. The ordered pair (2, 9) is said to *satisfy* the equation. The set of all such ordered pairs that satisfy an equation is called the **solution set** of the equation.

The equation $y = x + 7$ does not result in a true statement for all ordered pairs of real numbers. For example, (3, 5) is not a solution because $5 \neq 3 + 7$. Such equations are called **conditional equations.** On the other hand, if an equation such as

$$x + y = y + x$$

results in a true statement for every replacement of x and y, for which both

For $a, b, c, d \in R$:

Equality Axioms

1. $a = a$ Reflexive law
2. If $a = b$, then $b = a$ Symmetric law
3. If $a = b$ and $b = c$, then $a = c$ Transitive law
4. If $a = b$, then a may be substituted   Substitution law
   for b or b for a in any expression.

Order Axioms

1. Exactly one of the following is true: Trichotomy law
   $$a < b, \quad a = b, \quad \text{or} \quad a > b.$$
2. If $a < b$ and $b < c$, then $a < c$.   Transitive law
3. If $a, b > 0$, then   Closure for positive numbers
   $$a + b > 0 \quad \text{and} \quad a \cdot b > 0$$

Axioms for Operations

1. $a + b \in R$ Closure law for addition
2. $a + b = b + a$ Commutative law of addition
3. $(a + b) + c = a + (b + c)$ Associative law of addition
4. $a + 0 = 0 + a = a$ Identity law of addition
5. $a + (-a) = (-a) + a = 0$ Additive inverse law
6. $a \cdot b \in R$ Closure law for multiplication
7. $a \cdot b = b \cdot a$ Commutative law of multiplication
8. $(a \cdot b) \cdot c = a \cdot (b \cdot c)$ Associative law of multiplication
9. $a \cdot 1 = 1 \cdot a = a$ Identity law of multiplication
10. $a \cdot \dfrac{1}{a} = \dfrac{1}{a} \cdot a = 1 \quad (a \neq 0)$ Multiplicative inverse law
11. $a \cdot (b + c) = a \cdot b + a \cdot c$ Distributive law

Some Properties

1. $a \cdot 0 = 0$ 2. $-(-a) = a$
3. $-a = -1 \cdot a$ 4. If $a = b$, then $a + c = b + c$
5. If $a + c = b + c$, then $a = b$ 6. If $a = b$, then $a \cdot c = b \cdot c$
7. If $a \cdot c = b \cdot c$, then $a = b \quad (c \neq 0)$ 8. $a - b = a + (-b)$
9. $\dfrac{a}{b} = \dfrac{c}{d}$ if and only if $a \cdot d = b \cdot c \; (b, d \neq 0)$ 10. $\dfrac{a}{b} = \dfrac{a \cdot c}{b \cdot c} \quad (b, c \neq 0)$
11. $\dfrac{a}{c} + \dfrac{b}{c} = \dfrac{a + b}{c} \quad (c \neq 0)$ 12. $\dfrac{a}{b} \cdot \dfrac{c}{d} = \dfrac{a \cdot c}{b \cdot d} \quad (b, d \neq 0)$
13. $\dfrac{a}{b} = a \cdot \dfrac{1}{b} \quad (b \neq 0)$ 14. $\dfrac{a}{b} \div \dfrac{c}{d} = \dfrac{a}{b} \cdot \dfrac{d}{c} \quad (b, c, d \neq 0)$
15. $\dfrac{a}{b} = \dfrac{-a}{-b} = -\dfrac{a}{-b} = -\dfrac{-a}{b};$

   $\dfrac{-a}{b} = \dfrac{a}{-b} = -\dfrac{a}{b} = -\dfrac{-a}{-b} \quad (b \neq 0)$
16. $ab = 0$ if and only if $a = 0$ or $b = 0$

members have meaning, the equation is call an **identity.** The left-hand member and the right-hand member of identities are called **equivalent expressions.**

To show that an equation is not an identity, it is necessary to find only one replacement for the variables that will make the equation false. Such a procedure is referred to as showing a **counterexample.**

Example   We can show by counterexample that $x^2 - 3y = 3y - x^2$ is not an identity by substituting some arbitrary values for the variables. If we use 1 for x and 1 for y, the left-hand member of the equation equals $1^2 - 3 \cdot 1$, or -2, and the right-hand member equals $3 \cdot 1 - 1^2$, or 2. Hence,

$$x^2 - 3y \neq 3y - x^2$$

for the ordered pair $(1, 1)$, and the equation is not an identity.

Example   We can show that the equation

$$(x - y)^2 + 2xy = x^2 + y^2$$

is an identity by simplifying the left-hand member

$$(x - y)^2 + 2xy = x^2 - 2xy + y^2 + 2xy$$
$$= x^2 + y^2.$$

Thus, the left-hand member and the right-hand member are equivalent and the equation is an identity.

Sets of ordered pairs play an important role in mathematics and are given a special name, as follows:

Definition A.7   *A* **relation** *is a set of ordered pairs.*

The set of first components in a set of ordered pairs is called the **domain** of the relation and the set of second components is called the **range.**

Examples
   a. $\{(5, 8), (6, 9), (7, 10)\}$ is a relation with domain $\{5, 6, 7\}$ and range $\{8, 9, 10\}$.

   b. $\{(5, 8), (5, 9), (7, 10)\}$ is a relation with domain $\{5, 7\}$ and range $\{8, 9, 10\}$.

The set of ordered pairs exhibited in Example a is a special kind of relation in which each element in the domain is paired with only one element in the range. Such relations are given a special name.

Definition A.8   *A **function** is a relation in which each element in the domain is associated with only one element in the range.*

Note that the relation in Example b above is not a function because 5, an element in the domain, is associated with more than one element in the range.

The relations above were exhibited by listing the members in the set of ordered pairs. They can also be described by equations that specify which ordered pairs are in the relation and which are not; those that are in the relation are elements of the solution set. For example,

$$\{(x, y) \mid y = x + 3, \quad x \in \{5, 6, 7\}\}$$

also specifies the function in Example a above:

$$(5, 8), (6, 9), (7, 10)$$

The elements in the range, 8, 9, and 10, are obtained by replacing x in the equation $y = x + 3$ with 5, 6, and 7 in turn and solving for y. This function consists of three ordered pairs only.

Example   Given the function

$$\{(x, y) \mid y = x - 3, \quad x \in \{1, 2, 3\}\}:$$

1. The domain is $\{1, 2, 3\}$.
2. Replacing x with each element of the domain in turn, you can find the associated values of y to be -2, -1, and 0, respectively. Thus, the range is $\{-2, -1, 0\}$.
3. The function is $\{(1, -2), (2, -1), (3, 0)\}$.

If the domain of a function is the set of real numbers R, rather than a finite set, the function consists of an *infinite set* of ordered pairs and obviously not all its elements can be listed. *If the domain is not specified, we shall assume that it includes all real numbers for which both members of the defining equation are defined.*

Examples

a. The domain of the function defined by $y = x + 3$ is the set R because $x + 3$, and therefore y, are defined for all real number replacements for x.

b. The domain of the function defined by $y = \dfrac{3}{x - 4}$ is the set R with the restriction that $x \neq 4$, because $x - 4 = 0$ and y is undefined for $x = 4$.

Functions are sometimes designated by names (linear, quadratic, etc.) or by means of a single symbol, commonly the symbol f. If the discussion includes a consideration of more than one function, other letters such as g, h, F, P, etc., are used.

The symbol for the function can be used in conjunction with the variable representing an element in the domain to represent the associated element in the range. For example, $f(x)$ (read "f of x" or "the value of f at x") is the element in the range of f associated or paired with the element x in the domain. Thus, we can discuss functions defined by an equation such as

$$y = x - 2$$

$$f(x) = x - 2,$$

$$g(x) = x - 2,$$

$$\text{etc.}$$

where the symbols $f(x)$, $g(x)$, etc., are playing exactly the same role as y. The variable x represents an element in the domain, and y, $f(x)$, $g(x)$, etc., represent an element in the range.

The notation $f(x)$ is especially useful because, by replacing x with a specific real number a in the domain, the notation $f(a)$ will then denote the paired, or corresponding, element in the range of the function.

Example   If $f(x) = 2x - 5$,

$$f(3) = 2(3) - 5 = 6 - 5 = 1,$$

$$f(0) = 2(0) - 5 = 0 - 5 = -5,$$

$$f(-3) = 2(-3) - 5 = -6 - 5 = -11,$$

$$f(a) = 2(a) - 5 = 2a - 5.$$

A.4 Graphs of Relations and Functions

As you probably recall from your study of algebra, an association exists between ordered pairs of real numbers and the points on a geometric plane, sometimes called the **real plane.** If two number lines are drawn perpendicular to each other at their origins as in Figure A.4, they form a rectangular coordinate system. These perpendicular number lines are called **axes** (singular, axis). The horizontal line is usually called the **x-axis,** and the vertical line is usually called the **y-axis.** The x-axis and the y-axis divide the plane into four regions, called **quadrants,** as shown.

For a given rectangular coordinate system, any point in the plane can be associated with a unique ordered pair of real numbers. For example, the point A in Quadrant I, in Figure A.4, can be associated with the ordered

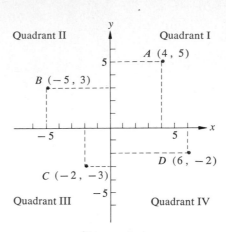

Figure A.4

pair (4, 5), where 4 represents the distance of the point to the right of the y-axis and 5 represents the distance of the point above the x-axis. The points B, C, and D correspond to the ordered pairs $(-5, 3), (-2, -3)$, and $(6, -2)$, respectively.

The *components* of an ordered pair associated with a point in the plane are called the **coordinates** of the point. The first component is called the **abscissa** of the point; the second component is called the **ordinate** of the point. The *point* is called the **graph** of the ordered pair. An ordered pair is sometimes used as a name for the corresponding point. Thus, we occasionally speak of the point (2, 3) or the point (4, −3) or, in general, the point (x, y).

Recall that in Section A.3, a function was defined in part as a set of ordered pairs. Consequently, a function—or at least a part of a function— can now be displayed as a set of points on the plane. The phrase "part of a function" is used here since the domain or range or both may be the infinite set of real numbers, which cannot be shown on a finite line. For example, consider the function

$$y = x - 4.$$

If arbitrary values are assigned to x, say −3, 0, 3, and 6, and the corresponding values for y are computed, the four ordered pairs

$$(-3, -7), \quad (0, -4), \quad (3, -1), \quad \text{and} \quad (6, 2)$$

are obtained, which are solutions to the equation. These points can be located on a coordinate system (see Figure A.5a on page 282). These points appear to lie on a straight line, and in fact they do. It can be shown that the coordinates of any point on the line (Figure A.5b) constitute a solution of the first-degree equation in two variables, $y = x - 4$ and that, conversely,

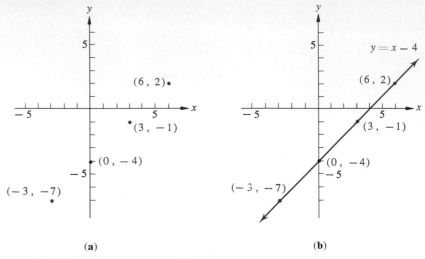

Figure A.5

every solution of $y = x - 4$ corresponds to a point on this line. The line is referred to as the *graph of the function* $y = x - 4$ or, alternatively, as the *graph of the solution set of the equation* or simply as the *graph of the equation*. Obviously, only part of the line can be displayed, so an arrowhead is placed at each end to indicate that the graph, or line, continues in both directions indefinitely.

Furthermore, since the graphs of first-degree equations

$$y = mx + b \quad \text{or} \quad ax + by = c$$

are straight lines, it is only necessary to find two ordered pairs in the solution set to sketch the graph. Two pairs that are easy to find are $(0, b)$ and $(a, 0)$, in which the numbers a and b are called the **intercepts.**

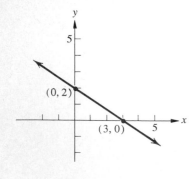

Example   The graph of the function $2x + 3y = 6$ can be obtained by first finding the x and y intercepts.

If $y = 0$, then $x = 3$ and $(3, 0)$ is a solution.

If $x = 0$, then $y = 2$ and $(0, 2)$ is a solution.

The x and y intercepts are 3 and 2, respectively. See the graph.

First-degree equations are also called **linear equations,** and the functions defined by these equations are called **linear functions.**

For any two points P_1 and P_2 on a line, the set of points containing P_1 and P_2 and all points lying between P_1 and P_2 is called a **line segment.** We shall

Figure A.6

(a)

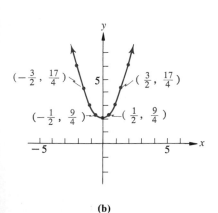

(b)

Figure A.7

designate the line segment by $\overline{P_1 P_2}$. A line segment has the important property of length. If we designate the endpoints of a line segment $\overline{P_1 P_2}$ with the pairs of coordinates (x_1, y_1) and (x_2, y_2), as in Figure A.6, the length, or measure, of the line segment, which we shall designate by $P_1 P_2$ or some variable such as d, can be determined by an application of the Pythagorean theorem (see Section A.8). From the right triangle shown, we observe that the lengths of the perpendicular sides are $y_2 - y_1$ and $x_2 - x_1$; the square of the distance between P_1 and P_2 is given by

$$d^2 = (x_2 - x_1)^2 + (y_2 - y_1)^2.$$

By considering only the positive square root of the right-hand member, we have that

$$d = \sqrt{(x_2 - x_1)^2 + (y_2 - y_1)^2}.$$

Example The distance between the pair of points $(1, 3)$ and $(5, 6)$ is given by

$$d = \sqrt{(5 - 1)^2 + (6 - 3)^2}$$
$$= \sqrt{16 + 9} = \sqrt{25} = 5.$$

As with first-degree equations in two variables, solutions of higher-degree equations in two variables are also ordered pairs, which can be found by arbitrarily assigning values to x and finding the associated values of y. For example, in the equation

$$y = x^2 + 2,$$

if $x = -2$, then

$$y = (-2)^2 + 2 = 6,$$

and $(-2, 6)$ is a solution. Similarly, by replacing x with $-1, 0, 1$, and 2, we obtain $(-1, 3), (0, 2), (1, 3)$, and $(2, 6)$ as additional solutions. Plotting these points on the plane, we have the graph in Figure A.7a. Clearly, these points do not lie on a straight line. By plotting additional solutions of $y = x^2 + 2$, such as

$$\left(-\frac{3}{2}, \frac{17}{4}\right), \quad \left(-\frac{1}{2}, \frac{9}{4}\right), \quad \left(\frac{1}{2}, \frac{9}{4}\right), \quad \text{and} \quad \left(\frac{3}{2}, \frac{17}{4}\right),$$

we have the additional points in Figure A.7b. These points can now be connected in sequence from left to right by a smooth curve as shown. We assume that this curve is a good approximation to the graph of $y = x^2 + 2$.

More generally, the graph of any equation of the form

$$y = ax^2 + bx + c \quad (a, b, c, x \in R, a \neq 0),$$

is called a **parabola.** Notice that such an equation defines a function, since with each x, an equation of this form will associate one and only one y. Such functions are called **quadratic functions.**

The graph of

$$x^2 + y^2 = r^2,$$

where r is a real number greater than zero, is a **circle** with center at the origin and radius of length r. However, although this equation defines a relation, it does not define a function.

Figure A.8

Whether or not a relation is also a function can be determined by inspecting its graph. For example, consider the graphs of equations $y = ax^2 + bx + c$ and $x^2 + y^2 = r^2$ in Figure A.8. Imagine a vertical line moving across each graph from left to right. Since the line cuts the graph of $y = ax^2 + bx + c$ at *only one point at each position, this equation defines a function;* that is, for each x in this relation, there is only one y. Since the line cuts the graph of $x^2 + y^2 = r^2$ at *more than one point at certain positions, the equation does not define a function,* although it does define a relation. There are ordered pairs in the relation which have the same first components and different second components.

A.5 Zeros of Functions; Equations in One Variable

In any function, the element(s) in the domain that are paired with the element 0 in the range are called **zeros of the function.**

Example   The zeros of

$$f = \{(-3, 1), (-2, 0), (-1, -2), (0, -4), (1, 0)\}$$

Figure A.9

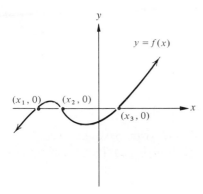

Figure A.10

are -2 and 1 because each of these first components is paired with a second component 0.

The graph of each pair containing a zero in the above example lies on the x-axis, as shown in Figure A.9.

For any function

$$y = f(x),$$

the zeros of the function are values of x (x_1, x_2, x_3 in Figure A.10) for which

$$y = f(x) = 0.$$

These values of x are the solutions of the equation in one variable, $f(x) = 0$, and the set of such numbers is the solution set of the equation.

Note that the ordered pairs $(x_1, 0), (x_2, 0)$, and $(x_3, 0)$ are the coordinates of the points of intersection of the graph of $y = f(x)$ and the x-axis.

Thus, there are three different names for a single idea:

1. The zeros of $y = f(x)$.
2. The x intercepts of the graph of $y = f(x)$.
3. Solutions of $f(x) = 0$.

In many simple cases, such as

$$x - 2 = 0,$$

the solution set of an equation in one variable is evident by inspection. However, this is not the case for most first-degree equations. Recall from your work in algebra that solutions of first-degree equations in one variable can usually be found by generating **equivalent equations** (a set of equations with the same solution set) until you obtain an equation whose solution set is evident by inspection.

In general, if each member of an equation is multiplied by the same nonzero number, or if the same quantity is added to both members, the result is an equation equivalent to the original equation.

Solutions of quadratic equations in one variable are obtained by various methods. For example if the left-hand member of a quadratic equation of the form

$$ax^2 + bx + c = 0 \quad (a \neq 0)$$

is factorable, the solutions can be obtained from the fact that the left-hand member will equal zero for values of x for which one or both of the factors equals zero. Such a procedure of solving a quadratic equation is called *solution by factoring*.

Example We can solve $x^2 + x = 30$, by first writing the equation equivalently as

$$x^2 + x - 30 = 0, \tag{1}$$

in which form the right-hand member is zero. Factoring the left-hand member yields

$$(x + 6)(x - 5) = 0. \tag{2}$$

The left-hand member equals zero for values of x for which

$$x + 6 = 0 \qquad \text{or} \qquad x - 5 = 0 \qquad \text{or both.}$$

By inspection, the solution set of (2) and, therefore, of (1) is $\{-6, 5\}$.

Quadratic equations of the form

$$x^2 = a$$

can be solved by a method called *extraction of roots*. If the equation has a solution, then x must be a square root of a (see Definition A.4). Since each real number a has two square roots, the solution set is $\{\sqrt{a}, -\sqrt{a}\}$. If $a = 0$, the solution set is $\{0\}$.

Example The method of extraction of roots can be used to find the solution set of the equation $4x^2 = 24$. Multiplying each member by $1/4$ yields

$$x^2 = 6$$

Hence, $x = \sqrt{6}$ or $x = -\sqrt{6}$ and the solution set is $\{\sqrt{6}, -\sqrt{6}\}$.

From the method of extraction of roots applied to the equation

$$(x - a)^2 = b$$

we obtain

$$x - a = \sqrt{b} \qquad \text{and} \qquad x - a = -\sqrt{b},$$

from which the solution set $\{a + \sqrt{b}, a - \sqrt{b}\}$ is obtained. If $b = 0$, the solution set is $\{a\}$. From this it follows that we can find the solution set of any quadratic equation by first rewriting the equation in the form

$$(x - a)^2 = b.$$

This procedure of solving a quadratic equation is called *completing the square*.

Applying this procedure to the general quadratic equation

$$ax^2 + bx + c = 0 \quad (a \neq 0)$$

results in the formula

$$x = \frac{-b \pm \sqrt{b^2 - 4ac}}{2a},$$

called the **quadratic formula,** which gives us the solutions of any quadratic equation in the form $ax^2 + bx + c = 0$ expressed in terms of a, b, and c. The solution set is

$$\left\{ \frac{-b + \sqrt{b^2 - 4ac}}{2a}, \frac{-b - \sqrt{b^2 - 4ac}}{2a} \right\}.$$

If a, b, $c \in R$, the elements in the solution set are real numbers if and only if $b^2 - 4ac \geq 0$.

Example   The equation $x^2 - 5x = 6$ can be solved by using the quadratic formula as follows. First, rewrite the equation as

$$x^2 - 5x - 6 = 0.$$

Then, replacing a with 1, b with -5, and c with -6 in the quadratic formula yields

$$x = \frac{-(-5) \pm \sqrt{(-5)^2 - 4(1)(-6)}}{2(1)},$$

from which

$$x = \frac{5 + \sqrt{25 + 24}}{2} = 6 \quad \text{or} \quad x = \frac{5 - \sqrt{25 + 24}}{2} = -1,$$

and the solution set is $\{6, -1\}$.

A.6 Inverse of a Function

If the components of each ordered pair in a given relation are interchanged, the resulting relation and the given relation are called **inverses** of each other.

Example $\{(1, 3), (5, 7), (11, 13)\}$ and $\{(3, 1), (7, 5), (13, 11)\}$ are inverse relations.

The inverse of a relation $\mathcal{R}$ is denoted by $\mathcal{R}^{-1}$ (read "$\mathcal{R}$ inverse" or "the inverse of $\mathcal{R}$"). It is evident from the above example that the domain and range of $\mathcal{R}^{-1}$ are the range and domain, respectively, of $\mathcal{R}$. We have used the script $\mathcal{R}$ for a relation so that it should not be confused with the R that represents the set of real numbers. Furthermore, it should be understood that -1 in the inverse notation symbol $\mathcal{R}^{-1}$ is not an exponent.

The inverse of the relation

$$\mathcal{R}: y = 3x - 5 \tag{1}$$

can be obtained by replacing x with y and y with x. Hence, the inverse relation of (1) is given by

$$\mathcal{R}^{-1}: x = 3y - 5 \tag{2}$$

or, when y is expressed in terms of x in the defining equation,

$$\mathcal{R}^{-1}: y = \frac{1}{3}(x + 5). \tag{2'}$$

Notice in the above example that $y = 3x - 5$ and $x = 3y - 5$ from (1) and (2) define inverse relations. The domains and ranges of the relations are interchanged because the variables have been interchanged. However, $x = 3y - 5$ from (2) and $y = \frac{1}{3}(x + 5)$ from (2') are equivalent equations and define the same relation.

The graphs of inverse relations are always located symmetrically with respect to the graph of the linear equation $y = x$. To see this, notice the graphs of the ordered pairs (a, b) and (b, a) in Figure A.11. They are the same distance from, but on opposite sides of, the line that is the graph of the equation $y = x$. The graph of $y = x$ serves as a reflecting line, or mirror, for these points.

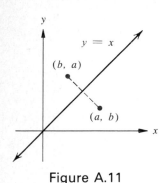

Figure A.11

Using the example above, Figure A.12 on page 289 shows the graphs of

$$y = 3x - 5$$

and its inverse,

$$x = 3y - 5 \qquad \text{or} \qquad y = \frac{1}{3}(x + 5),$$

together with the graph of $y = x$.

Because every function is a relation, *every function has an inverse, but the inverse is not always a function.* For example, consider the function

$$y = x^2. \tag{3}$$

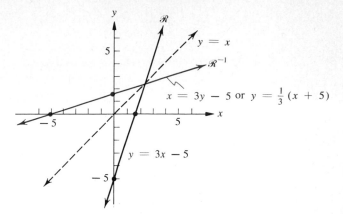

Figure A.12

Its graph and the graph of its inverse,

$$x = y^2 \qquad \text{or} \qquad y = \pm\sqrt{x}. \tag{4}$$

are shown in Figure A.13. Since for all but one value of x ($x = 0$), the inverse specified by (4) associates two different values of y (one positive and one negative) in the range, the inverse of (3) is not a function. Observe in Figure A.13 that the graphs of these relations are symmetric with respect to the graph of $y = x$.

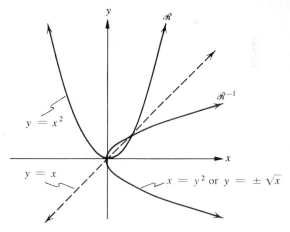

Figure A.13

In order for a function to have an inverse that is also a function, it must be a *one-to-one function;* that is, each element in the domain of the original function must be associated with one and only one element in its range, and each element in its range with one and only one element in its domain.

(a)

(b)

Figure A.14

Figure A.15

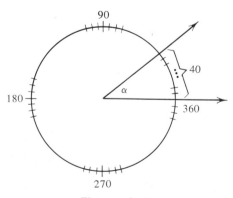

A.7 Angles and Their Measure

In geometry, a **ray** is a figure such as that shown in Figure A.14a, where the arrow indicates that the ray extends infinitely far in one direction. The point *A* (points are usually named with capital letters) in A.14a is the **endpoint** of that ray, and point *C* is the endpoint of the two rays in A.14b.

A ray is named by naming its endpoint and one other point on the ray. In Figure A.14a the ray is named by $\overrightarrow{AB}$ (read "ray *AB*"), and in A.14b the rays are named by $\overrightarrow{CD}$ and $\overrightarrow{CE}$.

An **angle** is the union of two rays having a common endpoint called the **vertex** of the angle. The rays are the **sides** of the angle. In Figure A.14b, an angle is formed by $\overrightarrow{CD}$ and $\overrightarrow{CE}$, where $\overrightarrow{CD}$ and $\overrightarrow{CE}$ are the sides, and point *C* is the vertex.

An angle can be named in several ways. For example, if a point is the vertex of only one angle, as in Figure A.15, then that letter can be used to name the angle; in this case, $\angle B$. Also, an angle can be named using a single letter placed "inside" the angle, such as β (beta), or by using three letters, where the middle letter names the vertex, such as $\angle ABC$ or $\angle CBA$.

If a point is the vertex of more than one angle, as in Figure A.16, each angle can also be named either by using three letters, such as $\angle DEF$, $\angle DEG$, or $\angle GEF$, or by using a single letter "inside" the angle, such as α (alpha), or γ (gamma). However, since three angles have the vertex *E*, "$\angle E$" cannot be used since it would not clearly name any one of them.

Angles can be measured by various units of measurement, but in geometry they are commonly measured in **degrees.** Such measures can be assigned to an angle by using a circle whose center is at the vertex of the angle. If a circle is divided into 360 arcs of equal length (Figure A.17 below), the number of arcs intercepted by a central angle α is the measure of the angle in degree units and is designated by $\alpha°$ (the measure of the angle is independent of the radius of the measuring circle). From Figure A.17, we have that $\alpha = 40°$.

Figure A.16

Figure A.17

Each degree is divided into sixty minutes (60′) and each minute of a degree is divided into sixty seconds (60″). To convert from tenths of degrees to minutes, you simply multiply the number of tenths by 60.

Examples

a. $0.4° = (0.4 \times 60)'$

$= 24'$

b. $35.2° = 35° + (0.2 \times 60)'$

$= 35° \ 12'$

Conversely, to convert from minutes to tenths of degrees, simply divide the number of minutes by 60.

Examples

a. $36' = \left(\dfrac{36}{60}\right)°$

$= 0.6°$

b. $75° \ 18' = 75° + \left(\dfrac{18}{60}\right)°$

$= 75.3°$

A **protractor,** usually in the shape of a semicircle, as shown in Figure A.18, with 180 equally spaced marks on the semicircle, can be used to measure angles. If an angle is placed with its vertex at the center of the semicircle with one side passing through the 0° mark, the measure of the angle can be read directly.

Figure A.18

Examples   In Figure A.18,

$$\angle ABC = 50°, \quad \angle ABE = 110°, \quad \text{and} \quad \angle DBE = 70°.$$

Also

$$\angle CBE = 110° - 50° = 60°.$$

If the measure of an angle is less than 90°, the angle is an **acute angle;** if more than 90° and less than 180°, it is an **obtuse angle;** if equal to 90°, it is a **right angle.** If two lines, or rays, form a right angle, the lines, or rays, are **perpendicular,** indicated by the symbol ⊥. Thus $\overrightarrow{DE} \perp \overrightarrow{FG}$ is read "ray DE is perpendicular to ray FG."

Examples   In the figure, α is an acute angle, β is an obtuse angle, and γ is a right angle. Also, $\overrightarrow{AC} \perp \overrightarrow{AB}$.

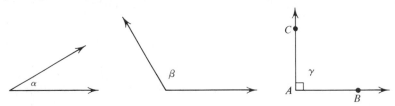

If the sum of the measures of two angles is 90°, the angles are **complementary;** if the sum is 180°, the angles are **supplementary.**

Examples

a. If $\angle D = 32°$, $\angle E = 148°$, and $\angle F = 58°$, then $\angle D$ and $\angle F$ are complementary, and $\angle D$ and $\angle E$ are supplementary.

b. In the figure, angles α and β are complementary, and angles γ and θ (theta) are supplementary.

Frequently, pairs of supplementary angles are introduced through the use of parallel lines. In Figure A.19, in which lines m and n are intersected by a transversal, line t, angles 1, 2, 7, and 8 are called **exterior angles** and angles 3, 4, 5, and 6 are called **interior angles.** Interior angles on opposite sides of the transversal are called **alternate interior angles** and exterior angles on opposite sides of the transversal are called **alternate exterior angles.** For example, pairs of angles 3 and 6, and 4 and 5, are alternate interior angles and pairs of angles 1 and 8, and 7 and 2 are alternate exterior angles.

Angles that are on the same side of the transversal and also in the same relative position in relation to the parallel lines are called **corresponding angles.** For example, pairs of angles 1 and 5; 2 and 6; 3 and 7; and 4 and 8 are corresponding angles.

If lines m and n are parallel, it follows that alternate interior angles are equal and alternate exterior angles are equal, as are corresponding angles. Also interior angles on the same side of the transversal are supplementary, as are exterior angles on the same side of the transversal.

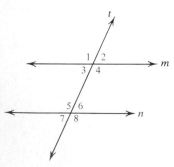

Figure A.19

A.8 Triangles

A **triangle** is a figure formed by three line segments joining three points that are not in the same line. The line segments are the **sides,** and the three points are the **vertices** (plural of **vertex**) of the triangle. The letters assigned to the vertices can be used to name a triangle.

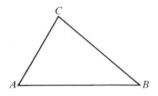

Example   In the figure, the points A, B, and C are the vertices of $\triangle ABC$ (read "triangle ABC"), which can also be named by $\triangle ACB$, $\triangle BAC$, $\triangle BCA$, $\triangle CAB$, or $\triangle CBA$. The sides are $\overline{AB}$, $\overline{BC}$, and $\overline{CA}$; their lengths (positive real numbers) can be represented by AB, BC, and CA, respectively.

If each angle of a triangle is acute, the triangle is an **acute triangle;** if one angle is obtuse, it is an **obtuse triangle;** if one angle is a right angle, it is a **right triangle.** In a right triangle, the side opposite the right angle is the **hypotenuse,** and the other sides are the **legs.**

Examples   In the figure, $\triangle DEF$ is an acute triangle, $\triangle KGH$ is an obtuse triangle, and $\triangle PQR$ is a right triangle. In $\triangle PQR$, $\overline{PQ}$ and $\overline{PR}$ are the legs, and $\overline{RQ}$ is the hypotenuse.

Triangles also have names related to the relative lengths of the sides. If the three sides are of different lengths, the triangle is a **scalene triangle;** if two sides have equal lengths, it is an **isosceles triangle;** if all three sides have equal lengths, it is an **equilateral triangle.**

Examples   In the figure, $\triangle ABC$ is scalene, $\triangle DEF$ is isosceles ($FD = FE$), and $\triangle HKL$ is equilateral ($HK = KL = LH$). Note, further, that $\triangle ABC$ is an obtuse triangle, whereas $\triangle DEF$ and $\triangle HKL$ are both acute triangles.

An **altitude** of a triangle is a segment from any vertex of a triangle perpendicular to the opposite side. Sometimes the length of the segment is called the altitude.

Examples   In the figure, in $\triangle ABC$, h, or $\overline{CD}$, is the altitude to $\overline{AB}$. In $\triangle EFG$, $\overline{GH}$ is the altitude from G, and $\overline{FL}$ is the altitude from F.

The following properties hold for any triangle (see Figure A.20).

1. The perimeter is $P = a + b + c$.
2. The area is $\mathscr{A} = \frac{1}{2}bh$.
3. $\angle A + \angle B + \angle C = 180°$.

As in Properties 1 and 2, lowercase letters are often used to indicate the lengths of the sides opposite the angles named by the corresponding capital letters.

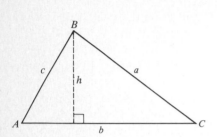

Figure A.20

Examples

a. The perimeter of $\triangle ABC$ is

$$P = 3 + 5 + 6 = 14.$$

b. If $h = 6$ and $CD = 10$, the area of $\triangle CDE$ is

$$\mathscr{A} = \frac{1}{2}(10)(6) = 30.$$

a.

b.

c. If $\angle G = 70°$ and $\angle K = 50°$,

$$70° + 50° + \angle H = 180°,$$

from which we have that $\angle H = 60°$.

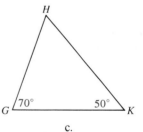

c.

In addition to Properties 1, 2, and 3, the following properties also apply to certain special triangles.

4. In an isosceles triangle, the angles opposite the equal sides are equal.

Example   In triangle BCD, if $BD = CD$, then $\angle B = \angle C$.

5. If a triangle is equilateral, then (see Figure A.21)

$$\angle A = \angle B = \angle C = 60°, \quad h = \frac{s}{2}\sqrt{3}, \quad \text{and} \quad \mathscr{A} = \frac{\sqrt{3}}{4}s^2.$$

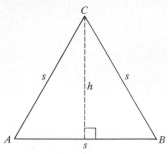

Figure A.21

Example   If $s = 12$, then

$$h = \frac{12}{2}\sqrt{3} = 6\sqrt{3} \quad \text{and} \quad \mathscr{A} = \frac{\sqrt{3}}{4}(12)^2 = 36\sqrt{3}.$$

6. In a right triangle ($\angle C = 90°$),

    (i) $a^2 + b^2 = c^2$ (the Pythagorean theorem).

    (ii) If $a = b$, then $\angle A = \angle B = 45°$, and $c = a\sqrt{2}$.

    (iii) If one acute angle is 30°, and the length of the opposite side is a, then the other acute angle is 60°, the length of the hypotenuse is $2a$, and the length of the longer leg is $a\sqrt{3}$.

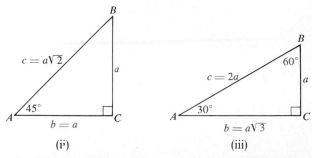

Examples   Using the measurements of the appropriate triangle:

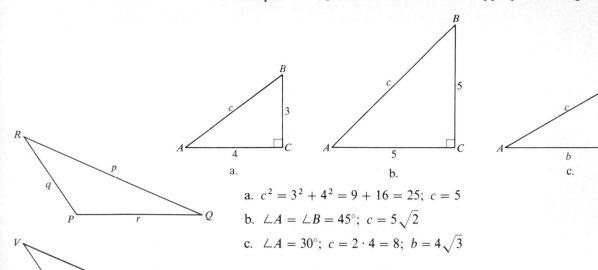

a.   b.   c.

a. $c^2 = 3^2 + 4^2 = 9 + 16 = 25$; $c = 5$

b. $\angle A = \angle B = 45°$; $c = 5\sqrt{2}$

c. $\angle A = 30°$; $c = 2 \cdot 4 = 8$; $b = 4\sqrt{3}$

A.9 Congruent and Similar Triangles

If two triangles are the same size and shape, they are said to be **congruent triangles**. In Figure A.22, the triangles are congruent. This is indicated by use of the symbol $\cong$ (read "is congruent to"). Thus, in Figure A.22,

Figure A.22

$\triangle PQR \cong \triangle STV$. From this it follows that

$$\angle P = \angle S, \quad p = s,$$
$$\angle Q = \angle T, \quad q = t,$$
$$\angle R = \angle V, \quad r = v.$$

If two triangles are the same shape, but not necessarily the same size, they are said to be **similar triangles.** In Figure A.23, the triangles are similar. This is written $\triangle ABC \sim \triangle DEF$, where "$\sim$" is read "is similar to." In two similar triangles, the corresponding angles are equal—that is, $\angle A = \angle D$, $\angle B = \angle E$, $\angle C = \angle F$; and the corresponding sides are proportional—that is,

$$\frac{AC}{DF} = \frac{AB}{DE} = \frac{BC}{EF}.$$

Example Given that $\triangle RST \sim \triangle ADF$.

 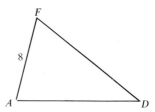

Figure A.23

We can find AD as follows. Since the triangles are similar, we have

$$\frac{AD}{RS} = \frac{AF}{RT},$$

$$\frac{AD}{6} = \frac{8}{4},$$

from which

$$AD = 12.$$

The following properties can be used to prove that two triangles are similar:

1. two angles of one triangle are equal, respectively, to two angles of the other, or
2. one angle of one triangle is equal to one angle of the other, and the including sides are proportional, or
3. three sides of one triangle are proportional to three sides of the other.

Examples Using the measurements in the figures on page 297:

a. $\angle P = \angle M$ and $\angle Q = \angle N$;

therefore, by Property 1, $\triangle PQR \sim \triangle MNL$.

b. $\angle A = \angle D$, and $\dfrac{AB}{DE} = \dfrac{AC}{DF}$ because $\dfrac{9}{6} = \dfrac{6}{4}$;

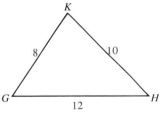

therefore, by Property 2, $\triangle ABC \sim \triangle DEF$.

c. $\dfrac{GK}{SW} = \dfrac{KH}{WT} = \dfrac{GH}{ST}$ because $\dfrac{8}{4} = \dfrac{10}{5} = \dfrac{12}{6}$;

therefore, by Property 3, $\triangle GHK \sim \triangle STW$.

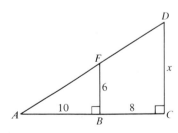

Example   To find the length x in the figure, we proceed as follows. $\angle A = \angle A$ and $\angle ABF = \angle ACD$; hence, by Property 1, we have $\triangle ABF \sim \triangle ACD$. Also, $AC = AB + BC = 18$. Because the triangles are similar, the corresponding sides are proportional. Thus,

$$\frac{DC}{FB} = \frac{AC}{AB} \quad \text{and} \quad \frac{x}{6} = \frac{18}{10}, \quad \text{from which}$$

$$x = \frac{6 \cdot 18}{10} = \frac{54}{5}.$$

A list of symbols introduced in Appendix A is shown inside the front cover.

B ROUNDING-OFF PROCEDURES

Figure B.1 shows the names of selected place values for the decimal number system: *whole number places* (to the left of the decimal point) are named to thousands; *decimal places* (to the right of the decimal point) are named to ten-thousandths. Familiarity with these place values is sufficient for our work.

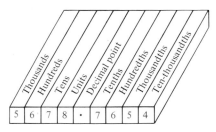

Figure B.1

Frequently the result of a computation includes more digits than may be needed for a particular purpose. For example, the result of a calculation that involves the cost of an item might be $32.527. The actual cost, to the nearest cent, is $32.53. We say that 32.527 has been "rounded off to the nearest hundredth" (or "to the nearest cent"). Note that 32.527 is "nearer" to 32.53 than it is to 32.52.

The process shown below indicates the rounding-off process most commonly used. In this procedure, the *round-off digit* is the last digit to be retained; the *test digit* is the digit immediately to the right of the round-off digit. For example, to round off 318.04591 to the nearest *hundredth*, underline the round-off digit 4 in the hundredths place, then proceed as follows:

round-off digit

318.04591

test digit

If the test digit is 5 or more, the round-off digit is increased by 1, but if the test digit is less than 5, the round-off digit is not changed. Here the test digit is 5, hence 1 is added to the round-off digit:

$$\overset{5}{318.0\not{4}591}$$

The round-off digit is in a decimal place, so all digits to the right of the round-off digit are dropped.

$$\overset{5}{318.0\not{4}\not{5}\not{9}\not{1}} \longrightarrow 318.05$$

Thus, 318.04591 is written as 318.05 to the nearest hundredth.

To round off 318.04591 to the nearest *hundred*, we have

round-off digit

318.04591

test digit

Here the test digit is less than 5, so the round-off digit is not changed. Also, because the round-off digit is to the left of the decimal point, the digits following are replaced with zeros, and all digits to the right of the decimal point are dropped. Thus, to the nearest hundred, we have

$$318.04591 \longrightarrow \overset{00}{3\not{1}\not{8}.\not{0}\not{4}\not{5}\not{9}\not{1}} \longrightarrow 300$$

Examples   In each case the round-off digit is underlined.

a. Round off to the nearest ten:

$$31\underline{8}.04591 \longrightarrow \overset{20}{3\not{1}\underline{8}.\not{0}\not{4}\not{5}\not{9}\not{1}} \longrightarrow 320$$

b. Round off to the nearest tenth:

$$318.\underline{0}4591 \longrightarrow 318.\underline{0}\not{4}\not{5}\not{9}\not{1} \longrightarrow 318.0$$

c. Round off to the nearest thousandth:

$$318.04\underline{5}91 \longrightarrow 318.04\overset{6}{\cancel{591}} \longrightarrow 318.046$$

Round-off instructions sometimes specify a given number of decimal places.

Examples

a. Round off to two decimal places:

$$21.5\underline{9}71 \longrightarrow 21.\overset{60}{\cancel{5971}} \longrightarrow 21.60$$

b. Round off to three decimal places:

$$21.59\underline{7}1 \longrightarrow 21.59\underline{7}\cancel{1} \longrightarrow 21.597$$

Exact and Approximate Numbers

Any number arrived at by a counting process is an **exact number.** For example, if we say "60 students are enrolled in a class," the number 60 is exact. Exact numbers may also arise by definition. In the statement, "One hour equals 60 minutes," the number 60 is exact. On the other hand, a number that is arrived at by some process of measurement is an **approximate number.** Thus, if the distance between two cities, measured by driving from one city to the other, is given as 60 miles, this distance is an approximation to the actual distance, and the number 60 is an approximate number. The actual distance may be a little more or a little less than 60 miles.

Examples

a. "One dime is equal to 10 cents," the number 10 is *exact*; 10 is defined to be the number of cents in a dime.

b. "This sheet of postage stamps contains 10 stamps," the number 10 is *exact*; 10 is associated with the concept of counting.

c. "The reading shown on a voltmeter is 10 volts," the number 10 is *approximate*; 10 is a *measure* of voltage.

Significant Digits

When working with approximate numbers, the term **significant digits** is used to refer to those digits in a number that have meaning. The following rules can be used to decide which digits of an approximate number are significant.

SIGNIFICANT DIGITS OF APPROXIMATE NUMBERS

1. Nonzero digits (1, 2, 3, 4, 5, 6, 7, 8, 9) are always significant.
2. Zeros that are preceded and followed by nonzero significant digits are always significant.
3. When a number has no nonzero digits on the left of the decimal point, zeros between the decimal point and the first nonzero digit are not significant.
4. Final zeros on the right of a decimal point are significant.
5. Final zeros on a whole number are assumed to be not significant unless further information is available.

In the examples illustrating each rule, the significant digits are underlined.

Examples

a. 62.18 has four significant digits.   (Rule 1)

b. 0.042 has two significant digits.   (Rules 1 and 3)

c. 23.0017 has six significant digits.   (Rule 2)

d. 4005 has four significant digits.   (Rule 2)

e. 7.200 has four significant digits.   (Rule 4)

f. 0.0160 has three significant digits.   (Rules 3 and 4)

g. In the statement "Hawaii is 2400 miles west of Los Angeles," without further information on how the measurement was made, assume that 2400 has only two significant digits.   (Rule 5)

h. If a measurement is given as 2400 miles to *the nearest mile*, then the number 2400 has four significant digits.   (Rule 5)

Numbers can be rounded off to a specified number of significant digits by using the same procedure that is used to round off a number to a specified place value.

Examples

a. Round off to two significant digits:

$$58.603 \longrightarrow 58.\overset{9}{\cancel{6}}\cancel{0}\cancel{3} \longrightarrow 59$$

b. Round off to four significant digits:

$$58.603 \longrightarrow 58.60\cancel{3} \longrightarrow 58.60$$

C USING TABLES

In previous sections we obtained approximations to powers, square roots, reciprocals, and trigonometric function values using a scientific calculator. Approximations for these values can also be obtained by using tables that have been prepared for this purpose.

Table I in this appendix lists approximations to powers, square roots, and reciprocals of some natural numbers and can be used if a calculator is not available. Tables II and III list trigonometric function values.

Function Values for Acute Angles

Table II lists rational number approximations for function values of selected angles over the interval $0° \leq \alpha \leq 90°$. The table is graduated in intervals of six minutes (6'), where one minute is equal to one-sixtieth of a degree, and in intervals of tenths of a degree. Observe that the table reads from top to bottom in the left margin with function values identified at the tops of the columns. It reads from bottom to top in the right margin with function values identified at the bottoms of the columns. (See extract of Table II.)

PART OF TABLE II

θ deg	deg min		$\sin \theta$	$\cos \theta$	$\tan \theta$	$\csc \theta$	$\sec \theta$	$\cot \theta$			
30.0	30	0	0.5000	0.8660	0.5774	2.0000	1.1547	1.7321	60	0	60.0
30.1	30	6	0.5015	0.8652	0.5797	1.9940	1.1559	1.7251	59	54	59.9
30.2	30	12	0.5030	0.8643	0.5820	1.9880	1.1570	1.7182	59	48	59.8
30.3	30	18	0.5045	0.8634	0.5844	1.9821	1.1582	1.7113	59	42	59.7
30.4	30	24	0.5060	0.8625	0.5867	1.9762	1.1594	1.7045	59	36	59.6
30.5	30	30	0.5075	0.8616	0.5890	1.9703	1.1606	1.6977	59	30	59.5
30.6	30	36	0.5090	0.8607	0.5914	1.9645	1.1618	1.6909	59	24	59.4
30.7	30	42	0.5105	0.8599	0.5938	1.9587	1.1630	1.6842	59	18	59.3
30.8	30	48	0.5120	0.8590	0.5961	1.9530	1.1642	1.6775	59	12	59.2
			$\cos \theta$	$\sin \theta$	$\cot \theta$	$\sec \theta$	$\csc \theta$	$\tan \theta$	deg	min	θ deg

Function values are, in general, irrational numbers, and the four-place decimals shown in the table are mostly rational number approximations.

Examples Find approximations for

a. sin 30° 12′   b. cot 59.5°

Solutions From Table II,

a. sin 30° 12′ ≈ 0.5030   b. cot 59.5° ≈ 0.5890

Table III lists function values for angles whose measures are given in radian units in intervals of 0.01^R. (See extract of Table III.)

PART OF TABLE III

θ radians or Real Number x	θ degrees	sin θ or sin x	cos θ or cos x	tan θ or tan x	csc θ or csc x	sec θ or sec x	cot θ or cot x
0.45	25° 47′	0.4350	0.9004	0.4831	2.299	1.111	2.070
.46	26° 21′	.4439	.8961	.4954	2.253	1.116	2.018
.47	26° 56′	.4529	.8916	.5080	2.208	1.122	1.969
.48	27° 30′	.4618	.8870	.5206	2.166	1.127	1.921
.49	28° 04′	.4706	.8823	.5334	2.125	1.133	1.875
0.50	28° 39′	0.4794	0.8776	0.5463	2.086	1.139	1.830
.51	29° 13′	.4882	.8727	.5594	2.048	1.146	1.788
.52	29° 48′	.4969	.8678	.5726	2.013	1.152	1.747
.53	30° 22′	.5055	.8628	.5859	1.978	1.159	1.707
.54	30° 56′	.5141	.8577	.5994	1.945	1.166	1.668

Examples Find approximations for

a. cos 0.47^R b. cot 0.53^R

Solutions From Table III,

a. cos 0.47^R ≈ 0.8916   b. cot 0.53^R ≈ 1.707

Notice that in both tables the function values sin θ, tan θ, and sec θ increase, while cos θ, cot θ, and csc θ decrease, for increasing measures of θ where $0° \le \theta \le 90°$.

Tables II and III consist of function values for angles with the specified measures. To approximate function values for angle measures not listed in the tables, we can read the entry in the table for that angle measure nearest the angle measure with which we are concerned. Thus, from Table III,

$$\sin 0.514^R \approx \sin 0.51^R \approx 0.4882.$$

Function Values for All Angles

Because the trigonometric tables are restricted to acute angles, we have to use a reference angle in conjunction with these tables to find approximations for function values for angles that are not acute. We will use the same procedure that we used to find function values for integral multiples of 30°, 45°, and 60° in Section 1.4.

Examples   Find an approximation for each function value.

a. csc 334° 36′   b. sec 128° 12′.

Solutions   A sketch showing the reference angle is helpful.

$\alpha = 334° \ 36′$

$\tilde{\alpha} = 25° \ 24′ = 25.4°$

a. $\tilde{\alpha} = 360° - 334° \ 36′ = 25° \ 24′.$

Also, since α is in Quadrant IV, csc 334° 36′ is negative. Therefore, using Table II, we obtain

$$\text{csc } 334° \ 36′ = -\text{csc } \tilde{\alpha}$$
$$= -\text{csc } 25° \ 24′ \approx -2.3314.$$

b. $\tilde{\alpha} = 180° - 128° \ 12′ = 51° \ 48′.$

$\alpha = 128° \ 12′$

$\tilde{\alpha} = 51° \ 48′ = 51.8°$

Since 128° 12′ is in Quadrant II, sec 128° 12′ is negative. Therefore, using Table II,

$$\text{sec } 128° \ 12′ = -\text{sec } 51° \ 48′ = -1.6171.$$

When the measure of an angle is given in radians, and a reference angle is needed, we have to use an approximation for π. We shall use 3.14.

Examples   Find an approximation for each function value.

a. 1.95^R b. cot 5.36^R

Solutions   A sketch showing the reference angle is helpful.

  a. Using $\pi = 3.14$,

$\tilde{\alpha}$

$\alpha = 1.95^R$

$$\tilde{\alpha} \approx (3.14 - 1.95)^R = 1.19^R.$$

Since α is in Quadrant II, cos 1.95^R is negative. Therefore, using Table III, we obtain

$$\cos 1.95^R = -\cos \tilde{\alpha}$$
$$\approx -\cos 1.19^R \approx -0.3717.$$

$\alpha = 5.36^R$

$\tilde{\alpha} \approx 0.92^R$

  b. Using 6.28 for 2π,

$$\tilde{\alpha} \approx 6.28^R - 5.36^R = 0.92^R.$$

Since 5.36^R is in Quadrant IV, cot 5.36^R is negative. Hence, using Table III,

$$\cot 5.36^R \approx -\cot 0.92^R \approx -0.7615.$$

The answers in the last example were obtained using $\pi \approx 3.14$. Different results will be obtained with other approximations for π such as 3.142 or 3.1416. In fact, if a calculator is used to find an approximation in Example a above, we obtain

$$\cos 1.95^R \approx -0.3701808 \approx -0.3702,$$

a significant difference from the value -0.3717 that we obtained using the table with $\pi \approx 3.14$. The calculator value is a more accurate approximation because it is programmed for a more accurate approximation to π.

Function values for angles with negative measures can also be found using the procedures we used for angles with positive measures.

Examples   Find an approximation for each function value.

a. $\sin(-34° \ 12')$ b. $\cos(-422° \ 30')$

Solutions

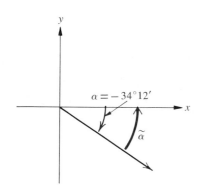

a. The reference angle is

$$\tilde{\alpha} = 34° \ 12'.$$

Because $-34° \ 12'$ is in Quadrant IV, we have, using Table II,

$$\sin(-34° \ 12') = -\sin 34° \ 12'$$
$$= -\sin 34.2° \approx -0.5621.$$

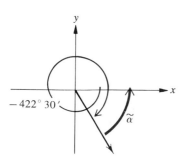

b. The reference angle is

$$\tilde{\alpha} = |360° - 422° \ 30'| = 62° \ 30'.$$

Because $-422° \ 30'$ is in Quadrant IV, $\cos(-422° \ 30')$ is positive, and we have, using Table II,

$$\cos(-422° \ 30') = \cos 62° \ 30' \approx 0.4617.$$

TABLE I Powers, roots, and reciprocals

N	N^2	$\sqrt{N}$	$\sqrt{10N}$	$1/N$	N	N^2	$\sqrt{N}$	$\sqrt{10N}$	$1/N$
1	1	1.0 00	3.1 62	1.00 00	51	2 601	7.1 41	22.5 83	.019 61
2	4	1.4 14	4.4 72	.500 00	52	2 704	7.2 11	22.8 04	.019 23
3	9	1.7 32	5.4 77	.333 33	53	2 809	7.2 80	23.0 22	.018 87
4	16	2.0 00	6.3 25	.250 00	54	2 916	7.3 48	23.2 38	.018 52
5	25	2.2 36	7.0 71	.200 00	55	3 025	7.4 16	23.4 52	.018 18
6	36	2.4 49	7.7 46	.166 67	56	3 136	7.4 83	23.6 64	.017 86
7	49	2.6 46	8.3 67	.142 86	57	3 249	7.5 50	23.8 75	.017 54
8	64	2.8 28	8.9 44	.125 00	58	3 364	7.6 16	24.0 83	.017 24
9	81	3.0 00	9.4 87	.111 11	59	3 481	7.6 81	24.2 90	.016 95
10	100	3.1 62	10.0 00	.100 00	60	3 600	7.7 46	24.4 95	.016 67
11	121	3.3 17	10.4 88	.090 91	61	3 721	7.8 10	24.6 98	.016 39
12	144	3.4 64	10.9 54	.083 33	62	3 844	7.8 74	24.9 00	.016 13
13	169	3.6 06	11.4 02	.076 92	63	3 969	7.9 37	25.1 00	.015 87
14	196	3.7 42	11.8 32	.071 43	64	4 096	8.0 00	25.2 98	.015 62
15	225	3.8 73	12.2 47	.066 67	65	4 225	8.0 62	25.4 95	.015 38
16	256	4.0 00	12.6 49	.062 50	66	4 356	8.1 24	25.6 90	.015 15
17	289	4.1 23	13.0 38	.058 82	67	4 489	8.1 85	25.8 84	.014 93
18	324	4.2 43	13.4 16	.055 56	68	4 624	8.2 46	26.0 77	.014 71
19	361	4.3 59	13.7 84	.052 63	69	4 761	8.3 07	26.2 68	.014 49
20	400	4.4 72	14.1 42	.050 00	70	4 900	8.3 67	26.4 58	.014 29
21	441	4.5 83	14.4 91	.047 62	71	5 041	8.4 26	26.6 46	.014 08
22	484	4.6 90	14.8 32	.045 45	72	5 184	8.4 85	26.8 33	.013 89
23	529	4.7 96	15.1 66	.043 48	73	5 329	8.5 44	27.0 19	.013 70
24	576	4.8 99	15.4 92	.041 67	74	5 476	8.6 02	27.2 03	.013 51
25	625	5.0 00	15.8 11	.040 00	75	5 625	8.6 60	27.3 86	.013 33
26	676	5.0 99	16.1 25	.038 46	76	5 776	8.7 18	27.5 68	.013 16
27	729	5.1 96	16.4 32	.037 04	77	5 929	8.7 75	27.7 49	.012 99
28	784	5.2 92	16.7 33	.035 71	78	6 084	8.8 32	27.9 28	.012 82
29	841	5.3 85	17.0 29	.034 48	79	6 241	8.8 88	28.1 07	.012 66
30	900	5.4 77	17.3 21	.033 33	80	6 400	8.9 44	28.2 84	.012 50
31	961	5.5 68	17.6 07	.032 26	81	6 561	9.0 00	28.4 60	.012 35
32	1 024	5.6 57	17.8 89	.031 25	82	6 724	9.0 55	28.6 36	.012 20
33	1 089	5.7 45	18.1 66	.030 30	83	6 889	9.1 10	28.8 10	.012 05
34	1 156	5.8 31	18.4 39	.029 41	84	7 056	9.1 65	28.9 83	.011 90
35	1 225	5.9 16	18.7 08	.028 57	85	7 225	9.2 20	29.1 55	.011 76
36	1 296	6.0 00	18.9 74	.027 78	86	7 396	9.2 74	29.3 26	.011 63
37	1 369	6.0 83	19.2 35	.027 03	87	7 569	9.3 27	29.4 96	.011 49
38	1 444	6.1 64	19.4 94	.026 32	88	7 744	9.3 81	29.6 65	.011 36
39	1 521	6.2 45	19.7 48	.025 64	89	7 921	9.4 34	29.8 33	.011 24
40	1 600	6.3 25	20.0 00	.025 00	90	8 100	9.4 87	30.0 00	.011 11
41	1 681	6.4 03	20.2 48	.024 39	91	8 281	9.5 39	30.1 66	.010 99
42	1 764	6.4 81	20.4 94	.023 81	92	8 464	9.5 92	30.3 32	.010 87
43	1 849	6.5 57	20.7 36	.023 26	93	8 649	9.6 44	30.4 96	.010 75
44	1 936	6.6 33	20.9 76	.022 73	94	8 836	9.6 95	30.6 59	.010 64
45	2 025	6.7 08	21.2 13	.022 22	95	9 025	9.7 47	30.8 22	.010 53
46	2 116	6.7 82	21.4 48	.021 74	96	9 216	9.7 98	30.9 84	.010 42
47	2 209	6.8 56	21.6 79	.021 28	97	9 409	9.8 49	31.1 45	.010 31
48	2 304	6.9 28	21.9 09	.020 83	98	9 604	9.8 99	31.3 05	.010 20
49	2 401	7.0 00	22.1 36	.020 41	99	9 801	9.9 50	31.4 64	.010 10
50	2 500	7.0 71	22.3 61	.020 00	100	10 000	10.0 00	31.6 23	.010 00
N	N^2	$\sqrt{N}$	$\sqrt{10N}$	$1/N$	N	N^2	$\sqrt{N}$	$\sqrt{10N}$	$1/N$

TABLE II Values of trigonometric functions

θ deg	deg	min	$\sin \theta$	$\cos \theta$	$\tan \theta$	$\csc \theta$	$\sec \theta$	$\cot \theta$			
0.0	0	0	0.0000	1.0000	0.0000	No value	1.0000	No value	90	0	90.0
0.1	0	6	0.0017	1.0000	0.0017	572.96	1.0000	572.96	89	54	89.9
0.2	0	12	0.0035	1.0000	0.0035	286.48	1.0000	286.48	89	48	89.8
0.3	0	18	0.0052	1.0000	0.0052	190.99	1.0000	190.98	89	42	89.7
0.4	0	24	0.0070	1.0000	0.0070	143.24	1.0000	143.24	89	36	89.6
0.5	0	30	0.0087	1.0000	0.0087	114.59	1.0000	114.59	89	30	89.5
0.6	0	36	0.0105	0.9999	0.0105	95.495	1.0001	95.490	89	24	89.4
0.7	0	42	0.0122	0.9999	0.0122	81.853	1.0001	81.847	89	18	89.3
0.8	0	48	0.0140	0.9999	0.0140	71.622	1.0001	71.615	89	12	89.2
0.9	0	54	0.0157	0.9999	0.0157	63.665	1.0001	63.657	89	6	89.1
1.0	1	0	0.0175	0.9998	0.0175	57.299	1.0002	57.290	89	0	89.0
1.1	1	6	0.0192	0.9998	0.0192	52.090	1.0002	52.081	88	54	88.9
1.2	1	12	0.0209	0.9998	0.0209	47.750	1.0002	47.740	88	48	88.8
1.3	1	18	0.0227	0.9997	0.0227	44.077	1.0003	44.066	88	42	88.7
1.4	1	24	0.0244	0.9997	0.0244	40.930	1.0003	40.917	88	36	88.6
1.5	1	30	0.0262	0.9997	0.0262	38.202	1.0003	38.188	88	30	88.5
1.6	1	36	0.0279	0.9996	0.0279	35.815	1.0004	35.801	88	24	88.4
1.7	1	42	0.0297	0.9996	0.0297	33.708	1.0004	33.694	88	18	88.3
1.8	1	48	0.0314	0.9995	0.0314	31.836	1.0005	31.821	88	12	88.2
1.9	1	54	0.0332	0.9995	0.0332	30.161	1.0005	30.145	88	6	88.1
2.0	2	0	0.0349	0.9994	0.0349	28.654	1.0006	28.636	88	0	88.0
2.1	2	6	0.0366	0.9993	0.0367	27.290	1.0007	27.271	87	54	87.9
2.2	2	12	0.0384	0.9993	0.0384	26.050	1.0007	26.031	87	48	87.8
2.3	2	18	0.0401	0.9992	0.0402	24.918	1.0008	24.898	87	42	87.7
2.4	2	24	0.0419	0.9991	0.0419	23.880	1.0009	23.859	87	36	87.6
2.5	2	30	0.0436	0.9990	0.0437	22.926	1.0010	22.904	87	30	87.5
2.6	2	36	0.0454	0.9990	0.0454	22.044	1.0010	22.022	87	24	87.4
2.7	2	42	0.0471	0.9989	0.0472	21.229	1.0011	21.205	87	18	87.3
2.8	2	48	0.0488	0.9988	0.0489	20.471	1.0012	20.446	87	12	87.2
2.9	2	54	0.0506	0.9987	0.0507	19.766	1.0013	19.740	87	6	87.1
3.0	3	0	0.0523	0.9986	0.0524	19.107	1.0014	19.081	87	0	87.0
3.1	3	6	0.0541	0.9985	0.0542	18.492	1.0015	18.464	86	54	86.9
3.2	3	12	0.0558	0.9984	0.0559	17.914	1.0016	17.886	86	48	86.8
3.3	3	18	0.0576	0.9983	0.0577	17.372	1.0017	17.343	86	42	86.7
3.4	3	24	0.0593	0.9982	0.0594	16.862	1.0018	16.832	86	36	86.6
3.5	3	30	0.0610	0.9981	0.0612	16.380	1.0019	16.350	86	30	86.5
3.6	3	36	0.0628	0.9980	0.0629	15.926	1.0020	15.895	86	24	86.4
3.7	3	42	0.0645	0.9979	0.0647	15.496	1.0021	15.464	86	18	86.3
3.8	3	48	0.0663	0.9978	0.0664	15.089	1.0022	15.056	86	12	86.2
3.9	3	54	0.0680	0.9977	0.0682	14.703	1.0023	14.669	86	6	86.1
4.0	4	0	0.0698	0.9976	0.0699	14.336	1.0024	14.301	86	0	86.0
4.1	4	6	0.0715	0.9974	0.0717	13.987	1.0026	13.951	85	54	85.9
4.2	4	12	0.0732	0.9973	0.0734	13.654	1.0027	13.617	85	48	85.8
4.3	4	18	0.0750	0.9972	0.0752	13.337	1.0028	13.300	85	42	85.7
4.4	4	24	0.0767	0.9971	0.0769	13.035	1.0030	12.996	85	36	85.6
4.5	4	30	0.0785	0.9969	0.0787	12.746	1.0031	12.706	85	30	85.5
4.6	4	36	0.0802	0.9968	0.0805	12.469	1.0032	12.429	85	24	85.4
4.7	4	42	0.0819	0.9966	0.0822	12.204	1.0034	12.163	85	18	85.3
4.8	4	48	0.0837	0.9965	0.0840	11.951	1.0035	11.909	85	12	85.2
4.9	4	54	0.0854	0.9963	0.0857	11.707	1.0037	11.665	85	6	85.1
			$\sin \theta$	$\cot \theta$	$\sec \theta$	$\csc \theta$	$\tan \theta$	deg min		θ deg

TABLE II (*continued*)

θ deg	deg	min	sin θ	cos θ	tan θ	csc θ	sec θ	cot θ			
5.0	5	0	0.0872	0.9962	0.0875	11.474	1.0038	11.430	85	0	85.0
5.1	5	6	0.0889	0.9960	0.0892	11.249	1.0040	11.205	84	54	84.9
5.2	5	12	0.0906	0.9959	0.0910	11.034	1.0041	10.988	84	48	84.8
5.3	5	18	0.0924	0.9957	0.0928	10.826	1.0043	10.780	84	42	84.7
5.4	5	24	0.0941	0.9956	0.0945	10.626	1.0045	10.579	84	36	84.6
5.5	5	30	0.0958	0.9954	0.0963	10.433	1.0046	10.385	84	30	84.5
5.6	5	36	0.0976	0.9952	0.0981	10.248	1.0048	10.199	84	24	84.4
5.7	5	42	0.0993	0.9951	0.0998	10.069	1.0050	10.019	84	18	84.3
5.8	5	48	0.1011	0.9949	0.1016	9.8955	1.0051	9.8448	84	12	84.2
5.9	5	54	0.1028	0.9947	0.1033	9.7283	1.0053	9.6768	84	6	84.1
6.0	6	0	0.1045	0.9945	0.1051	9.5668	1.0055	9.5144	84	0	84.0
6.1	6	6	0.1063	0.9943	0.1069	9.4105	1.0057	9.3573	83	54	83.9
6.2	6	12	0.1080	0.9942	0.1086	9.2593	1.0059	9.2052	83	48	83.8
6.3	6	18	0.1097	0.9940	0.1104	9.1129	1.0061	9.0579	83	42	83.7
6.4	6	24	0.1115	0.9938	0.1122	8.9711	1.0063	8.9152	83	36	83.6
6.5	6	30	0.1132	0.9936	0.1139	8.8337	1.0065	8.7769	83	30	83.5
6.6	6	36	0.1149	0.9934	0.1157	8.7004	1.0067	8.6428	83	24	83.4
6.7	6	42	0.1167	0.9932	0.1175	8.5711	1.0069	8.5126	83	18	83.3
6.8	6	48	0.1184	0.9930	0.1192	8.4457	1.0071	8.3863	83	12	83.2
6.9	6	54	0.1201	0.9928	0.1210	8.3238	1.0073	8.2636	83	6	83.1
7.0	7	0	0.1219	0.9925	0.1228	8.2055	1.0075	8.1444	83	0	83.0
7.1	7	6	0.1236	0.9923	0.1246	8.0905	1.0077	8.0285	82	54	82.9
7.2	7	12	0.1253	0.9921	0.1263	7.9787	1.0079	7.9158	82	48	82.8
7.3	7	18	0.1271	0.9919	0.1281	7.8700	1.0082	7.8062	82	42	82.7
7.4	7	24	0.1288	0.9917	0.1299	7.7642	1.0084	7.6996	82	36	82.6
7.5	7	30	0.1305	0.9914	0.1317	7.6613	1.0086	7.5958	82	30	82.5
7.6	7	36	0.1323	0.9912	0.1334	7.5611	1.0089	7.4947	82	24	82.4
7.7	7	42	0.1340	0.9910	0.1352	7.4635	1.0091	7.3962	82	18	82.3
7.8	7	48	0.1357	0.9907	0.1370	7.3684	1.0093	7.3002	82	12	82.2
7.9	7	54	0.1374	0.9905	0.1388	7.2757	1.0096	7.2066	82	6	82.1
8.0	8	0	0.1392	0.9903	0.1405	7.1853	1.0098	7.1154	82	0	82.0
8.1	8	6	0.1409	0.9900	0.1423	7.0972	1.0101	7.0264	81	54	81.9
8.2	8	12	0.1426	0.9898	0.1441	7.0112	1.0103	6.9395	81	48	81.8
8.3	8	18	0.1444	0.9895	0.1459	6.9273	1.0106	6.8548	81	42	81.7
8.4	8	24	0.1461	0.9893	0.1477	6.8454	1.0108	6.7720	81	36	81.6
8.5	8	30	0.1478	0.9890	0.1495	6.7655	1.0111	6.6912	81	30	81.5
8.6	8	36	0.1495	0.9888	0.1512	6.6874	1.0114	6.6122	81	24	81.4
8.7	8	42	0.1513	0.9885	0.1530	6.6111	1.0116	6.5350	81	18	81.3
8.8	8	48	0.1530	0.9882	0.1548	6.5366	1.0119	6.4596	81	12	81.2
8.9	8	54	0.1547	0.9880	0.1566	6.4637	1.0122	6.3859	81	6	81.1
9.0	9	0	0.1564	0.9877	0.1584	6.3925	1.0125	6.3138	81	0	81.0
9.1	9	6	0.1582	0.9874	0.1602	6.3228	1.0127	6.2432	80	54	80.9
9.2	9	12	0.1599	0.9871	0.1620	6.2547	1.0130	6.1742	80	48	80.8
9.3	9	18	0.1616	0.9869	0.1638	6.1880	1.0133	6.1066	80	42	80.7
9.4	9	24	0.1633	0.9866	0.1655	6.1227	1.0136	6.0405	80	36	80.6
9.5	9	30	0.1650	0.9863	0.1673	6.0589	1.0139	5.9758	80	30	80.5
9.6	9	36	0.1668	0.9860	0.1691	5.9963	1.0142	5.9124	80	24	80.4
9.7	9	42	0.1685	0.9857	0.1709	5.9351	1.0145	5.8502	80	18	80.3
9.8	9	48	0.1702	0.9854	0.1727	5.8751	1.0148	5.7894	80	12	80.2
9.9	9	54	0.1719	0.9851	0.1745	5.8164	1.0151	5.7297	80	6	80.1
			cos θ	sin θ	cot θ	sec θ	csc θ	tan θ	deg	min	θ deg

TABLE II   (*continued*)

θ deg	deg min		sin θ	cos θ	tan θ	csc θ	sec θ	cot θ			
10.0	10	0	0.1736	0.9848	0.1763	5.7588	1.0154	5.6713	80	0	80.0
10.1	10	6	0.1754	0.9845	0.1781	5.7023	1.0157	5.6140	79	54	79.9
10.2	10	12	0.1771	0.9842	0.1799	5.6470	1.0161	5.5578	79	48	79.8
10.3	10	18	0.1788	0.9839	0.1817	5.5928	1.0164	5.5027	79	42	79.7
10.4	10	24	0.1805	0.9836	0.1835	5.5396	1.0167	5.4486	79	36	79.6
10.5	10	30	0.1822	0.9833	0.1853	5.4874	1.0170	5.3955	79	30	79.5
10.6	10	36	0.1840	0.9829	0.1871	5.4362	1.0174	5.3435	79	24	79.4
10.7	10	42	0.1857	0.9826	0.1890	5.3860	1.0177	5.2924	79	18	79.3
10.8	10	48	0.1874	0.9823	0.1908	5.3367	1.0180	5.2422	79	12	79.2
10.9	10	54	0.1891	0.9820	0.1926	5.2883	1.0184	5.1929	79	6	79.1
11.0	11	0	0.1908	0.9816	0.1944	5.2408	1.0187	5.1446	79	0	79.0
11.1	11	6	0.1925	0.9813	0.1962	5.1942	1.0191	5.0970	78	54	78.9
11.2	11	12	0.1942	9.9810	0.1980	5.1484	1.0194	5.0504	78	48	78.8
11.3	11	18	0.1959	0.9806	0.1998	5.1034	1.0198	5.0045	78	42	78.7
11.4	11	24	0.1977	0.9803	0.2016	5.0593	1.0201	4.9595	78	36	78.6
11.5	11	30	0.1994	0.9799	0.2035	5.0159	1.0205	4.9152	78	30	78.5
11.6	11	36	0.2011	0.9796	0.2053	4.9732	1.0209	4.8716	78	24	78.4
11.7	11	42	0.2028	0.9792	0.2071	4.9313	1.0212	4.8288	78	18	78.3
11.8	11	48	0.2045	0.9789	0.2089	4.8901	1.0216	4.7867	78	12	78.2
11.9	11	54	0.2062	0.9785	0.2107	4.8496	1.0220	4.7453	78	6	78.1
12.0	12	0	0.2079	0.9781	0.2126	4.8097	1.0223	4.7046	78	0	78.0
12.1	12	6	0.2096	0.9778	0.2144	4.7706	1.0227	4.6646	77	54	77.9
12.2	12	12	0.2113	0.9774	0.2162	4.7321	1.0231	4.6252	77	48	77.8
12.3	12	18	0.2130	0.9770	0.2180	4.6942	1.0235	4.5864	77	42	77.7
12.4	12	24	0.2147	0.9767	0.2199	4.6569	1.0239	4.5483	77	36	77.6
12.5	12	30	0.2164	0.9763	0.2217	4.6202	1.0243	4.5107	77	30	77.5
12.6	12	36	0.2181	0.9759	0.2235	4.5841	1.0247	4.4737	77	24	77.4
12.7	12	42	0.2198	0.9755	0.2254	4.5486	1.0251	4.4374	77	18	77.3
12.8	12	48	0.2215	0.9751	0.2272	4.5137	1.0255	4.4015	77	12	77.2
12.9	12	54	0.2232	0.9748	0.2290	4.4793	1.0259	4.3662	77	6	77.1
13.0	13	0	0.2250	0.9744	0.2309	4.4454	1.0263	4.3315	77	0	77.0
13.1	13	6	0.2267	0.9740	0.2327	4.4121	1.0267	4.2972	76	54	76.9
13.2	13	12	0.2284	0.9736	0.2345	4.3792	1.0271	4.2635	76	48	76.8
13.3	13	18	0.2300	0.9732	0.2364	4.3469	1.0276	4.2303	76	42	76.7
13.4	13	24	0.2317	0.9728	0.2382	4.3150	1.0280	4.1976	76	36	76.6
13.5	13	30	0.2334	0.9724	0.2401	4.2837	1.0284	4.1653	76	30	76.5
13.6	13	36	0.2351	0.9720	0.2419	4.2528	1.0288	4.1335	76	24	76.4
13.7	13	42	0.2368	0.9715	0.2438	4.2223	1.0293	4.1022	76	18	76.3
13.8	13	48	0.2385	0.9711	0.2456	4.1923	1.0297	4.0713	76	12	76.2
13.9	13	54	0.2402	0.9707	0.2475	4.1627	1.0302	4.0408	76	6	76.1
14.0	14	0	0.2419	0.9703	0.2493	4.1336	1.0306	4.0108	76	0	76.0
14.1	14	6	0.2436	0.9699	0.2512	4.1048	1.0311	3.9812	75	54	75.9
14.2	14	12	0.2453	0.9694	0.2530	4.0765	1.0315	3.9520	75	48	75.8
14.3	14	18	0.2470	0.9690	0.2549	4.0486	1.0320	3.9232	75	42	75.7
14.4	14	24	0.2487	0.9686	0.2568	4.0211	1.0324	3.8947	75	36	75.6
14.5	14	30	0.2504	0.9681	0.2586	3.9939	1.0329	3.8667	75	30	75.5
14.6	14	36	0.2521	0.9677	0.2605	3.9672	1.0334	3.8391	75	24	75.4
14.7	14	42	0.2538	0.9673	0.2623	3.9408	1.0338	3.8118	75	18	75.3
14.8	14	48	0.2554	0.9668	0.2642	3.9147	1.0343	3.7849	75	12	75.2
14.9	14	54	0.2571	0.9664	0.2661	3.8890	1.0348	3.7583	75	6	75.1
			cos θ	sin θ	cot θ	sec θ	csc θ	tan θ	deg min		θ deg

II

TABLE II (*continued*)

θ deg	deg min	sin θ	cos θ	tan θ	csc θ	sec θ	cot θ		
15.0	15 0	0.2588	0.9659	0.2679	3.8637	1.0353	3.7321	75 0	75.0
15.1	15 6	0.2605	0.9655	0.2698	3.8387	1.0358	3.7062	74 54	74.9
15.2	15 12	0.2622	0.9650	0.2717	3.8140	1.0363	3.6806	74 48	74.8
15.3	15 18	0.2639	0.9646	0.2736	3.7897	1.0367	3.6554	74 42	74.7
15.4	15 24	0.2656	0.9641	0.2754	3.7657	1.0372	3.6305	74 36	74.6
15.5	15 30	0.2672	0.9636	0.2773	3.7420	1.0377	3.6059	74 30	74.5
15.6	15 36	0.2689	0.9632	0.2792	3.7186	1.0382	3.5816	74 24	74.4
15.7	15 42	0.2706	0.9627	0.2811	3.6955	1.0388	3.5576	74 18	74.3
15.8	15 48	0.2723	0.9622	0.2830	3.6727	1.0393	3.5339	74 12	74.2
15.9	15 54	0.2740	0.9617	0.2849	3.6502	1.0398	3.5105	74 6	74.1
16.0	16 0	0.2756	0.9613	0.2867	3.6280	1.0403	3.4874	74 0	74.0
16.1	16 6	0.2773	0.9608	0.2886	3.6060	1.0408	3.4646	73 54	73.9
16.2	16 12	0.2790	0.9603	0.2905	3.5843	1.0413	3.4420	73 48	73.8
16.3	16 18	0.2807	0.9598	0.2924	3.5629	1.0419	3.4197	73 42	73.7
16.4	16 24	0.2823	0.9593	0.2943	3.5418	1.0424	3.3977	73 36	73.6
16.5	16 30	0.2840	0.9588	0.2962	3.5209	1.0429	3.3759	73 30	73.5
16.6	16 36	0.2857	0.9583	0.2981	3.5003	1.0435	3.3544	73 24	73.4
16.7	16 42	0.2874	0.9578	0.3000	3.4800	1.0440	3.3332	73 18	73.3
16.8	16 48	0.2890	0.9573	0.3019	3.4598	1.0446	3.3122	73 12	73.2
16.9	16 54	0.2907	0.9568	0.3038	3.4399	1.0451	3.2914	73 6	73.1
17.0	17 0	0.2924	0.9563	0.3057	3.4203	1.0457	3.2709	73 0	73.0
17.1	17 6	0.2940	0.9558	0.3076	3.4009	1.0463	3.2506	72 54	72.9
17.2	17 12	0.2957	0.9553	0.3096	3.3817	1.0468	3.2305	72 48	72.8
17.3	17 18	0.2974	0.9548	0.3115	3.3628	1.0474	3.2106	72 42	72.7
17.4	17 24	0.2990	0.9542	0.3134	3.3440	1.0480	3.1910	72 36	72.6
17.5	17 30	0.3007	0.9537	0.3153	3.3255	1.0485	3.1716	72 30	72.5
17.6	17 36	0.3024	0.9532	0.3172	3.3072	1.0491	3.1524	72 24	72.4
17.7	17 42	0.3040	0.9527	0.3191	3.2891	1.0497	3.1334	72 18	72.3
17.8	17 48	0.3057	0.9521	0.3211	3.2712	1.0503	3.1146	72 12	72.2
17.9	17 54	0.3074	0.9516	0.3230	3.2536	1.0509	3.0961	72 6	72.1
18.0	18 0	0.3090	0.9511	0.3249	3.2361	1.0515	3.0777	72 0	72.0
18.1	18 6	0.3107	0.9505	0.3268	3.2188	1.0521	3.0595	71 54	71.9
18.2	18 12	0.3123	0.9500	0.3288	3.2017	1.0527	3.0415	71 48	71.8
18.3	18 18	0.3140	0.9494	0.3307	3.1848	1.0533	3.0237	71 42	71.7
18.4	18 24	0.3156	0.9489	0.3327	3.1681	1.0539	3.0061	71 36	71.6
18.5	18 30	0.3173	0.9483	0.3346	3.1515	1.0545	2.9887	71 30	71.5
18.6	18 36	0.3190	0.9478	0.3365	3.1352	1.0551	2.9714	71 24	71.4
18.7	18 42	0.3206	0.9472	0.3385	3.1190	1.0557	2.9544	71 18	71.3
18.8	18 48	0.3223	0.9466	0.3404	3.1030	1.0564	2.9375	71 12	71.2
18.9	18 54	0.3239	0.9461	0.3424	3.0872	1.0570	2.9208	71 6	71.1
19.0	19 0	0.3256	0.9455	0.3443	2.0716	1.0576	2.9042	71 0	71.0
19.1	19 6	0.3272	0.9449	0.3463	3.0561	1.0583	2.8878	70 54	70.9
19.2	19 12	0.3289	0.9444	0.3482	3.0407	1.0589	2.8716	70 48	70.8
19.3	19 18	0.3305	0.9438	0.3502	3.0256	1.0595	2.8556	70 42	70.7
19.4	19 24	0.3322	0.9432	0.3522	3.0106	1.0602	2.8397	70 36	70.6
19.5	19 30	0.3338	0.9426	0.3541	2.9957	1.0608	2.8239	70 30	70.5
19.6	19 36	0.3355	0.9421	0.3561	2.9811	1.0615	2.8083	70 24	70.4
19.7	19 42	0.3371	0.9415	0.3581	2.9665	1.0622	2.7929	70 18	70.3
19.8	19 48	0.3387	0.9409	0.3600	2.9521	1.0628	2.7776	70 12	70.2
19.9	19 54	0.3404	0.9403	0.3620	2.9379	1.0635	2.7625	70 6	70.1
		cos θ	sin θ	cot θ	sec θ	csc θ	tan θ	deg min	θ deg

TABLE II *(continued)*

θ deg	deg	min	$\sin \theta$	$\cos \theta$	$\tan \theta$	$\csc \theta$	$\sec \theta$	$\cot \theta$			
20.0	20	0	0.3420	0.9397	0.3640	2.9238	1.0642	2.7475	70	0	70.0
20.1	20	6	0.3437	0.9391	0.3659	2.9099	1.0649	2.7326	69	54	69.9
20.2	20	12	0.3453	0.9385	0.3679	2.8960	1.0655	2.7179	69	48	69.8
20.3	20	18	0.3469	0.9379	0.3699	2.8824	1.0662	2.7034	69	42	69.7
20.4	20	24	0.3486	0.9373	0.3719	2.8688	1.0669	2.6889	69	36	69.6
20.5	20	30	0.3502	0.9367	0.3739	2.8555	1.0676	2.6746	69	30	69.5
20.6	20	36	0.3518	0.9361	0.3759	2.8422	1.0683	2.6605	69	24	69.4
20.7	20	42	0.3535	0.9354	0.3779	2.8291	1.0690	2.6464	69	18	69.3
20.8	20	48	0.3551	0.9348	0.3799	2.8161	1.0697	2.6325	69	12	69.2
20.9	20	54	0.3567	0.9342	0.3819	2.8032	1.0704	2.6187	69	6	69.1
21.0	21	0	0.3584	0.9336	0.3839	2.7904	1.0711	2.6051	69	0	69.0
21.1	21	6	0.3600	0.9330	0.3859	2.7778	1.0719	2.5916	68	54	68.9
21.2	21	12	0.3616	0.9323	0.3879	2.7653	1.0726	2.5782	68	48	68.8
21.3	21	18	0.3633	0.9317	0.3899	2.7529	1.0733	2.5649	68	42	68.7
21.4	21	24	0.3649	0.9311	0.3919	2.7407	1.0740	2.5517	68	36	68.6
21.5	21	30	0.3665	0.9304	0.3939	2.7285	1.0748	2.5386	68	30	68.5
21.6	21	36	0.3681	0.9298	0.3959	2.7165	1.0755	2.5257	68	24	68.4
21.7	21	42	0.3697	0.9291	0.3979	2.7046	1.0763	2.5129	68	18	68.3
21.8	21	48	0.3714	0.9285	0.4000	2.6927	1.0770	2.5002	68	12	68.2
21.9	21	54	0.3730	0.9278	0.4020	2.6811	1.0778	2.4876	68	6	68.1
22.0	22	0	0.3746	0.9272	0.4040	2.6695	1.0785	2.4751	68	0	68.0
22.1	22	6	0.3762	0.9265	0.4061	2.6580	1.0793	2.4627	67	54	67.9
22.2	22	12	0.3778	0.9259	0.4081	2.6466	1.0801	2.4504	67	48	67.8
22.3	22	18	0.3795	0.9252	0.4101	2.6354	1.0808	2.4383	67	42	67.7
22.4	22	24	0.3811	0.9245	0.4122	2.6242	1.0816	2.4262	67	36	67.6
22.5	22	30	0.3827	0.9239	0.4142	2.6131	1.0824	2.4142	67	30	67.5
22.6	22	36	0.3843	0.9232	0.4163	2.6022	1.0832	2.4023	67	24	67.4
22.7	22	42	0.3859	0.9225	0.4183	2.5913	1.0840	2.3906	67	18	67.3
22.8	22	48	0.3875	0.9219	0.4204	2.5805	1.0848	2.3789	67	12	67.2
22.9	22	54	0.3891	0.9212	0.4224	2.5699	1.0856	2.3673	67	6	67.1
23.0	23	0	0.3907	0.9205	0.4245	2.5593	1.0864	2.3559	67	0	67.0
23.1	23	6	0.3923	0.9198	0.4265	2.5488	1.0872	2.3445	66	54	66.9
23.3	23	12	0.3939	0.9191	0.4286	2.5384	1.0880	2.3332	66	48	66.8
23.3	23	18	0.3955	0.9184	0.4307	2.5282	1.0888	2.3220	66	42	66.7
23.4	23	24	0.3971	0.9178	0.4327	2.5180	1.0896	2.3109	66	36	66.6
23.5	23	30	0.3987	0.9171	0.4348	2.5078	1.0904	2.2998	66	30	66.5
23.6	23	36	0.4003	0.9164	0.4369	2.4978	1.0913	2.2889	66	24	66.4
23.7	23	42	0.4019	0.9157	0.4390	2.4879	1.0921	2.2781	66	18	66.3
23.8	23	48	0.4035	0.9150	0.4411	2.4780	1.0929	2.2673	66	12	66.2
23.9	23	54	0.4051	0.9143	0.4431	2.4683	1.0938	2.2566	66	6	66.1
24.0	24	0	0.4067	0.9135	0.4452	2.4586	1.0946	2.2460	66	0	66.0
24.1	24	6	0.4083	0.9128	0.4473	2.4490	1.0955	2.2355	65	54	65.9
24.2	24	12	0.4099	0.9121	0.4494	2.4395	1.0963	2.2251	65	48	65.8
24.3	24	18	0.4115	0.9114	0.4515	2.4301	1.0972	2.2148	65	42	65.7
24.4	24	24	0.4131	0.9107	0.4536	2.4207	1.0981	2.2045	65	36	65.6
24.5	24	30	0.4147	0.9100	0.4557	2.4114	1.0989	2.1943	65	30	65.5
24.6	24	36	0.4163	0.9092	0.4578	2.4022	1.0998	2.1842	65	24	65.4
24.7	24	42	0.4179	0.9085	0.4599	2.3931	1.1007	2.1742	65	18	65.3
24.8	24	48	0.4195	0.9078	0.4621	2.3841	1.1016	2.1642	65	12	65.2
24.9	24	54	0.4210	0.9070	0.4642	2.3751	1.1025	2.1543	65	6	65.1
			$\cos \theta$	$\sin \theta$	$\cot \theta$	$\sec \theta$	$\csc \theta$	$\tan \theta$	deg	min	θ deg

TABLE II   (*continued*)

θ deg	deg	min	sin θ	cos θ	tan θ	csc θ	sec θ	cot θ			
25.0	25	0	0.4226	0.9063	0.4663	2.3662	1.1034	2.1445	65	0	65.0
25.1	25	6	0.4242	0.9056	0.4684	2.3574	1.1043	2.1348	64	54	64.9
25.2	25	12	0.4258	0.9048	0.4706	2.3486	1.1052	2.1251	64	48	64.8
25.3	25	18	0.4274	0.9041	0.4727	2.3400	1.1061	2.1155	64	42	64.7
25.4	25	24	0.4289	0.9033	0.4748	2.3314	1.1070	2.1060	64	36	64.6
25.5	25	30	0.4305	0.9026	0.4770	2.3228	1.1079	2.0965	64	30	64.5
25.6	25	36	0.4321	0.9018	0.4791	2.3144	1.1089	2.0872	64	24	64.4
25.7	25	42	0.4337	0.9011	0.4813	2.3060	1.1098	2.0778	64	18	64.3
25.8	25	48	0.4352	0.9003	0.4834	2.2976	1.1107	2.0686	64	12	64.2
25.9	25	54	0.4368	0.8996	0.4856	2.2894	1.1117	2.0594	64	6	64.1
26.0	26	0	0.4384	0.8988	0.4877	2.2812	1.1126	2.0503	64	0	64.0
26.1	26	6	0.4399	0.8980	0.4899	2.2730	1.1136	2.0413	63	54	63.9
26.2	26	12	0.4415	0.8973	0.4921	2.2650	1.1145	2.0323	63	48	63.8
26.3	26	18	0.4431	0.8965	0.4942	2.2570	1.1155	2.0233	63	42	63.7
26.4	26	24	0.4446	0.8957	0.4964	2.2490	1.1164	2.0145	63	36	63.6
26.5	26	30	0.4462	0.8949	0.4986	2.2412	1.1174	2.0057	63	30	63.5
26.6	26	36	0.4478	0.8942	0.5008	2.2333	1.1184	1.9970	63	24	63.4
26.7	26	42	0.4493	0.8934	0.5029	2.2256	1.1194	1.9883	63	18	63.3
26.8	26	48	0.4509	0.8926	0.5051	2.2179	1.1203	1.9797	63	12	63.2
26.9	26	54	0.4524	0.8918	0.5073	2.2103	1.1213	1.9711	63	6	63.1
27.0	27	0	0.4540	0.8910	0.5095	2.2027	1.1223	1.9626	63	0	63.0
27.1	27	6	0.4555	0.8902	0.5117	2.1952	1.1233	1.9542	62	54	62.9
27.2	27	12	0.4571	0.8894	0.5139	2.1877	1.1243	1.9458	62	48	62.8
27.3	27	18	0.4586	0.8886	0.5161	2.1803	1.1253	1.9375	62	42	62.7
27.4	27	24	0.4602	0.8878	0.5184	2.1730	1.1264	1.9292	62	36	62.6
27.5	27	30	0.4617	0.8870	0.5206	2.1657	1.1274	1.9210	62	30	62.5
27.6	27	36	0.4633	0.8862	0.5228	2.1584	1.1284	1.9128	62	24	62.4
27.7	27	42	0.4648	0.8854	0.5250	2.1513	1.1294	1.9047	62	18	62.3
27.8	27	48	0.4664	0.8846	0.5272	2.1441	1.1305	1.8967	62	12	62.2
27.9	27	54	0.4679	0.8838	0.5295	2.1371	1.1315	1.8887	62	6	62.1
28.0	28	0	0.4695	0.8829	0.5317	2.1301	1.1326	1,8807	62	0	62.0
28.1	28	6	0.4710	0.8821	0.5339	2.1231	1.1336	1.8728	61	54	61.9
28.2	28	12	0.4726	0.8813	0.5362	2.1162	1.1347	1.8650	61	48	61.8
28.3	28	18	0.4741	0.8805	0.5384	2.1093	1.1357	1.8572	61	42	61.7
28.4	28	24	0.4756	0.8796	0.5407	2.1025	1.1368	1.8495	61	36	61.6
28.5	28	30	0.4772	0.8788	0.5430	2.0957	1.1379	1.8418	61	30	61.5
28.6	28	36	0.4787	0.8780	0.5452	2.0890	1.1390	1.8341	61	24	61.4
28.7	28	42	0.4802	0.8771	0.5475	2.0824	1.1401	1.8265	61	18	61.3
28.8	28	48	0.4818	0.8763	0.5498	2.0758	1.1412	1.8190	61	12	61.2
28.9	28	54	0.4833	0.8755	0.5520	2.0692	1.1423	1.8115	61	6	61.1
29.0	29	0	0.4848	0.8746	0.5543	2.0627	1.1434	1.8040	61	0	61.0
29.1	29	6	0.4863	0.8738	0.5566	2.0562	1.1445	1.7966	60	54	60.9
29.2	29	12	0.4879	0.8729	0.5589	2.0598	1.1456	1.7893	60	48	60.8
29.3	29	18	0.4894	0.8721	0.5612	2.0434	1.1467	1.7820	60	42	60.7
29.4	29	24	0.4909	0.8712	0.5635	2.0371	1.1478	1.7747	60	36	60.6
29.5	29	30	0.4924	0.8704	0.5658	2.0308	1.1490	1.7675	60	30	60.5
29.6	29	36	0.4939	0.8695	0.5681	2.0245	1.1501	1.7603	60	24	60.4
29.7	29	42	0.4955	0.8686	0.5704	2.0183	1.1512	1.7532	60	18	60.3
29.8	29	48	0.4970	0.8678	0.5727	2.0122	1.1524	1.7461	60	12	60.2
29.9	29	54	0.4985	0.8669	0.5750	2.0061	1.1535	1.7391	60	6	60.1
			cos θ	sin θ	cot θ	sec θ	csc θ	tan θ	deg	min	θ deg

TABLE II (*continued*)

θ deg	deg	min	sin θ	cos θ	tan θ	csc θ	sec θ	cot θ			
30.0	30	0	0.5000	0.8660	0.5774	2.0000	1.1547	1.7321	60	0	60.0
30.1	30	6	0.5015	0.8652	0.5797	1.9940	1.1559	1.7251	59	54	59.9
30.2	30	12	0.5030	0.8643	0.5820	1.9880	1.1570	1.7182	59	48	59.8
30.3	30	18	0.5045	0.8634	0.5844	1.9821	1.1582	1.7113	59	42	59.7
30.4	30	24	0.5060	0.8625	0.5867	1.9762	1.1594	1.7045	59	36	59.6
30.5	30	30	0.5075	0.8616	0.5890	1.9703	1.1606	1.6977	59	30	59.5
30.6	30	36	0.5090	0.8607	0.5914	1.9645	1.1618	1.6909	59	24	59.4
30.7	30	42	0.5105	0.8599	0.5938	1.9587	1.1630	1.6842	59	18	59.3
30.8	30	48	0.5120	0.8590	0.5961	1.9530	1.1642	1.6775	59	12	59.2
30.9	30	54	0.5135	0.8581	0.5985	1.9473	1.1654	1.6709	59	6	59.1
31.0	31	0	0.5150	0.8572	0.6009	1.9416	1.1666	1.6643	59	0	59.0
31.1	31	6	0.5165	0.8563	0.6032	1.9360	1.1679	1.6577	58	54	58.9
31.2	31	12	0.5180	0.8554	0.6056	1.9304	1.1691	1.6512	58	48	58.8
31.3	31	18	0.5195	0.8545	0.6080	1.9249	1.1703	1.6447	58	42	58.7
31.4	31	24	0.5210	0.8536	0.6104	1.9194	1.1716	1.6383	58	36	58.6
31.5	31	30	0.5225	0.8526	0.6128	1.9139	1.1728	1.6319	58	30	58.5
31.6	31	36	0.5240	0.8517	0.6152	1.9084	1.1741	1.6255	58	24	58.4
31.7	31	42	0.5255	0.8508	0.6176	1.9031	1.1753	1.6191	58	18	58.3
31.8	31	48	0.5270	0.8499	0.6200	1.8977	1.1766	1.6128	58	12	58.2
31.9	31	54	0.5284	0.8490	0.6224	1.8924	1.1779	1.6066	58	6	58.1
32.0	32	0	0.5299	0.8480	0.6249	1.8871	1.1792	1.6003	58	0	58.0
32.1	32	6	0.5314	0.8471	0.6273	1.8818	1.1805	1.5941	57	54	57.9
32.2	32	12	0.5329	0.8462	0.6297	1.8766	1.1818	1.5880	57	48	57.8
32.3	32	18	0.5344	0.8453	0.6322	1.8714	1.1831	1.5818	57	42	57.7
32.4	32	24	0.5358	0.8443	0.6346	1.8663	1.1844	1.5757	57	36	57.6
32.5	32	30	0.5373	0.8434	0.6371	1.8612	1.1857	1.5697	57	30	57.5
32.6	32	36	0.5388	0.8425	0.6395	1.8561	1.1870	1.5637	57	24	57.4
32.7	32	42	0.5402	0.8415	0.6420	1.8510	1.1883	1.5577	57	18	57.3
32.8	32	48	0.5417	0.8406	0.6445	1.8460	1.1897	1.5517	57	12	57.2
32.9	32	54	0.5432	0.8396	0.6469	1.8410	1.1910	1.5458	57	6	57.1
33.0	33	0	0.5446	0.8387	0.6494	1.8361	1.1924	1.5399	57	0	57.0
33.1	33	6	0.5461	0.8377	0.6519	1.8312	1.1937	1.5340	56	54	56.9
33.2	33	12	0.5476	0.8368	0.6544	1.8263	1.1951	1.5282	56	48	56.8
33.3	33	18	0.5490	0.8358	0.6569	1.8214	1.1964	1.5224	56	42	56.7
33.4	33	24	0.5505	0.8348	0.6594	1.8166	1.1978	1.5166	56	36	56.6
33.5	33	30	0.5519	0.8339	0.6619	1.8118	1.1992	1.5108	56	30	56.5
33.6	33	36	0.5534	0.8329	0.6644	1.8070	1.2006	1.5051	56	24	56.4
33.7	33	42	0.5548	0.8320	0.6669	1.8023	1.2020	1.4994	56	18	56.3
33.8	33	48	0.5563	0.8310	0.6694	1.7976	1.2034	1.4938	56	12	56.2
33.9	33	54	0.5577	0.8300	0.6720	1.7929	1.2048	1.4882	56	6	56.1
34.0	34	0	0.5592	0.8290	0.6745	1.7883	1.2062	1.4826	56	0	56.0
34.1	34	6	0.5606	0.8281	0.6771	1.7837	1.2076	1.4770	55	54	55.9
34.2	34	12	0.5621	0.8271	0.6796	1.7791	1.2091	1.4715	55	48	55.8
34.3	34	18	0.5635	0.8261	0.6822	1.7745	1.2105	1.4659	55	42	55.7
34.4	34	24	0.5650	0.8251	0.6847	1.7700	1.2120	1.4605	55	36	55.6
34.5	34	30	0.5664	0.8241	0.6873	1.7655	1.2134	1.4550	55	30	55.5
34.6	34	36	0.5678	0.8231	0.6899	1.7610	1.2149	1.4496	55	24	55.4
34.7	34	42	0.5693	0.8221	0.6924	1.7566	1.2163	1.4442	55	18	55.3
34.8	34	48	0.5707	0.8211	0.6950	1.7522	1.2178	1.4388	55	12	55.2
34.9	34	54	0.5721	0.8202	0.6976	1.7478	1.2193	1.4335	55	6	55.1
			cos θ	sin θ	cot θ	sec θ	csc θ	tan θ	deg	min	θ deg

TABLE II (*continued*)

θ deg	deg min		sin θ	cos θ	tan θ	csc θ	sec θ	cot θ			
35.0	35	0	0.5736	0.8192	0.7002	1.7434	1.2208	1.4281	55	0	55.0
35.1	35	6	0.5750	0.8181	0.7028	1.7391	1.2223	1.4229	54	54	54.9
35.2	35	12	0.5764	0.8171	0.7054	1.7348	1.2238	1.4176	54	48	54.8
35.3	35	18	0.5779	0.8161	0.7080	1.7305	1.2253	1.4124	54	42	54.7
35.4	35	24	0.5793	0.8151	0.7107	1.7263	1.2268	1.4071	54	36	54.6
35.5	35	30	0.5807	0.8141	0.7133	1.7221	1.2283	1.4019	54	30	54.5
35.6	35	36	0.5821	0.8131	0.7159	1.7179	1.2299	1.3968	54	24	54.4
35.7	35	42	0.5835	0.8121	0.7186	1.7137	1.2314	1.3916	54	18	54.3
35.8	35	48	0.5850	0.8111	0.7212	1.7095	1.2329	1.3865	54	12	54.2
35.9	35	54	0.5864	0.8100	0.7239	1.7054	1.2345	1.3814	54	6	54.1
36.0	36	0	0.5878	0.8090	0.7265	1.7013	1.2361	1.3764	54	0	54.0
36.1	36	6	0.5892	0.8080	0.7292	1.6972	1.2376	1.3713	53	54	53.9
36.2	36	12	0.5906	0.8070	0.7319	1.6932	1.2392	1.3663	53	48	53.8
36.3	36	18	0.5920	0.8059	0.7346	1.6892	1.2408	1.3613	53	42	53.7
36.4	36	24	0.5934	0.8049	0.7373	1.6852	1.2424	1.3564	53	36	53.6
36.5	36	30	0.5948	0.8039	0.7400	1.6812	1.2440	1.3514	53	30	53.5
36.6	36	36	0.5962	0.8028	0.7427	1.6772	1.2456	1.3465	53	24	53.4
36.7	36	42	0.5976	0.8018	0.7454	1.6733	1.2472	1.3416	53	18	53.3
36.8	36	48	0.5990	0.8007	0.7481	1.6694	1.2489	1.3367	53	12	53.2
36.9	36	54	0.6004	0.7997	0.7508	1.6655	1.2505	1.3319	53	6	53.1
37.0	37	0	0.6018	0.7986	0.7536	1.6616	1.2521	1.3270	53	0	53.0
37.1	37	6	0.6032	0.7976	0.7563	1.6578	1.2538	1.3222	52	54	52.9
37.2	37	12	0.6046	0.7965	0.7590	1.6540	1.2554	1.3175	52	48	52.8
37.3	37	18	0.6060	0.7955	0.7618	1.6502	1.2571	1.3127	52	42	52.7
37.4	37	24	0.6074	0.7944	0.7646	1.6464	1.2588	1.3079	52	36	52.6
37.5	37	30	0.6088	0.7934	0.7673	1.6427	1.2605	1.3032	52	30	52.5
37.6	37	36	0.6101	0.7923	0.7701	1.6390	1.2622	1.2985	52	24	52.4
37.7	37	42	0.6115	0.7912	0.7729	1.6353	1.2639	1.2938	52	18	52.3
37.8	37	48	0.6129	0.7902	0.7757	1.6316	1.2656	1.2892	52	12	52.2
37.9	37	54	0.6143	0.7891	0.7785	1.6279	1.2673	1.2846	52	6	52.1
38.0	38	0	0.6157	0.7880	0.7813	1.6243	1.2690	1.2799	52	0	52.0
38.1	38	6	0.6170	0.7869	0.7841	1.6207	1.2708	1.2753	51	54	51.9
38.2	38	12	0.6184	0.7859	0.7869	1.6171	1.2725	1.2708	51	48	51.8
38.3	38	18	0.6198	0.7848	0.7898	1.6135	1.2742	1.2662	51	42	51.7
38.4	38	24	0.6211	0.7837	0.7926	1.6099	1.2760	1.2617	51	36	51.6
38.5	38	30	0.6225	0.7826	0.7954	1.6064	1.2778	1.2572	51	30	51.5
38.6	38	36	0.6239	0.7815	0.7983	1.6029	1.2796	1.2527	51	24	51.4
38.7	38	42	0.6252	0.7804	0.8012	1.5994	1.2813	1.2482	51	18	51.3
38.8	38	48	0.6266	0.7793	0.8040	1.5959	1.2831	1.2437	51	12	51.2
38.9	38	54	0.6280	0.7782	0.8069	1.5925	1.2849	1.2393	51	6	51.1
39.0	39	0	0.6293	0.7771	0.8098	1.5890	1.2868	1.2349	51	0	51.0
39.1	39	6	0.6307	0.7760	0.8127	1.5856	1.2886	1.2305	50	54	50.9
39.2	39	12	0.6320	0.7749	0.8156	1.5822	1.2904	1.2261	50	48	50.8
39.3	39	18	0.6334	0.7738	0.8185	1.5788	1.2923	1.2218	50	42	50.7
39.4	39	24	0.6347	0.7727	0.8214	1.5755	1.2941	1.2174	50	36	50.6
39.5	39	30	0.6361	0.7716	0.8243	1.5721	1.2960	1.2131	50	30	50.5
39.6	39	36	0.6374	0.7705	0.8273	1.5688	1.2978	1.2088	50	24	50.4
39.7	39	42	0.6388	0.7694	0.8302	1.5655	1.2997	1.2045	50	18	50.3
39.8	39	48	0.6401	0.7683	0.8332	1.5622	1.3016	1.2002	50	12	50.2
39.9	39	54	0.6414	0.7672	0.8361	1.5590	1.3035	1.1960	50	6	50.1
			cos θ	sin θ	cot θ	sec θ	csc θ	tan θ	deg min		θ deg

TABLE II (continued)

θ deg	deg	min	sin θ	cos θ	tan θ	csc θ	sec θ	cot θ			
40.0	40	0	0.6428	0.7660	0.8391	1.5557	1.3054	1.1918	50	0	50.0
40.1	40	6	0.6441	0.7649	0.8421	1.5525	1.3073	1.1875	49	54	49.9
40.2	40	12	0.6455	0.7638	0.8451	1.5493	1.3092	1.1833	49	48	49.8
40.3	40	18	0.6468	0.7627	0.8481	1.5461	1.3112	1.1792	49	42	49.7
40.4	40	24	0.6481	0.7615	0.8511	1.5429	1.3131	1.1750	49	36	49.6
40.5	40	30	0.6494	0.7604	0.8541	1.5398	1.3151	1.1708	49	30	49.5
40.6	40	36	0.6508	0.7593	0.8571	1.5366	1.3171	1.1667	49	24	49.4
40.7	40	42	0.6521	0.7581	0.8601	1.5335	1.3190	1.1626	49	18	49.3
40.8	40	48	0.6534	0.7570	0.8632	1.5304	1.3210	1.1585	49	12	49.2
40.9	40	54	0.6547	0.7559	0.8662	1.5273	1.3230	1.1544	49	6	49.1
41.0	41	0	0.6561	0.7547	0.8693	1.5243	1.3250	1.1504	49	0	49.0
41.1	41	6	0.6574	0.7536	0.8724	1.5212	1.3270	1.1463	48	54	48.9
41.2	41	12	0.6587	0.7524	0.8754	1.5182	1.3291	1.1423	48	48	48.8
41.3	41	18	0.6600	0.7513	0.8785	1.5151	1.3311	1.1383	48	42	48.7
41.4	41	24	0.6613	0.7501	0.8816	1.5121	1.3331	1.1343	48	36	48.6
41.5	41	30	0.6626	0.7490	0.8847	1.5092	1.3352	1.1303	48	30	48.5
41.6	41	36	0.6639	0.7478	0.8878	1.5062	1.3373	1.1263	48	24	48.4
41.7	41	42	0.6652	0.7466	0.8910	1.5032	1.3393	1.1224	48	18	48.3
41.8	41	48	0.6665	0.7455	0.8941	1.5003	1.3414	1.1184	48	12	48.2
41.9	41	54	0.6678	0.7443	0.8972	1.4974	1.3435	1.1145	48	6	48.1
42.0	42	0	0.6691	0.7431	0.9004	1.4945	1.3456	1.1106	48	0	48.0
42.1	42	6	0.6704	0.7420	0.9036	1.4916	1.3478	1.1067	47	54	47.9
42.2	42	12	0.6717	0.7408	0.9067	1.4887	1.3499	1.1028	47	48	47.8
42.3	42	18	0.6730	0.7396	0.9099	1.4859	1.3520	1.0990	47	42	47.7
42.4	42	24	0.6743	0.7385	0.9131	1.4830	1.3542	1.0951	47	36	47.6
42.5	42	30	0.6756	0.7373	0.9163	1.4802	1.3563	1.0913	47	30	47.5
42.6	42	36	0.6769	0.7361	0.9195	1.4774	1.3585	1.0875	47	24	47.4
42.7	42	42	0.6782	0.7349	0.9228	1.4746	1.3607	1.0837	47	18	47.3
42.8	42	48	0.6794	0.7337	0.9260	1.4718	1.3629	1.0799	47	12	47.2
42.9	42	54	0.6807	0.7325	0.9293	1.4690	1.3651	1.0761	47	6	47.1
43.0	43	0	0.6820	0.7314	0.9325	1.4663	1.3673	1.0724	47	0	47.0
43.1	43	6	0.6833	0.7302	0.9358	1.4635	1.3696	1.0686	46	54	46.9
43.2	43	12	0.6845	0.7290	0.9391	1.4608	1.3718	1.0649	46	48	46.8
43.3	43	18	0.6858	0.7278	0.9424	1.4581	1.3741	1.0612	46	42	46.7
43.4	43	24	0.6871	0.7266	0.9457	1.4554	1.3763	1.0575	46	36	46.6
43.5	43	30	0.6884	0.7254	0.9490	1.4527	1.3786	1.0538	46	30	46.5
43.6	43	36	0.6896	0.7242	0.9523	1.4501	1.3809	1.0501	46	24	46.4
43.7	43	42	0.6909	0.7230	0.9556	1.4474	1.3832	1.0464	46	18	46.3
43.8	43	48	0.6921	0.7218	0.9590	1.4448	1.3855	1.0428	46	12	46.2
43.9	43	54	0.6934	0.7206	0.9623	1.4422	1.3878	1.0392	46	6	46.1
44.0	44	0	0.6947	0.7193	0.9657	1.4396	1.3902	1.0355	46	0	46.0
44.1	44	6	0.6959	0.7181	0.9691	1.4370	1.3925	1.0319	45	54	45.9
44.2	44	12	0.6972	0.7169	0.9725	1.4344	1.3949	1.0283	45	48	45.8
44.3	44	18	0.6984	0.7157	0.9759	1.4318	1.3972	1.0247	45	42	45.7
44.4	44	24	0.6997	0.7145	0.9793	1.4293	1.3996	1.0212	45	36	45.6
44.5	44	30	0.7009	0.7133	0.9827	1.4267	1.4020	1.0176	45	30	45.5
44.6	44	36	0.7022	0.7120	0.9861	1.4242	1.4044	1.0141	45	24	45.4
44.7	44	42	0.7034	0.7108	0.9896	1.4217	1.4069	1.0105	45	18	45.3
44.8	44	48	0.7046	0.7096	0.9930	1.4192	1.4093	1.0070	45	12	45.2
44.9	44	54	0.7059	0.7083	0.9965	1.4167	1.4118	1.0035	45	6	45.1
45.0	45	0	0.7071	0.7071	1.0000	1.4142	1.4142	1.0000	45	0	45.0
			cos θ	sin θ	cot θ	sec θ	csc θ	tan θ	deg	min	θ deg

II

TABLE III Values of trigonometric functions (radians)

θ radians or Real Number x	θ degrees	$\sin \theta$ or $\sin x$	$\cos \theta$ or $\cos x$	$\tan \theta$ or $\tan x$	$\csc \theta$ or $\csc x$	$\sec \theta$ or $\sec x$	$\cot \theta$ or $\cot x$
0.00	0° 00′	0.0000	1.000	0.0000	No value	1.000	No value
.01	0° 34′	.0100	1.000	.0100	100.0	1.000	100.0
.02	1° 09′	.0200	0.9998	.0200	50.00	1.000	49.99
.03	1° 43′	.0300	0.9996	.0300	33.34	1.000	33.32
.04	2° 18′	.0400	0.9992	.0400	25.01	1.001	24.99
0.05	2° 52′	0.0500	0.9988	0.0500	20.01	1.001	19.98
.06	3° 26′	.0600	.9982	.0601	16.68	1.002	16.65
.07	4° 01′	.0699	.9976	.0701	14.30	1.002	14.26
.08	4° 35′	.0799	.9968	.0802	12.51	1.003	12.47
.09	5° 09′	.0899	.9960	.0902	11.13	1.004	11.08
0.10	5° 44′	0.0998	0.9950	0.1003	10.02	1.005	9.967
.11	6° 18′	.1098	.9940	.1104	9.109	1.006	9.054
.12	6° 53′	.1197	.9928	.1206	8.353	1.007	8.293
.13	7° 27′	.1296	.9916	.1307	7.714	1.009	7.649
.14	8° 01′	.1395	.9902	.1409	7.166	1.010	7.096
0.15	8° 36′	0.1494	0.9888	0.1511	6.692	1.011	6.617
.16	9° 10′	.1593	.9872	.1614	6.277	1.013	6.197
.17	9° 44′	.1692	.9856	.1717	5.911	1.015	5.826
.18	10° 19′	.1790	.9838	.1820	5.586	1.016	5.495
.19	10° 53′	.1889	.9820	.1923	5.295	1.018	5.200
0.20	11° 28′	0.1987	0.9801	0.2027	5.033	1.020	4.933
.21	12° 02′	.2085	.9780	.2131	4.797	1.022	4.692
.22	12° 36′	.2182	.9759	.2236	4.582	1.025	4.472
.23	13° 11′	.2280	.9737	.2341	4.386	1.027	4.271
.24	13° 45′	.2377	.9713	.2447	4.207	1.030	4.086
0.25	14° 19′	0.2474	0.9689	0.2553	4.042	1.032	3.916
.26	14° 54′	.2571	.9664	.2660	3.890	1.035	3.759
.27	15° 28′	.2667	.9638	.2768	3.749	1.038	3.613
.28	16° 03′	.2764	.9611	.2876	3.619	1.041	3.478
.29	16° 37′	.2860	.9582	.2984	3.497	1.044	3.351
0.30	17° 11′	0.2955	0.9553	0.3093	3.384	1.047	3.233
.31	17° 46′	.3051	.9523	.3203	3.278	1.050	3.122
.32	18° 20′	.3146	.9492	.3314	3.179	1.053	3.018
.33	18° 54′	.3240	.9460	.3425	3.086	1.057	2.920
.34	19° 29′	.3335	.9428	.3537	2.999	1.061	1.827
0.35	20° 03′	0.3429	0.9394	0.3650	2.916	1.065	2.740
.36	20° 38′	.3523	.9359	.3764	2.839	1.068	2.657
.37	21° 12′	.3616	.9323	.3879	2.765	1.073	2.578
.38	21° 46′	.3709	.9287	.3994	2.696	1.077	2.504
.39	22° 21′	.3802	.9249	.4111	2.630	1.081	2.433
0.40	22° 55′	0.3894	0.9211	0.4228	2.568	1.086	2.365
.41	23° 29′	.3986	.9171	.4346	2.509	1.090	2.301
.42	24° 04′	.4078	.9131	.4466	2.452	1.095	2.239
.43	24° 38′	.4169	.9090	.4586	2.399	1.100	2.180
.44	25° 13′	.4259	.9048	.4708	2.348	1.105	2.124
0.45	25° 47′	0.4350	0.9004	0.4831	2.299	1.111	2.070

TABLE III (*continued*)

θ radians or Real Number x	θ degrees	sin θ or sin x	cos θ or cos x	tan θ or tan x	csc θ or csc x	sec θ or sec x	cot θ or cot x
0.45	25° 47′	0.4350	0.9004	0.4831	2.299	1.111	2.070
.46	26° 21′	.4439	.8961	.4954	2·253	1.116	2.018
.47	26° 56′	.4529	.8916	.5080	2.208	1.122	1.969
.48	27° 30′	.4618	.8870	.5206	2.166	1.127	1.921
.49	28° 04′	.4706	.8823	.5334	2.125	1.133	1.875
0.50	28° 39′	0.4794	0.8776	0.5463	2.086	1.139	1.830
.51	29° 13′	.4882	.8727	.5594	2.048	1.146	1.788
.52	29° 48′	.4969	.8678	.5726	2.013	1.152	1.747
.53	30° 22′	.5055	.8628	.5859	1.978	1.159	1.707
.54	30° 56′	.5141	.8577	.5994	1.945	1.166	1.668
0.55	31° 31′	0.5227	0.8525	0.6131	1.913	1.173	1.631
.56	32° 05′	.5312	.8473	.6269	1.883	1.180	1.595
.57	32° 40′	.5396	.8419	.6410	1.853	1.188	1.560
.58	33° 14′	.5480	.8365	.6552	1.825	1.196	1.526
.59	33° 48′	.5564	.8309	.6696	1.797	1.203	1.494
0.60	34° 23′	0.5646	0.8253	0.6841	1.771	1.212	1.462
.61	34° 57′	.5729	.8196	.6989	1.746	1.220	1.431
.62	35° 31′	.5810	.8139	.7139	1.721	1.229	1.401
.63	36° 06′	.5891	.8080	.7291	1.697	1.238	1.372
.64	36° 40′	.5972	.8021	.7445	1.674	1.247	1.343
0.65	37° 15′	0.6052	0.7961	0.7602	1.652	1.256	1.315
.66	37° 49′	.6131	.7900	.7761	1.631	1.266	1.288
.67	38° 23′	.6210	.7838	.7923	1.610	1.276	1.262
.68	38° 58′	.6288	.7776	.8087	1.590	1.286	1.237
.69	39° 32′	.6365	.7712	.8253	1.571	1.297	1.212
0.70	40° 06′	0.6442	0.7648	0.8423	1.552	1.307	1.187
.71	40° 41′	.6518	.7584	.8595	1.534	1.319	1.163
.72	41° 15′	·6594	.7518	.8771	1.517	1.330	1.140
.73	41° 50′	.6669	.7452	.8949	1.500	1.342	1.117
.74	42° 24′	.6743	.7385	.9131	1.483	1.354	1.095
0.75	42° 58′	0.6816	0.7317	0.9316	1.467	1.367	1.073
.76	43° 33′	.6889	.7248	.9505	1.452	1.380	1.052
.77	44° 07′	.6961	.7179	.9697	1.436	1.393	1.031
.78	44° 41′	.7033	.7109	.9893	1.422	1.407	1.011
.79	45° 16′	.7104	.7038	1.009	1.408	1.421	0.9908
0.80	45° 50′	0.7174	0.6967	1.030	1.394	1.435	0.9712
.81	46° 25′	.7243	.6895	1.050	1.381	1.450	.9520
.82	46° 59′	.7311	.6822	1.072	1.368	1.466	.9331
.83	47° 33′	.7379	.6749	1.093	1.355	1.482	.9146
.84	48° 08′	.7446	.6675	1.116	1.343	1.498	.8964
0.85	48° 42′	0.7513	0.6600	1.138	1.331	1.515	0.8785
.86	49° 16′	.7578	.6524	1.162	1.320	1.533	.8609
.87	49° 51′	.7643	.6448	1.185	1.308	1.551	.8437
.88	50° 25′	.7707	.6372	1.210	1.297	1.569	.8267
.89	51° 00′	.7771	.6294	1.235	1.287	1.589	.8100
0.90	51° 34′	0.7833	0.6216	1.260	1.277	1.609	0.7936
.91	52° 08′	.7895	.6137	1.286	1.267	1.629	.7774
.92	52° 43′	.7956	.6058	1.313	1.257	1.651	.7615
.93	53° 17′	.8016	.5978	1.341	1.247	1.673	.7458
.94	53° 51′	.8076	.5898	1.369	1.238	1.696	.7303
0.95	54° 26′	0.8134	0.5817	1.398	1.229	1.719	0.7151

TABLE III (continued)

θ radians or Real Number x	θ degrees	sin θ or sin x	cos θ or cos x	tan θ or tan x	csc θ or csc x	sec θ or sec x	cot θ or cot x
0.95	54° 26′	0.8134	0.5817	1.398	1.229	1.719	0.7151
.96	55° 00′	.8192	.5735	1.428	1.221	1.744	.7001
.97	55° 35′	.8249	.5653	1.459	1.212	1.769	.6853
.98	56° 09′	.8305	.5570	1.491	1.204	1.795	.6707
.99	56° 43′	.8360	·5487	1.524	1.196	1.823	.6563
1.00	57° 18′	0.8415	0.5403	1.557	1.188	1.851	0.6421
1.01	57° 52′	.8468	.5319	1.592	1.181	1.880	.6281
1.02	58° 27′	.8521	.5234	1.628	1.174	1.911	.6142
1.03	59° 01′	.8573	.5148	1.665	1.166	1.942	.6005
1.04	59° 35′	.8624	.5062	1.704	1.160	1.975	.5870
1.05	60° 10′	0.8674	0.4976	1.743	1.153	2.010	0.5736
1.06	60° 44′	.8724	.4889	1.784	1.146	2.046	.5604
1.07	61° 18′	.8772	.4801	1.827	1.140	2.083	.5473
1.08	61° 53′	.8820	.4713	1.871	1.134	2.122	.5344
1.09	62° 27′	.8866	.4625	1.917	1.128	2.162	.5216
1.10	63° 02′	0.8912	0.4536	1.965	1.122	2.205	0.5090
1.11	63° 36′	.8957	.4447	2.014	1.116	2.249	.4964
1.12	64° 10′	.9001	.4357	2.066	1.111	2.295	.4840
1.13	64° 45′	.9044	.4267	2.120	1.106	2.344	.4718
1.14	65° 19′	.9086	.4176	2.176	1.101	2.395	.4596
1.15	65° 53′	0.9128	0.4085	2.234	1.096	2.448	0.4475
1.16	66° 28′	.9168	.3993	2.296	1.091	2.504	.4356
1.17	67° 02′	.9208	.3902	2.360	1.086	2.563	.4237
1.18	67° 37′	.9246	.3809	2.427	1.082	2.625	.4120
1.19	68° 11′	.9284	.3717	2.498	1.077	2.691	.4003
1.20	68° 45′	0.9320	0.3624	2.572	1.073	2.760	0.3888
1.21	69° 20′	.9356	.3530	2.650	1.069	2.833	.3773
1.22	69° 54′	.9391	.3436	2.733	1.065	2.910	.3659
1.23	70° 28′	.9425	.3342	2.820	1.061	2.992	.3546
1.24	71° 03′	.9458	.3248	2.912	1.057	3.079	.3434
1.25	71° 37′	0.9490	0.3153	3.010	1.054	3.171	0.3323
1.26	72° 12′	.9521	.3058	3.113	1.050	3.270	.3212
1.27	72° 46′	.9551	.2963	3.224	1.047	3.375	.3102
1.28	73° 20′	.9580	.2867	3.341	1.044	3.488	.2993
1.29	73° 55′	.9608	.2771	3.467	1.041	3.609	.2884
1.30	74° 29′	0.9636	0.2675	3.602	1.038	3.738	0.2776
1.31	75° 03′	.9662	.2579	3.747	1.035	3.878	.2669
1.32	75° 38′	.9687	.2482	3.903	1.032	4.029	.2562
1.33	76° 12′	.9711	.2385	4.072	1.030	4.193	.2456
1.34	76° 47′	.9735	.2288	4.256	1.027	4.372	.2350
1.35	77° 21′	0.9757	0.2190	4.455	1.025	4.566	0.2245
1.36	77° 55′	.9779	.2092	4.673	1.023	4.779	.2140
1.37	78° 30′	.9799	.1994	4.913	1.021	5.014	.2035
1.38	79° 04′	.9819	.1896	5.177	1.018	5.273	.1931
1.39	79° 38′	.9837	.1798	5.471	1.017	5.561	.1828
1.40	80° 13′	0.9854	0.1700	5.798	1.015	5.883	0.1725
1.41	80° 47′	.9871	.1601	6.165	1.013	6.246	.1622
1.42	81° 22′	.9887	.1502	6.581	1.011	6.657	.1519
1.43	81° 56′	.9901	.1403	7.055	1.010	7.126	.1417
1.44	82° 30′	.9915	.1304	7.602	1.009	7.667	.1315
1.45	83° 05′	0.9927	0.1205	8.238	1.007	8.299	0.1214

TABLE III *(continued)*

θ radians or Real Number x	θ degrees	sin θ or sin x	cos θ or cos x	tan θ or tan x	csc θ or csc x	sec θ or sec x	cot θ or cot x
1.45	83° 05′	0.9927	0.1205	8.238	1.007	8.299	0.1214
1.46	83° 39′	.9939	.1106	8.989	1.006	9.044	.1113
1.47	84° 13′	.9949	.1006	9.887	1.005	9.938	.1001
1.48	84° 48′	.9959	.0907	10.98	1.004	11.03	.0910
1.49	85° 22′	.9967	.0807	12.35	1.003	12.39	.0810
1.50	85° 57′	0.9975	0.0707	14.10	1.003	14.14	0.0709
1.51	86° 31′	.9982	.0608	16.43	1.002	16.46	.0609
1.52	87° 05′	.9987	.0508	19.67	1.001	19.69	.0508
1.53	87° 40′	.9992	.0408	24.50	1.001	24.52	.0408
1.54	88° 14′	.9995	.0308	32.46	1.000	32.48	.0308
1.55	88° 49′	0.9998	0.0208	48.08	1.000	48.09	0.0208
1.56	89° 23′	.9999	.0108	92.62	1.000	92.63	.0108
1.57	89° 57′	1.000	.0008	1256	1.000	1256	.0008

ANSWERS TO ODD-NUMBERED EXERCISES

Exercise Set 1.1

1.

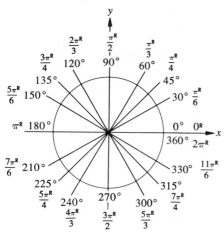

3. 60.0°; Quad. I

5. 840.0°; Quad. II

7. −288.0°; Quad. I

9. −405.0°; Quad. IV

11. 10.3°; Quad. I

13. 295.7°; Quad. IV

15. −15.5°; Quad. IV

17. −231.5°; Quad. II

19. 0.52^R; Quad. I

21. 1.31^R; Quad. I

23. 2.50^R; Quad. II

25. 7.01^R; Quad. I

27. −0.65^R; Quad. IV

29. −1.40^R; Quad. IV

31. −4.19^R; Quad. II

33. −8.90^R; Quad. III

35. 0.44^R

37. 19°

39. 1.45^R

41. $288°$

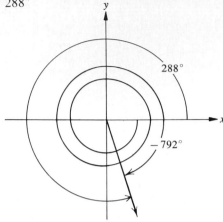

43. $(21 + k \cdot 360)°, \quad k \in J$

47. $(\pi/6 + k \cdot 2\pi)^R, \quad k \in J$

45. $(242 + k \cdot 360)°, \quad k \in J$

49. $(5\pi/3 + k \cdot 2\pi)^R, \quad k \in J$

Exercise Set 1.2

1.

$$\sin \alpha = 3/5, \qquad \csc \alpha = 5/3$$
$$\cos \alpha = 4/5, \qquad \sec \alpha = 5/4$$
$$\tan \alpha = 3/4, \qquad \cot \alpha = 4/3$$

3.

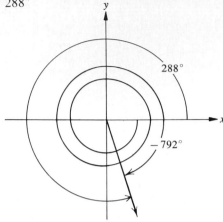

$$\sin \alpha = -15/17, \qquad \csc \alpha = -17/15$$
$$\cos \alpha = 8/17, \qquad \sec \alpha = 17/8$$
$$\tan \alpha = -15/8, \qquad \cot \alpha = -8/15$$

5.

$$\sin \alpha = 3/\sqrt{10}, \qquad \csc \alpha = \sqrt{10}/3$$
$$\cos \alpha = -1/\sqrt{10}, \qquad \sec \alpha = -\sqrt{10}$$
$$\tan \alpha = -3, \qquad \cot \alpha = -1/3$$

7.

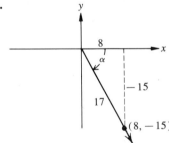

$$\sin \alpha = 1/\sqrt{2}, \qquad \csc \alpha = \sqrt{2}$$
$$\cos \alpha = -1/\sqrt{2}, \qquad \sec \alpha = -\sqrt{2}$$
$$\tan \alpha = -1, \qquad \cot \alpha = -1$$

9.

$$\sin \alpha = -\sqrt{3}/2, \qquad \csc \alpha = -2/\sqrt{3}$$
$$\cos \alpha = -1/2, \qquad \sec \alpha = -2$$
$$\tan \alpha = \sqrt{3}, \qquad \cot \alpha = 1/\sqrt{3}$$

11.

$$\sin \alpha = -1, \qquad \csc \alpha = -1$$
$$\cos \alpha = 0, \qquad \sec \alpha \text{ is not defined}$$
$$\tan \alpha \text{ is not defined}, \quad \cot \alpha = 0$$

13. $\sin \alpha = 4/5, \qquad \csc \alpha = 5/4$
$\cos \alpha = 3/5, \qquad \sec \alpha = 5/3$
$\tan \alpha = 4/3, \qquad \cot \alpha = 3/4$

15. $\sin \alpha = 2/5, \qquad \csc \alpha = 5/2$
$\cos \alpha = \sqrt{21}/5, \quad \sec \alpha = 5/\sqrt{21}$
$\tan \alpha = 2/\sqrt{21}, \qquad \cot \alpha = \sqrt{21}/2$

17. $\sin \alpha = 2\sqrt{10}/7, \qquad \csc \alpha = 7/2\sqrt{10}$
$\cos \alpha = 3/7, \qquad \sec \alpha = 7/3$
$\tan \alpha = 2\sqrt{10}/3, \qquad \cot \alpha = 3/2\sqrt{10}$

19. $\sin \alpha = \sqrt{3}/2, \qquad \csc \alpha = 2/\sqrt{3}$
$\cos \alpha = 1/2, \qquad \sec \alpha = 2$
$\tan \alpha = \sqrt{3}, \qquad \cot \alpha = 1/\sqrt{3}$

21. $\sin \alpha = 1/4, \qquad \csc \alpha = 4$
$\cos \alpha = \sqrt{15}/4, \qquad \sec \alpha = 4/\sqrt{15}$
$\tan \alpha = 1/\sqrt{15}, \qquad \cot \alpha = \sqrt{15}$

23. $\sin \alpha = 1/\sqrt{2}, \qquad \csc \alpha = \sqrt{2}$
$\cos \alpha = 1/\sqrt{2}, \qquad \sec \alpha = \sqrt{2}$
$\tan \alpha = 1, \qquad \cot \alpha = 1$

25. Domain: $\{\alpha \mid \alpha \in A\}$

27. Domain: $\{\alpha \mid \alpha \in A, \quad \alpha \neq (90 + k \cdot 180)°, \quad k \in J\}$

29. Domain: $\{\alpha \mid \alpha \in A, \quad \alpha \neq (90 + k \cdot 180)°, \quad k \in J\}$

31. In the figure, corresponding sides in similar triangles OAB and $OA'B'$ are proportional. Thus,

$$\frac{x_1}{y_1} = \frac{x_2}{y_2}, \quad \frac{x_1}{r_1} = \frac{x_2}{r_2}, \quad \text{and} \quad \frac{y_1}{r_1} = \frac{y_2}{r_2}.$$

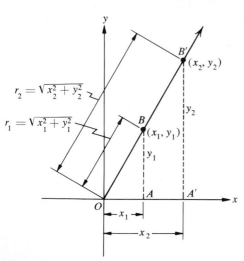

33. Dividing both members of $\cos^2 \alpha + \sin^2 \alpha = 1$ by $\cos^2 \alpha$ gives

$$\frac{\cos^2 \alpha}{\cos^2 \alpha} + \frac{\sin^2 \alpha}{\cos^2 \alpha} = \frac{1}{\cos^2 \alpha}.$$

Substituting $\tan^2 \alpha$ for $\sin^2 \alpha / \cos^2 \alpha$ and $\sec^2 \alpha$ for $1/\cos^2 \alpha$ gives

$$1 + \tan^2 \alpha = \sec^2 \alpha.$$

Exercise Set 1.3

1. $\sqrt{3}/2$ **3.** $1/\sqrt{2}$ **5.** $\sqrt{3}/2$ **7.** $1/\sqrt{3}$ **9.** $\sqrt{2}$
11. $2/\sqrt{3}$ **13.** 1   **15.** 0   **17.** Undefined   **19.** -1
21. 0   **23.** Undefined   **25.** 3/4   **27.** 1/2   **29.** -3
31. 1/12   **33.** 3/2   **35.** 0   **37.** $3\sqrt{3}/2\sqrt{2}$ **39.** 2
41. Undefined   **43.** 1   **45.** $-1/2$ **47.** Undefined   **49.** 0.2823
51. 2.6325   **53.** 0.9466   **55.** 0.1086   **57.** 3.9232   **59.** 4.6569
61. 0.7895   **63.** 0.4466   **65.** 1.238   **67.** 3.1218

69. Undefined for $\alpha = (90 + k \cdot 180)°$, $k \in J$;
equal to zero for $\alpha = (0 + k \cdot 180)°$, $k \in J$

71. Undefined for $\alpha = (90 + k \cdot 180)°$, $k \in J$; no secant function values are equal to zero

73. $\sin \alpha = y/1$, $\csc \alpha = 1/y$ **75.** 0.0100   **77.** 0.8415   **79.** 0.5646
     $\cos \alpha = \sqrt{1 - y^2}$, $\sec \alpha = 1/\sqrt{1 - y^2}$
     $\tan \alpha = y/\sqrt{1 - y^2}$, $\cot \alpha = \sqrt{1 - y^2}/y$

81. Consider a point $P(x, y)$ on the terminal side of an angle α. Construct a perpendicular from P to the x-axis to form a right triangle. Since the sum of the acute angles in a right triangle equals $90°$, $\alpha + \beta = 90°$. Hence,

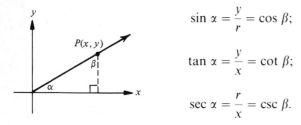

$$\sin \alpha = \frac{y}{r} = \cos \beta;$$

$$\tan \alpha = \frac{y}{x} = \cot \beta;$$

$$\sec \alpha = \frac{r}{x} = \csc \beta.$$

Therefore, sine and cosine are cofunctions; tangent and cotangent are cofunctions; secant and cosecant are cofunctions.

Exercise Set 1.4

1. Quad. IV   **3.** Quad. I   **5.** Quad. III   **7.** Quad. II

9. $\sin \alpha = 5/13$, $\csc \alpha = 13/5$ **11.** $\sin \alpha = -2/3$, $\csc \alpha = -3/2$
     $\cos \alpha = -12/13$, $\sec \alpha = -13/12$ $\cos \alpha = -\sqrt{5}/3$, $\sec \alpha = -3/\sqrt{5}$
     $\tan \alpha = -5/12$, $\cot \alpha = -12/5$ $\tan \alpha = 2/\sqrt{5}$, $\cot \alpha = \sqrt{5}/2$

13. $\sin \alpha = -3/5,$ $\csc \alpha = -5/3$
   $\cos \alpha = 4/5,$ $\sec \alpha = 5/4$
   $\tan \alpha = -3/4,$ $\cot \alpha = -4/3$

15.

17.

19.

21.

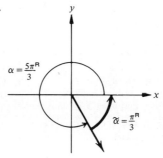

23. $-1/2$ **25.** 1 **27.** $-1/2$ **29.** 1
31. $-1/\sqrt{2}$ **33.** $-\sqrt{3}$ **35.** $-2/\sqrt{3}$

Where required, the answer was obtained by first finding a reference angle.

37. $\sqrt{3}$ **39.** $1/\sqrt{2}$ **41.** $\sqrt{3}$ **43.** 0
45. $-1/\sqrt{2}$ **47.** $\sqrt{3}$ **49.** 0.6428 **51.** 8.6428
53. -3.1716 **55.** -1.6390 **57.** -1.704 **59.** -0.0508
61. 6.246 **63.** -0.9780

Exercise Set 1.5

1. $\alpha = \text{Tan}^{-1} 1/\sqrt{3}$ **3.** $\alpha = \text{Sec}^{-1} \sqrt{2}$ **5.** $\alpha = \text{Cos}^{-1} \sqrt{3}/2$
7. $\alpha = \text{Sin}^{-1} 0.2169$ **9.** $\alpha = \text{Cot}^{-1} 2.0941$ **11.** $\alpha = \text{Sec}^{-1} 2.2013$
13. $\cot \alpha = \sqrt{3}$ **15.** $\sec \alpha = 1.4214$ **17.** $\sin \alpha = 0.4112$
19. $60°$ **21.** $30°$ **23.** $60°$
25. $21.2°$ **27.** $26.0°$ **29.** $55.5°$
31. $35.2°$ **33.** 0.27^R **35.** 0.49^R
37. 1.31^R **39.** 0.83^R **41.** 20.7^R
43. $63.7°$

Review Exercises, Chapter 1

1. $72.0°$ **2.** $138.1°$ **3.** 1.31^R **4.** 4.48^R **5.** $218°$ **6.** 0.95^R

7. $(131 + k \cdot 360)°, \quad k \in J$ **8.** $(7\pi/4 + k \cdot 2\pi)^R, \quad k \in J$

9. $\sin \alpha = 3/5, \qquad \csc \alpha = 5/3$
  $\cos \alpha = -4/5, \qquad \sec \alpha = -5/4$
  $\tan \alpha = -3/4, \qquad \cot \alpha = -4/3$

10. $\sin \alpha = 2/\sqrt{5}, \qquad \csc \alpha = \sqrt{5}/2$
  $\cos \alpha = -1/\sqrt{5}, \qquad \sec \alpha = -\sqrt{5}$
  $\tan \alpha = -2, \qquad \cot \alpha = -1/2$

11. $\sin \alpha = -1/2, \qquad \csc \alpha = -2$
  $\cos \alpha = \sqrt{3}/2, \qquad \sec \alpha = 2/\sqrt{3}$
  $\tan \alpha = -1/\sqrt{3}, \qquad \cot \alpha = -\sqrt{3}$

12. $\sin \alpha = -15/17, \qquad \csc \alpha = -17/15$
  $\cos \alpha = -8/17, \qquad \sec \alpha = -17/8$
  $\tan \alpha = 15/8, \qquad \cot \alpha = 8/15$

13. $\sin \alpha = 4/5, \qquad \csc \alpha = 5/4$
  $\cos \alpha = 3/5, \qquad \sec \alpha = 5/3$
  $\tan \alpha = 4/3, \qquad \cot \alpha = 3/4$

14. $\sin \alpha = 4/\sqrt{65}, \qquad \csc \alpha = \sqrt{65}/4$
  $\cos \alpha = 7/\sqrt{65}, \qquad \sec \alpha = \sqrt{65}/7$
  $\tan \alpha = 4/7, \qquad \cot \alpha = 7/4$

15. $\sin \alpha = 1/3, \qquad \csc \alpha = 3$
  $\cos \alpha = 2\sqrt{2}/3, \qquad \sec \alpha = 3/2\sqrt{2}$
  $\tan \alpha = 1/2\sqrt{2}, \qquad \cot \alpha = 2\sqrt{2}$

16. $\sin \alpha = \sqrt{21}/5, \qquad \csc \alpha = 5/\sqrt{21}$
  $\cos \alpha = 2/5, \qquad \sec \alpha = 5/2$
  $\tan \alpha = \sqrt{21}/2, \qquad \cot \alpha = 2/\sqrt{21}$

17. $2/\sqrt{3}$ **18.** $\sqrt{2}$ **19.** Undefined   **20.** 0 **21.** $4/3$

22. -1 **23.** $\sqrt{2}/6$ **24.** $13/2$ **25.** 5.3955 **26.** 1.3972

27. 2.9986 **28.** 0.7452 **29.** Quad. III   **30.** Quad. III

31. $\sin \alpha = 2/5, \qquad \csc \alpha = 5/2$
  $\cos \alpha = -\sqrt{21}/5, \qquad \sec \alpha = -5/\sqrt{21}$
  $\tan \alpha = -2/\sqrt{21}, \qquad \cot \alpha = -\sqrt{21}/2$

32. $\sin \alpha = -3/5, \qquad \csc \alpha = -5/3$
  $\cos \alpha = -4/5, \qquad \sec \alpha = -5/4$
  $\tan \alpha = 3/4, \qquad \cot \alpha = 4/3$

33.

34.

35.

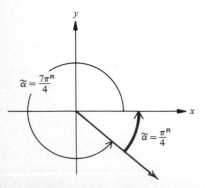

36.

37. $1/2$

38. $-1/\sqrt{3}$

39. $-1/\sqrt{2}$

40. $-\sqrt{2}$

41. 0.2924

42. -1.8650

43. 0.1291

44. -2.4023

45. $\alpha = \text{Tan}^{-1}\ 1.4321$

46. $\cos \alpha = 0.4232$

47. $55.9°$

48. $52.7°$

49. 0.69^R

50. 0.22^R

Exercise Set 2.1

1. $\alpha \approx 33.7°,\ \beta \approx 56.3°,\ c \approx 3.61$

3. $\alpha \approx 61.9°,\ \beta \approx 28.1°,\ b \approx 8.00$

5. $\alpha \approx 39.0°,\ a \approx 14.8,\ b \approx 18.3$

7. $\alpha \approx 67.8°,\ b \approx 4.49,\ c \approx 11.9$

9. $\beta \approx 39.7°,\ a \approx 18.1,\ c \approx 23.5$

11. $\alpha \approx 48.4°,\ a \approx 33.8,\ c \approx 45.2$

13. 60.0 square units

15. 24.695 square units

17. 360 meters

19. 53° 6′

21. 27.1 decimeters

23. 239 meters

25. 61.4 feet

27. 12,300 feet

29. 5.88 centimeters

31. $4\sqrt{3}$

33. $\cos \alpha = \dfrac{l-h}{l}$, from which $l \cos \alpha = l - h$ and $h = l(1 - \cos \alpha)$.

Exercise Set 2.2

1. $\gamma = 85.0°,\ b \approx 14.1,\ c \approx 14.6$

3. $\alpha = 42.8°,\ b \approx 52.0,\ c \approx 84.7$

5. $\beta = 52.9°,\ a \approx 0.307,\ c \approx 0.526$

7. $\gamma = 77.5°,\ a \approx 7.05,\ b \approx 13.3$

9. One triangle: $\beta \approx 20.9°,\ \gamma \approx 129.1°,\ c \approx 10.9$

11. Two triangles: $\gamma \approx 50.7°,\ \beta \approx 97.1°,\ b \approx 7.82$ or $\gamma' \approx 129.3°,\ \beta' \approx 18.5°,\ b' \approx 2.50$

13. One right triangle: $\alpha = 60.0°,\ \gamma = 90.0°,\ a \approx 27.7$

15. Two triangles: $\alpha \approx 60.4°,\ \gamma \approx 76.9°,\ c \approx 5.60$ or $\alpha' \approx 119.6°,\ \gamma' \approx 17.7°,\ c' \approx 1.75$

17. 1.51

19. 3,049 feet

21. 7.90 centimeters; 14.8 centimeters

23. 78.4 meters

25. 57.8 feet

27. 64.3 meters

29. From the figure, $\tan \gamma = \dfrac{c}{AC}$, or $AC = \dfrac{c}{\tan \gamma}$.

Using the Law of Sines, $\dfrac{AC}{\sin \beta} = \dfrac{a}{\sin[180° - (\alpha + \beta)]} = \dfrac{a}{\sin(\alpha + \beta)}$.

Hence, $AC = \dfrac{a \sin \beta}{\sin(\alpha + \beta)}$, from which $\dfrac{c}{\tan \gamma} = \dfrac{a \sin \beta}{\sin(\alpha + \beta)}$; $c = \dfrac{a \sin \beta \tan \gamma}{\sin(\alpha + \beta)}$

31. From the Law of Sines, $\dfrac{a}{b} = \dfrac{\sin \alpha}{\sin \beta}$. Adding 1 to both members gives $\dfrac{a}{b} + 1 = \dfrac{\sin \alpha}{\sin \beta} + 1$,

from which $\dfrac{a + b}{b} = \dfrac{\sin \alpha + \sin \beta}{\sin \beta}$.

33. From the results of Exercises 31 and 32 and dividing left members and right members respectively gives

$$\dfrac{\dfrac{a - b}{b}}{\dfrac{a + b}{b}} = \dfrac{\dfrac{\sin \alpha - \sin \beta}{\sin \beta}}{\dfrac{\sin \alpha + \sin \beta}{\sin \beta}},\quad \text{from which}\quad \dfrac{a - b}{a + b} = \dfrac{\sin \alpha - \sin \beta}{\sin \alpha + \sin \beta}.$$

Exercise Set 2.3

1. $a \approx 4.37$, $\beta \approx 59.9°$, $\gamma \approx 49.1°$ **3.** $c \approx 3.39$, $\alpha \approx 65.9°$, $\beta \approx 38.8°$
5. $b \approx 1.42$, $\alpha \approx 17.9°$, $\gamma \approx 20.6°$ **7.** $a \approx 3.83$, $\beta \approx 24.3°$, $\gamma \approx 55.3°$
9. $\alpha \approx 95.7°$, $\beta \approx 50.7°$, $\gamma \approx 33.6°$ **11.** $\alpha \approx 4.0°$, $\beta \approx 31.6°$, $\gamma \approx 144.4°$
13. $\alpha \approx 26.4°$, $\beta \approx 36.4$, $\gamma \approx 117.2°$ **15.** $\alpha \approx 41.4°$, $\beta \approx 55.8°$, $\gamma \approx 82.8°$
17. $42.7°$ **19.** $82.5°$ **21.** 10.9 **23.** 5.33 kilometers
25. 11.7 centimeters   **27.** $47.8°$ **29.** $1,240$ feet   **31.** 875 kilometers

33. Adding 1 to both members of $\cos \alpha = \dfrac{b^2 + c^2 - a^2}{2bc}$ gives

$$1 + \cos \alpha = 1 + \frac{b^2 + c^2 - a^2}{2bc}$$

$$= \frac{2bc + b^2 + c^2 - a^2}{2bc} = \frac{b^2 + 2bc + c^2 - a^2}{2bc}$$

$$= \frac{(b + c)^2 - a^2}{2bc} = \frac{(b + c + a)(b + c - a)}{2bc}$$

35. $a + b + c = 2s$ and $b + c - a = 2s - 2a = 2(s - a)$. Substituting $2s$ for $a + b + c$ and $2(s - a)$ for $b + c - a$ in the results of Exercise 33 gives

$$1 + \cos \alpha = \frac{(2s)(2)(s - a)}{2bc}, \quad \text{from which} \quad \frac{1 + \cos \alpha}{2} = \frac{s(s - a)}{bc}.$$

37. Multiplying left members and right members, respectively, of the result of Exercise 35 and 36 gives

$$\left(\frac{1 + \cos \alpha}{2}\right)\left(\frac{1 - \cos \alpha}{2}\right) = \left[\frac{s(s - a)}{bc}\right]\left[\frac{(s - b)(s - c)}{bc}\right],$$

$$\frac{1 - \cos^2 \alpha}{4} = \frac{s(s - a)(s - b)(s - c)}{b^2 c^2},$$

$$\frac{\sin^2 \alpha}{4} = \frac{s(s - a)(s - b)(s - c)}{b^2 c^2}.$$

Taking the square root of both members gives

$$\frac{1}{2} \sin \alpha = \frac{\sqrt{s(s - a)(s - b)(s - c)}}{bc},$$

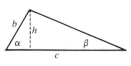

$$\frac{1}{2} bc \sin \alpha = \sqrt{s(s - a)(s - b)(s - c)}.$$

Because $h = b \sin \alpha$ (see the figure), the area $\mathscr{A}$ of a triangle is $\frac{1}{2}$(base $\times$ altitude), or $\frac{1}{2}c(b \sin \alpha)$. Thus,

$$\mathscr{A} = \sqrt{s(s - a)(s - b)(s - c)}.$$

39. 3.14

41. Noting in the figure that the radius r is the altitude of the three inner triangles, we have that the total area is $\frac{1}{2}(ra + rb + rc) = \frac{1}{2}(a + b + c)(r) = sr$. From the results of Exercise 37, we have

$$\mathscr{A} = \sqrt{s(s - a)(s - b)(s - c)} = sr,$$

from which $\qquad r = \sqrt{\dfrac{s(s - a)(s - b)(s - c)}{s^2}} = \sqrt{\dfrac{(s - a)(s - b)(s - c)}{s}}.$

Exercise Set 2.4

1. 6.2 meters

3. $5\pi^2/6 \approx 8.22$ decimeters

5. $-9\pi/4 \approx -7.07$ feet

7. 126.9°

9. 280.7°

11. 3.6 meters per second

13. 3,000 radians per minute

15. $-54\pi \approx -169.65$ decimeters per second

17. $-12/\pi \approx -3.82$ degrees per second

19. 104.72 meters

21. 1,745 miles

23. 3.98 centimeters

25. 5024 meters per minute

27. 9.39 radians per minute

29. 7.17×10^{-4} radians per hour

31. 2.24×10^3 miles per hour

33. Area of the sector is to the area of the circle as the central angle θ is to 2π radians. Thus,

$$\frac{\mathscr{A}(\text{of sector})}{\pi r^2} = \frac{\theta^R}{2\pi}, \quad \text{from which} \quad \mathscr{A} = \frac{\pi r^2 (\theta^R)}{2\pi} = \frac{1}{2}r^2\theta^R.$$

35. 1.92×10^{13} miles

37. 4690^R

Exercise Set 2.5

1. $\cos\left(\dfrac{3\pi}{4} + 2\pi\right); \dfrac{-1}{\sqrt{2}}$

3. $\sin\left(\dfrac{\pi}{4} + \pi\right); \dfrac{-1}{\sqrt{2}}$

5. $\sin\left(\dfrac{5\pi}{4} + 2\pi\right); \dfrac{-1}{\sqrt{2}}$

7. $\cos\left(\dfrac{3\pi}{4} + 3\pi\right); \dfrac{1}{\sqrt{2}}$

9. $\sin\left(\dfrac{7\pi}{6} + 2\pi\right); \dfrac{-1}{2}$

11. $\cos\left(\dfrac{\pi}{3} + 4\pi\right); \dfrac{1}{2}$

13. 0.9687

15. -0.1900

17. -0.0584

19. 10 centimeters; -10 centimeters

21. -8 centimeters; 0 centimeter

23. 22 centimeters

25. $40/\sqrt{3}$ centimeters

27. -106 volts

29. -66.7 volts

31. 85.2 volts

33. -0.0295 ampere

35. 0.002 ampere

37. 0.00993 ampere

39. -5.89 dynes per square centimeter

41. 0 dyne per square centimeter

43. 0.242 foot

Review Exercises, Chapter 2

1. $\alpha = 39.0°$, $a \approx 8.81$, $b \approx 10.9$

2. $\alpha \approx 53.7°$, $\beta \approx 36.3°$, $b \approx 3.97$

3. $\beta \approx 57.4°$, $b \approx 20.2$, $c \approx 24.0$

4. $\beta = 18.7°$, $a \approx 28.7$, $c \approx 30.3$

5. $\gamma = 77.5°$, $a \approx 7.03$, $b \approx 13.3$

6. $\beta = 10.0°$, $a \approx 1.73$, $b \approx 0.425$

7. One triangle: $c \approx 8.1$, $\alpha \approx 28.3°$, $\gamma \approx 130.0°$

8. One triangle: $\beta \approx 30.6°$, $\gamma \approx 13.9°$, $c \approx 17.5$

9. $\alpha \approx 51.3$, $\beta \approx 70.5°$, $c \approx 2.07$

10. $\beta \approx 14.1°$, $\gamma \approx 142.8°$, $a = 3.7$

11. $\alpha \approx 35.2°$, $\beta \approx 42.2°$, $\gamma \approx 102.6°$

12. $\alpha \approx 30.8°$, $\beta \approx 56.0°$, $\gamma \approx 93.2°$

13. 2.86 meters

14. 0.19^R

15. 12 decimeters per minute

16. 0.39 radians per second

17. $\cos\left(\dfrac{2\pi}{3} + 2\pi\right)$; $-\dfrac{1}{2}$

18. $\sin\left(\dfrac{\pi}{4} + 4\pi\right)$; $\dfrac{1}{\sqrt{2}}$

19. 0.9521

20. 0.7756

21. 4 centimeters

22. 0 centimeter

23. 0 volt

24. 0.0259

Exercise Set 3.1

1.

3.

5.

7. Zeros: $\{x \mid x = k\pi,\ k \in J\}$;
amplitude: 3

9. Zeros: $\{x \mid x = \pi/2 + k\pi,\ k \in J\}$;
amplitude: 1/2

11. Zeros: $\{x \mid x = k\pi,\ k \in J\}$;
amplitude: 1

13. Zeros: $\{x \mid x = k\pi/2,\ k \in J\}$;
amplitude: 1

15. Zeros: $\{x \mid x = \pi + k \cdot 2\pi, \ k \in J\}$;
amplitude: 1

17. Zeros: $\{x \mid x = \pi/2 + k\pi, \ k \in J\}$;
amplitude: 1

19. Zeros: $\{-3, -1, 1, 3\}$;
amplitude: 1

21. Zeros: $\{-2, -1, 0, 1, 2\}$;
amplitude: 1

3

23.

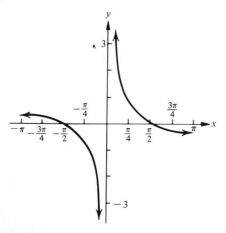

As x approaches zero from the left or right, $\dfrac{\sin x}{x}$ approaches a value of 1. The graph is not periodic because there is no number $a \ (a \neq 0)$ such that $\dfrac{\sin(x - a)}{x + a} = \dfrac{\sin x}{x}$.

25.

As x approaches zero from the left, $\dfrac{\cos x}{x}$ is negative and its absolute value becomes greater. As x approaches zero from right, $\dfrac{\cos x}{x}$ is positive and its absolute value becomes greater. The graph is not periodic because there is no number $a \ (a \neq 0)$ such that $\dfrac{\cos(x + a)}{x + a} = \dfrac{\cos x}{x}$.

27. Period is 1.

29. Period is 1/2.

Exercise Set 3.2

1. Zeros: $\{-2\pi, -\pi, 0, \pi, 2\pi\}$;
amplitude: 3

3. Zeros: $\left\{-\dfrac{3\pi}{2}, -\dfrac{\pi}{2}, \dfrac{\pi}{2}, \dfrac{3\pi}{2}\right\}$;
amplitude: 2

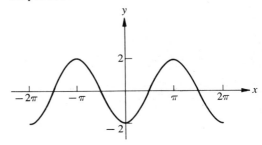

5. Zeros: $\{-2\pi, -\pi, 0, \pi, 2\pi\}$;
amplitude: $\dfrac{1}{2}$

7. Zeros: $\left\{-\dfrac{7\pi}{4}, -\dfrac{5\pi}{4}, -\dfrac{3\pi}{4}, -\dfrac{\pi}{4}, \dfrac{\pi}{4}, \dfrac{3\pi}{4}, \dfrac{5\pi}{4}, \dfrac{7\pi}{4}\right\}$;
amplitude: 1

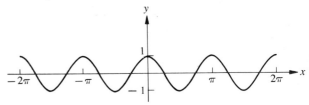

9. Zeros: $\{x \mid -2\pi, 0, 2\pi\}$;
amplitude: 1

11. Zeros: $\left\{-2\pi, -\dfrac{3\pi}{2}, -\pi, -\dfrac{\pi}{2}, 0, \dfrac{\pi}{2}, \pi, \dfrac{3\pi}{2}, 2\pi\right\}$;
amplitude: 2

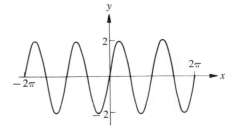

13. Zeros: $\left\{-\dfrac{3\pi}{2}, -\dfrac{\pi}{2}, \dfrac{\pi}{2}, \dfrac{3\pi}{2}\right\}$;
amplitude: 1;   phase shift: π

15. Zeros: $\left\{-\dfrac{7\pi}{4}, -\dfrac{3\pi}{4}, \dfrac{\pi}{4}, \dfrac{5\pi}{4}\right\}$;
amplitude: 2;   phase shift: $\dfrac{\pi}{4}$

17. Zeros: $\left\{-\dfrac{4\pi}{3}, -\dfrac{\pi}{3}, \dfrac{2\pi}{3}, \dfrac{5\pi}{3}\right\}$;
amplitude: $\dfrac{1}{2}$;   phase shift: $\dfrac{\pi}{3}$

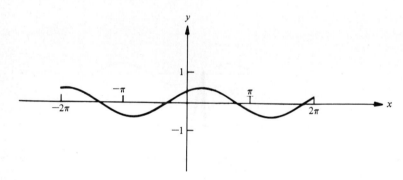

19. Zeros: $\left\{-2\pi, -\dfrac{3\pi}{2}, -\pi, -\dfrac{\pi}{2}, 0, \dfrac{\pi}{2}, \pi, \dfrac{3\pi}{2}, 2\pi\right\}$;
amplitude: 2;   phase shift: π

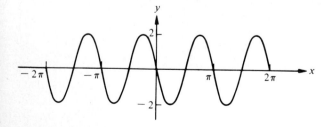

21. Zeros: $\{-\pi, \pi\}$;
amplitude: $\dfrac{1}{2}$;   phase shift: π

3

23. Zeros: $\left\{-\dfrac{\pi}{2}\right\}$; amplitude: 3; phase shift: π

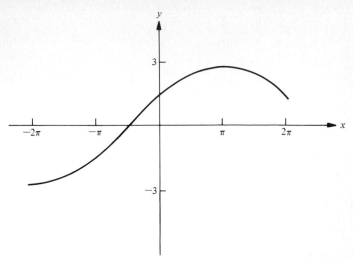

25. Zeros: $\left\{\dfrac{1}{2}, \dfrac{3}{2}, \dfrac{5}{2}, \dfrac{7}{2}\right\}$;
amplitude: 1

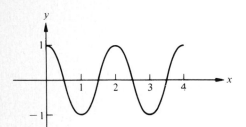

27. Zeros: $\{0, 3, 6, 9, 12\}$;
amplitude: $\dfrac{1}{2}$

29.

31.

33.

Exercise Set 3.3

1.

3.

5.

7.

9.

11.

13.

15.

17.

19.

21.

Exercise Set 3.4

1.

a. $\{y \mid y \in R\}$

b. $x = \dfrac{\pi}{6} + k \cdot \dfrac{\pi}{3}, \ k \in J$

c. $\left\{ x \mid x = k \cdot \dfrac{\pi}{3}, \ k \in J \right\}$

3.

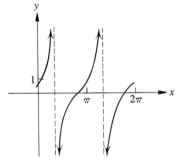

a. $\{y \mid y \in R\}$

b. $x = \pi/3 + k\pi, \ k \in J$

c. $\{x \mid x = 5\pi/6 + k\pi, \ k \in J\}$

5.

a. $\{y \mid y \in R\}$

b. $x = k \cdot 2\pi, \ k \in J$

c. $\{x \mid x = \pi + k \cdot 2\pi, \ k \in J\}$

7.

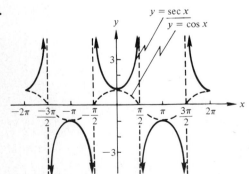

9. $x = \dfrac{\pi}{2} + k \cdot \pi, \ k \in J$

11.

13.

3

15.

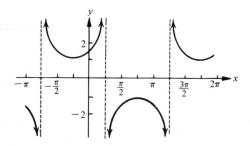

Exercise Set 3.5

1. $\{(0, 1), (6.3, 1)\}$

3. $\{(2.1, 0.5)\}$

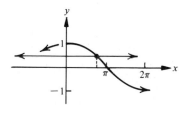

5. $\{(3\pi/4, -1), (7\pi/4, -1)\}$

7. $\{0, 1.3, \pi, 5.0, 2\pi\}$

9. $\{0.7, 2.6, 3.7, 5.6\}$

11. a.

b. $x^2 + y^2 = 1$

13. a.

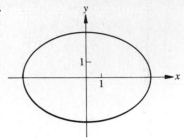

b. $\dfrac{x^2}{16} + \dfrac{y^2}{9} = 1$

15. a.

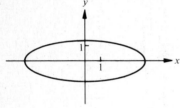

b. $\dfrac{x^2}{25} + \dfrac{y^2}{4} = 1$

17. a.

b. $\dfrac{x^2}{1} + \dfrac{y^2}{9} = 1$

19. $y = \dfrac{8}{x^2 + 4}$

21.

23. $xy = 1;\ |x| \geq 1, |y| \leq 1$

25. $\dfrac{(x - h)^2}{a^2} + \dfrac{(y - k)^2}{b^2} = 1$

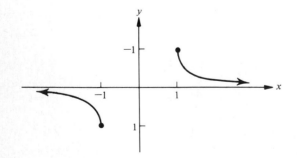

Exercise Set 3.6

1. $\cos y = \sqrt{3}/2$; $\pi/6$ **3.** $\sin y = 1$; $\pi/2$ **5.** $\cot y = 1/\sqrt{3}$; $\pi/3$ **7.** $\pi/4$

9. $\pi/4$ **11.** $\pi/3$ **13.** 0   **15.** $\pi/6$

17. $\pi/4$ **19.** $-\pi/6$ **21.** $-\pi/3$ **23.** $2\pi/3$

25. 0.52   **27.** 0.46   **29.** 0.36   **31.** 1.24

33. 1.45   **35.** 1.19   **37.** -0.50 **39.** 2.92

41. -1.45 **43.** $x = \mathrm{Cos}^{-1}\dfrac{3}{4}$, $0 \le x \le \pi$

45. $x = \dfrac{1}{3}\,\mathrm{Sin}^{-1}\dfrac{1}{4}$, $-\dfrac{\pi}{6} \le x \le \dfrac{\pi}{6}$ **47.** $x = \dfrac{1}{2}\,\mathrm{Tan}^{-1}\dfrac{5}{2}$, $-\dfrac{\pi}{4} < x < \dfrac{\pi}{4}$

49. $x = \dfrac{1}{3}\,\mathrm{Cos}^{-1}0.62$, $0 \le x \le \dfrac{\pi}{3}$ **51.** $x = \dfrac{1}{3}\,\mathrm{Tan}^{-1}\dfrac{y}{2}$, $-\dfrac{\pi}{6} < x < \dfrac{\pi}{6}$

53. $x = 2\,\mathrm{Sin}^{-1}\dfrac{y}{4}$, $-\pi \le x \le \pi$ **55.** $\sqrt{3}/2$

57. 0   **59.** $-1/\sqrt{2}$ **61.** $\pi/4$ **63.** $2\pi/3$ **65.** 0

67. 0   **69.** 0.7151   **71.** 2.4482   **73.** 1.0729   **75.** 4/5

77. 2/5   **79.** 3/4   **81.** $\dfrac{\sqrt{9-y^2}}{3}$ **83.** $\sqrt{1-x^2}$

Review Exercises

1.

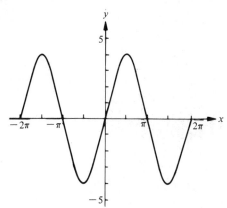

a. amplitude: 4
b. period: 2π
c. no phase shift

2.

a. amplitude: 1/2
b. period: π
c. no phase shift

3.

a. amplitude: 1
b. period: 2π
c. phase shift: $\pi/2$ to the left

4.

a. amplitude: 2
b. period: 4π
c. phase shift: $\pi/3$ to the right

5.

a. amplitude: 1
b. period: 8
c. no phase shift

6.

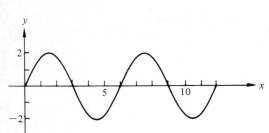

a. amplitude: 2
b. period: 6
c. no phase shift

7.

8.

9.

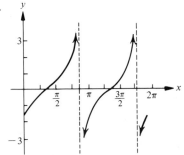

a. range: $\{y \mid y \in R\}$
b. asymptotes: $x = \dfrac{5\pi}{6} + k \cdot \pi, \ k \in J$
c. zeros: $\left\{ x \mid x = \dfrac{\pi}{3} + k \cdot \pi, \ k \in J \right\}$

10.

a. range: $\{y \mid y \in R\}$

b. asymptotes: $x = k \cdot \dfrac{\pi}{3}, \ k \in J$

c. zeros: $\left\{x \mid x = \dfrac{\pi}{6} + k \cdot \dfrac{\pi}{3}, \ k \in J\right\}$

11.

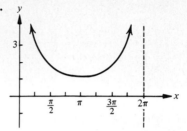

a. range: $\{y \mid y \le -1 \text{ or } y \ge 1\}$

b. asymptotes: $x = k \cdot 2\pi, \ k \in J$

c. no zeros

3

12.

a. range: $\{y \mid y \le -2 \text{ or } y \ge 2\}$

b. asymptotes: $x = \dfrac{\pi}{2} + k \cdot \pi, \ k \in J$

c. no zeros

13.

Solution set is $\left\{(1.2, \frac{1}{3}), (5.0, \frac{1}{3})\right\}$

14.

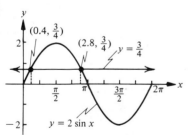

Solution set is $\left\{(0.4, \frac{3}{4}), (2.8, \frac{3}{4})\right\}$

15.

Solution set is
$\{(0.3, 0.5), (1.6, 0), (2.9, -0.5), (4.7, 0)\}$

16.

Solution set is
$\{(0.6, 0.3), (2.5, 0.3), (4.1, -0.3), (5.3, -0.3)\}$

17.

18.

19. $x^2 + y^2 = 9$ **20.** $\dfrac{x^2}{9} + \dfrac{y^2}{16} = 1$ **21.** $\cos y = 1/\sqrt{2};\ (\pi/4)^R$ **22.** $\cot y = -1;\ (-\pi/4)^R$

23. $\csc y = -2;\ (-\pi/6)^R$ **24.** $\sec y = 2/\sqrt{3};\ (\pi/6)^R$ **25.** $(\pi/2)^R$ **26.** $(-\pi/6)^R$

27. $(3\pi/4)^R$ **28.** $(\pi/6)^R$ **29.** 1.45^R **30.** 0.95^R

31. 0.47^R **32.** -1.02^R

33. $x = \text{Cos}^{-1}\dfrac{3}{5},\ 0 \le x \le \pi$ **34.** $x = \dfrac{1}{2}\,\text{Sin}^{-1}\dfrac{1}{5},\ -\dfrac{\pi}{4} \le x \le \dfrac{\pi}{4}$

35. $x = \dfrac{1}{2}\,\text{Tan}^{-1}\dfrac{y}{3},\ -\dfrac{\pi}{4} < x < \dfrac{\pi}{4}$ **36.** $x = 2\,\text{Cos}^{-1}\dfrac{y}{2},\ 0 \le x \le 2\pi$

37. $\sqrt{3}$ **38.** $\dfrac{\sqrt{3}}{2}$ **39.** $\dfrac{\pi}{2}$ **40.** $\dfrac{\pi}{4}$

Exercise Set 4.1

1. $\sec^2 \alpha$ **3.** $\sin^2 \alpha$ **5.** $-\sin^2 \alpha$ **7.** $\tan^3 \alpha$

9. $\dfrac{1}{\sin \alpha}$ **11.** $\dfrac{\cos \alpha}{\sin^2 \alpha}$ **13.** $\dfrac{\cos^2 \alpha - \sin^2 \alpha}{\cos^2 \alpha}$ **15.** $-\dfrac{\sin^2 \alpha}{\cos^2 \alpha}$

17. $\dfrac{1}{\sin \beta \cos \beta}$ **19.** $\dfrac{\cos \beta}{\sin^3 \beta}$ **21.** $\dfrac{1}{\sin \gamma}$ **23.** $\dfrac{\cos^2 \gamma}{\sin \gamma}$

25. $\cos^2 x$ **27.** $\dfrac{\sin^2 x}{\cos^4 x}$ **29.** $\dfrac{1}{\sin y \cos y}$ **31.** $\dfrac{\sin^4 \theta + \cos^4 \theta}{\cos^2 \theta \sin^2 \theta}$

33. $\dfrac{1}{\cos \theta \sin^2 \theta}$ **35.** $\dfrac{\cos^2 \alpha}{\sin^3 \alpha}$ **37.** $\dfrac{\sin^2 \alpha + \cos^2 \alpha}{\cos \alpha \sin^3 \alpha}$ **39.** $\dfrac{-1}{\cos \alpha}$

41. $\dfrac{-1}{\cos \beta \sin \beta}$ **43.** $\dfrac{-1}{\cos x}$ **45.** $\dfrac{-1}{\sin y \cos y}$ **47.** $\dfrac{1}{\cos \alpha - \sin \alpha}$

49. $\dfrac{\sin \alpha + \cos \alpha}{\sin \alpha}$

Exercise Set 4.2

51. If $x = \pi/4$, then $\sin\dfrac{\pi}{4} + \left(\cos\dfrac{\pi}{4}\right)\left(\tan\dfrac{\pi}{4}\right) = 2;\ \dfrac{1}{\sqrt{2}} + \left(\dfrac{1}{\sqrt{2}}\right) \cdot 1 \ne 2$

53. If $y = \pi/2$, then $2\left(\sin\dfrac{\pi}{2}\right)\left(\cos\dfrac{\pi}{2}\right) + \sin\dfrac{\pi}{2} = 0;\ 2(1)(0) + 1 \ne 0$

Exercise Set 4.3

1. $\dfrac{\sqrt{6}-\sqrt{2}}{4}$ **3.** $-\dfrac{\sqrt{2}+\sqrt{6}}{4}$ **5.** $\dfrac{\sqrt{2}+\sqrt{6}}{4}$ **13.** $\cos 9x$

15. $\cos 2x$ **17.** $\cos x$ **19.** $\cos 2x$ **21.** $16/65$

23. $13/\sqrt{170}$ **25.** $-\cos 60°$ **27.** $-\cos 30°$ **29.** $\cos 45°$

37. If $\alpha = 0°$ and $\beta = 60°$, then

$$\cos(\alpha + \beta) = \cos(0° + 60°) = \cos 60° = \frac{1}{2};$$

$$\cos \alpha + \cos \beta = \cos 0° + \cos 60° = 1 + \frac{1}{2} = \frac{3}{2} \neq \frac{1}{2}.$$

39. Using the difference formula and substituting $\pi/2$ for α and α for β, we obtain

$$\cos\left(\frac{\pi}{2} - \alpha\right) = \cos \frac{\pi}{2} \cos \alpha + \sin \frac{\pi}{2} \sin \alpha$$

$$= 0(\cos \alpha) + 1(\sin \alpha) = \sin \alpha.$$

41. 0   **43.** $\dfrac{\sqrt{3}+1}{2\sqrt{2}}$ **45.** $\dfrac{\sqrt{3}-1}{2\sqrt{2}}$

Exercise Set 4.4

1. $\dfrac{\sqrt{6}+\sqrt{2}}{4}$ **3.** $\dfrac{\sqrt{6}-\sqrt{2}}{4}$ **5.** $\dfrac{\sqrt{2}+\sqrt{6}}{4}$ **13.** $\sin 3x$

15. $\sin 4x$ **17.** $\sin 2x$ **19.** $\sin 2x$ **21.** $-44/125$

23. $5/\sqrt{754}$ **25.** $-\sin 60°$ **27.** $\sin 30°$ **29.** $-\sin 30°$

37. If $\alpha = 90°$ and $\beta = 60°$, then

$$\sin(\alpha - \beta) = \sin(90° - 60°) = \sin 30° = \frac{1}{2};$$

$$\sin \alpha - \sin \beta = \sin 90° - \sin 60° = 1 - \frac{\sqrt{3}}{2} \neq \frac{1}{2}.$$

39. 0   **41.** $-\dfrac{1}{2}$ **43.** $\dfrac{\sqrt{3}-1}{2\sqrt{2}}$

Exercise Set 4.5

1. $\dfrac{1+\sqrt{3}}{1-\sqrt{3}}$ **3.** $\dfrac{\sqrt{3}-1}{1+\sqrt{3}}$ **5.** $\dfrac{\sqrt{3}-1}{1+\sqrt{3}}$ **13.** $\tan 9x$ **15.** $\tan 2x$

17. $\tan x$ **19.** $-\tan 60°$ **21.** $\tan 30°$ **23.** $-\tan 45°$

Exercise Set 4.6

1. $\sin 2x$ **3.** $\cos 2x$ **5.** $\sin 4x$ **7.** $\frac{1}{2}\sin x$ **9.** $\cos 10\alpha$ **11.** $2\cos 12\alpha$

13. $\sin \alpha$ **15.** $\cos 2x$ **33.** $24/25$ **35.** $5/\sqrt{34}$ **37.** $-24/7$

Exercise Set 4.7

1. $\sqrt{\dfrac{1 - \sqrt{3}/2}{2}}$ **3.** $\dfrac{1 - \sqrt{3}/2}{1/2}$ **5.** $\sqrt{\dfrac{1 - \sqrt{3}/2}{2}}$ **7.** $\sqrt{\dfrac{1 + 1/\sqrt{2}}{2}}$

9. $\dfrac{1 + \sqrt{3}/2}{1/2}$ **11.** $-\sqrt{\dfrac{1 - \sqrt{3}/2}{2}}$ **19.** $\dfrac{1}{\sqrt{3}}$ **21.** $\sqrt{\dfrac{2}{3}}$

23. $-\sqrt{\dfrac{1 + 1/\sqrt{2}}{2}}$ **25.** $\frac{1}{2}(\sin 17° - \sin 7°)$ **27.** $\frac{1}{2}(\sin 3x - \sin x)$ **29.** $\frac{1}{2}(\cos 7x + \cos 3x)$

31. $2\cos 12° \cos 8°$ **33.** $2\sin 4\alpha \cos \alpha$ **35.** $2\cos 3\alpha \sin 2\alpha$

37. $\cos \alpha \cos \beta = \frac{1}{2}[\cos(\alpha + \beta) + \cos(\alpha - \beta)]$

$\qquad\qquad\quad = \frac{1}{2}[\cos \alpha \cos \beta - \sin \alpha \sin \beta + \cos \alpha \cos \beta + \sin \alpha \sin \beta]$

$\qquad\qquad\quad = \frac{1}{2}[2\cos \alpha \cos \beta] = \cos \alpha \cos \beta$

39. $\sin \alpha \cos \beta = \frac{1}{2}[\sin(\alpha + \beta) + \sin(\alpha - \beta)]$

$\qquad\qquad\quad = \frac{1}{2}(\sin \alpha \cos \beta + \cos \alpha \sin \beta + \sin \alpha \cos \beta - \cos \alpha \sin \beta)$

$\qquad\qquad\quad = \frac{1}{2}(2\sin \alpha \cos \beta) = \sin \alpha \cos \beta$

41. Let $\alpha = x + y$ and $\beta = x - y$. Then,

$$\cos \alpha + \cos \beta = \cos(x + y) + \cos(x - y)$$

$$= \cos x \cos y - \sin x \sin y + \cos x \cos y + \sin x \sin y = 2\cos x \cos y.$$

Solving $\alpha = x + y$ and $\beta = x - y$ for α and β, respectively, gives $x = \dfrac{\alpha + \beta}{2}$ and $y = \dfrac{\alpha - \beta}{2}$. Thus,

$$\cos \alpha + \cos \beta = 2\cos x \cos y = 2\cos \frac{\alpha + \beta}{2} \cos \frac{\alpha - \beta}{2}.$$

43. Let $\alpha = x + y$ and $\beta = x - y$. Then,

$$\sin \alpha + \sin \beta = \sin(x + y) + \sin(x - y)$$

$$= \sin x \cos y + \cos x \sin y + \sin x \cos y - \cos x \sin y = 2\sin x \cos y.$$

As in Exercise 41, $x = \dfrac{\alpha + \beta}{2}$ and $y = \dfrac{\alpha - \beta}{2}$. Thus,

$$\sin \alpha + \sin \beta = 2\sin x \cos y = 2\sin \frac{\alpha + \beta}{2} \cos \frac{\alpha - \beta}{2}.$$

Exercise Set 4.8

1a. $30°$ **b.** $\{30°, 330°\}$
5a. $30°$ **b.** $\{30°, 210°\}$
9a. $210°$ **b.** $\{210°, 330°\}$
13a. $135°$ **b.** $\{135°, 225°\}$
17a. $0°$ **b.** $\{0°, 180°\}$
21a. $0°$ **b.** $\{0°, 180°\}$
25a. $14°$ **b.** $\{14°, 166°\}$
29a. $14.3°$ **b.** $\{14.3°, 345.7°\}$
33a. $25.4°$ **b.** $\{25.4°, 205.4°\}$
37a. $213.2°$ **b.** $\{213.2°, 326.8°\}$
41a. $45.6°$ **b.** $\{45.6°, 225.6°\}$
45a. 0.23^R **b.** $\{0.23^R, 3.37^R\}$
49a. 2.89^R **b.** $\{2.89^R, 3.39^R\}$
53a. 3.63^R **b.** $\{3.63^R, 5.79^R\}$

3a. $150°$ **b.** $\{150°, 330°\}$
7a. $45°$ **b.** $\{45°, 315°\}$
11a. $210°$ **b.** $\{210°, 330°\}$
15a. $45°$ **b.** $\{45°, 135°\}$
19a. $150°$ **b.** $\{150°, 210°\}$
23a. $0°$ **b.** $\{0°, 180°\}$
27a. $111°$ **b.** $\{111°, 291°\}$
31a. $155.3°$ **b.** $\{155.3°, 204.7°\}$
35a. $231.4°$ **b.** $\{231.4°, 308.6°\}$
39a. $66.7°$ **b.** $\{66.7°, 293.3°\}$
43a. 0.70^R **b.** $\{0.70^R, 2.44^R\}$
47a. 0.98^R **b.** $\{0.98^R, 5.30^R\}$
51a. 3.26^R **b.** $\{3.26^R, 6.16^R\}$

55. $\{\alpha \mid \alpha \approx 16.2° + k \cdot 360° \text{ or } \alpha \approx 163.8° + k \cdot 360°\},\ k \in J$
57. $\{\alpha \mid \alpha \approx 142.0° + k \cdot 180°\},\ k \in J$
59. $\{\alpha \mid \alpha \approx 231.3° + k \cdot 360° \text{ or } 308.7° + k \cdot 360°\},\ k \in J$
61. $\{\alpha \mid \alpha \approx 16.8° + k \cdot 180°\},\ k \in J$

Exercise Set 4.9

1. $\{\pi/6\}$
7. $\{\pi/4, \pi/2\}$
13. $\{0\}$

3. $\{\pi/4\}$
9. $\{0, \pi/2\}$
15. $\{0, 2\pi/3, 4\pi/3\}$

5. $\{\pi/2\}$
11. $\{\pi/4, \pi/2\}$
17. $\{1.91, 4.37\}$

19. a. $\{270°\}$ **b.** $\{x \mid x = 270° + k \cdot 360°\},\ k \in J$

21. a. $\{60°, 180°, 300°\}$
b. $\{x \mid x = 60° + k \cdot 360°\} \cup \{x \mid x = 180° + k \cdot 360°\} \cup \{x \mid x = 300° + k \cdot 360°\},\ k \in J$

23. a. $\{30°, 90°, 150°\}$
b. $\{x \mid x = 30° + k \cdot 360°\} \cup \{x \mid x = 90° + k \cdot 360°\} \cup \{x \mid x = 150° + k \cdot 360°\},\ k \in J$

25. a. $\{218.2°, 321.8°\}$
b. $\{x \mid x = 218.2° + k \cdot 360°\} \cup \{x \mid x = 321.8° + k \cdot 360°\},\ k \in J$

27. a. $\{39.9°, 115.7°, 244.3°, 320.1°\}$
b. $\{x \mid x = 39.9° + k \cdot 360°\} \cup \{x \mid x = 115.7° + k \cdot 360°\} \cup \{x \mid x = 244.3° + k \cdot 360°\}$
$$\cup \{x \mid x = 320.1° + k \cdot 360°\},\ k \in J$$

29. $\left\{\dfrac{\pi}{3}, \pi\right\}$

31. $\left\{\dfrac{3\pi}{2}\right\}$

33. $\left\{x \mid x = \dfrac{3\pi}{4} + k \cdot \pi\right\},\ k \in J$

Exercise Set 4.10

1. $\{0, \pi/3\}$
11. $\{\pi/8\}$

3. $\{\pi/6, 5\pi/12\}$
13. $\{0\}$

5. $\{\pi/15\}$

7. $\{\pi/6\}$
15. $\{10°, 50°, 130°, 170°, 250°, 290°\}$

9. $\{0\}$

4

17. $\{15°, 75°, 135°, 195°, 225°, 315°\}$ **19.** $\{45°, 135°, 225°, 315°\}$
21. $\{15°, 45°, 75°, 105°, 135°, 165°, 195°, 225°, 255°, 285°, 315°, 345°\}$
23. $\{180°\}$ **25.** $\{60°, 180°, 300°\}$
27. $\{x \mid x = \pi/2 + k \cdot 2\pi\} \cup \{x \mid x = \pi/6 + k \cdot 2\pi\} \cup \{x \mid x = 5\pi/6 + k \cdot 2\pi\}, \ k \in J$
29. $\{x \mid x = k\pi\} \cup \{x \mid x = \pi/6 + k\pi/3\}, \ k \in J$
31. $\{x \mid x \approx 1.28 + k\pi/2\} \cup \{x \mid x = \pi/8 + k\pi/2\}, \ k \in J$
33. $\{x \mid x = \pi/12 + k(2\pi/3)\} \cup \{x \mid x = \pi/4 + k(2\pi/3)\}, \ k \in J$

35. $\{0.5\}$

Review Exercises

1. $1/\cos \alpha$ **2.** $1/\sin^2 \alpha$ **3.** $\cos^2 \alpha$ **4.** $1/\sin \alpha$ **5.** $-\cos \alpha$ **6.** $1/\cos \alpha$

23. If $\alpha = \dfrac{\pi}{4}$, then $\sin \dfrac{\pi}{4} \sec \dfrac{\pi}{4} = 2 \tan \dfrac{\pi}{4}$; $\dfrac{1}{\sqrt{2}} \cdot \sqrt{2} \neq 2 \cdot 1$

24. If $\alpha = \dfrac{\pi}{4}$, then $\dfrac{1 - \cos \pi/4}{\sin \pi/4} = \dfrac{1 + \sin \pi/4}{\cos \pi/4}$; $\dfrac{1 - 1/\sqrt{2}}{1/\sqrt{2}} \neq \dfrac{1 + 1\sqrt{2}}{1/\sqrt{2}}$; $\sqrt{2} - 1 \neq \sqrt{2} + 1$

25. $-\sqrt{2 - \sqrt{3}}/2$ **26.** $\sqrt{3} - 2$ **29.** $\frac{1}{2}(\cos 6x + \cos 4x)$ **30.** $2 \sin 5x \cos x$

31. a. $\{60°\}$
  b. $\{\alpha \mid \alpha = 60° + k \cdot 360°\} \cup \{\alpha \mid \alpha = 120° + k \cdot 360°\}, \ k \in J$

32. a. $\{135°\}$
  b. $\{\alpha \mid \alpha = 135° + k \cdot 180°\}, \ k \in J$

33. a. $\{118.0°\}$
  b. $\{\alpha \mid \alpha \approx 118.0° + k \cdot 360°\} \cup \{\alpha \mid \alpha \approx 242.0° + k \cdot 360°\}, \ k \in J$

34. a. $\{36.1°\}$
  b. $\{\alpha \mid \alpha \approx 36.1° + k \cdot 360°\} \cup \{\alpha \mid \alpha \approx 143.9° + k \cdot 360°\}, \ k \in J$

35. a. $\{0.369^R\}$
  b. $\{\alpha \mid \alpha \approx 0.369^R + k \cdot 2\pi^R\} \cup \{\alpha \mid \alpha \approx 5.915^R + k \cdot 2\pi^R\}, \ k \in J$

36. a. $\{1.810^R\}$
  b. $\{\alpha \mid \alpha \approx 1.810^R + k \cdot \pi^R\}, \ k \in J$

37. a. $\{0.316^R\}$
  b. $\{\alpha \mid \alpha \approx 0.316^R + k \cdot 2\pi^R\} \cup \{\alpha \mid \alpha \approx 2.826^R + k \cdot 2\pi^R\}, \ k \in J$

38. a. $\{0.459\}$

  b. $\{\alpha \mid \alpha \approx 0.459^R + k \cdot 2\pi^R\} \cup \{\alpha \mid \alpha \approx 5.824^R + k \cdot 2\pi^R\}, \; k \in J$

39. $\{\pi/3\}$ **40.** $\{\pi/6\}$ **41.** $\{\pi/4\}$ **42.** $\{\pi/2\}$

43. $\{30°, 150°, 210°, 330°\}$ **44.** $\{45°, 135°, 225°, 315°\}$

45. $\{15°, 105°, 135°, 225°, 255°, 345°\}$ **46.** $\{15°, 30°, 105°, 120°, 195°, 210°, 285°, 300°\}$

47. $\{0°, 45°, 135°, 180°, 225°, 315°\}$ **48.** $\{0°, 180°\}$

Exercise Set 5.1

1. $2.3; 90°$ **3.** $3.5; -60°$ **5.** $2.3; -90°$

7.

9.

11.

13.

15.

17.

19.

21.

23.

25.

27.

29.

31.

33.

35.

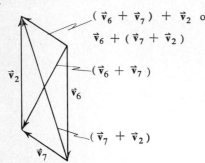

$(\vec{v}_6 + \vec{v}_7) + \vec{v}_2$ or

$\vec{v}_6 + (\vec{v}_7 + \vec{v}_2)$

$(\vec{v}_6 + \vec{v}_7)$

$(\vec{v}_7 + \vec{v}_2)$

37.

39.

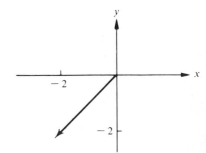

Exercise Set 5.2

1. $\|\vec{v}_x\| = 5$, $\|\vec{v}_y\| = 5$

3. $\|\vec{v}_x\| = 0$, $\|\vec{v}_y\| = 15$

5. $\|\vec{v}_x\| = 6\sqrt{3}$, $\|\vec{v}_y\| = 6$

7. $\|\vec{v}_x\| = 4$, $\|\vec{v}_y\| = 4\sqrt{3}$

9. $\|\vec{v}_3\| = 5$ pounds; 53.1° with 3-pound force, 36.9° with 4-pound force.

11. $\|\vec{v}_3\| = 27.9$ pounds; 43.8° with 15-pound force, 31.2° with 20-pound force.

13. 8.72 newtons, $\alpha \approx 36.6°$, $\beta \approx 83.4°$

15. 40 newtons, 36.9°

17. 19.7 pounds, 37.1 pounds

19. 511.3 pounds

21. 33.4°

23. $\|\vec{v}_x\| = 1.64$ meters/sec², $\|\vec{v}_y\| = 1.15$ meters/sec²

25. 0.967 kilometer per hour

27. Drift angle is 11.3° to the right; ground speed is 612 miles per hour; course is 146.3°

29. Drift angle is 12° to the right; wind direction is 318° measured from true north; wind speed is 106.5 miles per hour

31. Heading is 85.2°; air speed is 602 kilometers per hour

33. 34.6 nautical miles, 265.3°

35. 70.7 pounds

Exercise Set 5.3

1. $x = -3$, $y = 6$

3. $x = -7/2$, $y = -7$

5. $x = -2$, $y = -2$

7. a.

9. a.

11. a.

b. 5, 53° 6'

b. $\sqrt{2}$, 45°

b. $3\sqrt{2}$, 225°

13. a.

b. $4, 0°$

15. $(2, 2)$

17. $(0, 1)$

19. $\left(-\frac{5}{2}\sqrt{3}, -\frac{5}{2}\right)$

5

21.

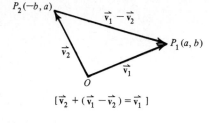

$[\vec{v_2} + (\vec{v_1} - \vec{v_2}) = \vec{v_1}]$

$$\mathbf{v_1} - \mathbf{v_2} = (a, b) - (-b, a) = (a + b, b - a);$$
$$\|\mathbf{v_1} - \mathbf{v_2}\|^2 = (a + b)^2 + (b - a)^2$$
$$= a^2 + 2ab + b^2 + b^2 - 2ab + a^2$$
$$= (a^2 + b^2) + (a^2 + b^2)$$
$$= \|\mathbf{v_1}\|^2 + \|\mathbf{v_2}\|^2$$

Hence, from the Pythagorean theorem, $\triangle P_1 O P_2$ is a right triangle, and $\angle P_1 O P_2 = 90°$.

Exercise Set 5.4

1. $(-4, 6)$

3. $(3, 1)$

5. $(-4, 1)$

7. $(-3, -6)$

9. $(-4, 2)$

11. $(6, 12)$

13. $(-13, -16)$

15. $(-9, 15)$

17. $(16, 10)$

19. $(7, -8)$

21. $(-7, -1)$

23. $(-27, 9)$

25. $(5/2, -1/2)$

27. $(19/6, 8/3)$

29. $(-5/2, 5)$

31. $3\mathbf{i} + 5\mathbf{j}$

33. $\mathbf{i} - 4\mathbf{j}$

35. $-3\mathbf{i}$

37. $4\sqrt{2}, -45°$

39. $\sqrt{5}, 26° \ 36'$

41. $\sqrt{10}, -108° \ 24'$

43. $2, 120°$

Exercise Set 5.5

1. 48   **3.** -2 **5.** 45   **7.** 13   **9.** $36\sqrt{2}$

11. 3/4   **13.** 156.8°   **15.** 90°   **17.** 24.0°

19. $20/\sqrt{29}$, 5   **21.** $7/\sqrt{41}$, $7/\sqrt{13}$ **23.** $-14/\sqrt{10}$, $-14/\sqrt{34}$

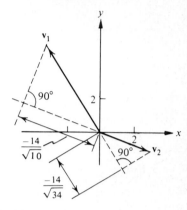

25. $0, 0$ **33.** 24   **35.** 36

37. $1000\sqrt{3}$ meter-newtons   **39.** 1,000,000 meter-newtons

41. Let $\mathbf{v}_1 = (a, b)$. Then $\|\mathbf{v}_1\| = \sqrt{a^2 + b^2}$ and $\|\mathbf{v}_1\|^2 = a^2 + b^2$.

Also, $\mathbf{v}_1 \cdot \mathbf{v}_1 = (a)(a) + (b)(b) = a^2 + b^2$. Thus, $\mathbf{v}_1 \cdot \mathbf{v}_1 = \|\mathbf{v}_1\|^2$.

43. Let $\mathbf{v}_1 = (a_1, b_1)$ and $\mathbf{v}_2 = (a_2, b_2)$. Then,

$$c(\mathbf{v}_1 \cdot \mathbf{v}_2) = c(a_1 a_2 + b_1 b_2) = (c a_1) a_2 + (c b_1) b_2$$
$$= (c a_1, c b_1) \cdot (a_2, b_2) = c(a_1, b_1) \cdot (a_2, b_2)$$
$$= (c \mathbf{v}_1) \cdot \mathbf{v}_2$$

Review Exercises

1. $(\vec{v}_1 + \vec{v}_2) + \vec{v}_3$

2.

3.

4.

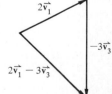

5. $4\sqrt{3}$; 4   **6.** 5; 36.9°   **7.** 120 newtons; 22.6°

8. 6.71 miles per hour; drift of 63.4°

9. Wind bearing is 357.7°; wind speed is 183 kilometers per hour

10. 36.9°   **11.** $x = -1, y = 3$ **12.** $x = 4, y = 1$

13. 5, 143.1°   **14.** $5\sqrt{2}, -45°$

15. $(-5\sqrt{2}, -5\sqrt{2})$ **16.** $(-\sqrt{3}, 1)$ **17.** $(12, -4)$ **18.** $(-10, 10)$

19. $(8, 2)$ **20.** $(12, 2)$ **21.** $(-1, 3)$ **22.** $(5/2\sqrt{13}, 2/\sqrt{13})$

23. $7\mathbf{i} - 9\mathbf{j}$ **24.** $-8\mathbf{i} - \mathbf{j}$ **25.** $2\sqrt{29}, 158° \ 12'$ **26.** $\sqrt{130}, -52° \ 6'$

27. -54 **28.** 75   **29.** $16\sqrt{3}$ **30.** $25\sqrt{3}$

31. 136° 48′   **32.** 56° 30′   **33.** $3/\sqrt{101}, 3/5\sqrt{2}$ **34.** $-82/\sqrt{89}, -82/\sqrt{85}$

35. $(-4)(4) + (-5)(-16/5) = -16 + 16 = 0$ **36.** $(7)(5) + (6)(-35/6) = 35 - 35 = 0$

Exercise Set 6.1

1. $x = 1, y = -3$ **3.** $x = -\dfrac{1}{2}, y = 8$ **5.** $1 + 10i$ **7.** $-10 - 11i$

9. $-1 + 6i$ **11.** $0 + 0i$ **13.** $12 - 12i$ **15.** $-36 - 12i$

17. $2 + 3i$ **19.** $-19 + 2i$ **21.** $62 - 39i$ **23.** $26 - 19i$

25. $a^2 + b^2$ **27.** i **29.** -1 **31.** -1

33. $-i$ **35.** $5i$ **37.** $2\sqrt{3}\, i$ **39.** $1 + 3i$

41. $3 - 2\sqrt{5}\, i$ **43.** $8 + 2i$ **45.** $-8 + 8i$ **47.** $-10\sqrt{5} + 2\sqrt{10}\, i$

49. $23 + 2i$ **51.** $9 + \sqrt{3}\, i$

53. If $x = 2i$, then $(2i)^2 + 4 = 4i^2 + 4$
$$= -4 + 4 = 0;$$
if $x = -2i$, then $(-2i)^2 + 4 = -4i^2 + 4$
$$= -4 + 4 = 0$$

55. a. $\left(-\dfrac{1}{2} + i\dfrac{\sqrt{3}}{2}\right)^2 = \dfrac{1}{4} - i\dfrac{\sqrt{3}}{2} + i^2\left(\dfrac{3}{4}\right) = -\dfrac{1}{2} - i\dfrac{\sqrt{3}}{2};$

$\left(-\dfrac{1}{2} + i\dfrac{\sqrt{3}}{2}\right)^3 = \left(-\dfrac{1}{2} + i\dfrac{\sqrt{3}}{2}\right)^2 \left(-\dfrac{1}{2} + i\dfrac{\sqrt{3}}{2}\right)$

$= \left(-\dfrac{1}{2} - i\dfrac{\sqrt{3}}{2}\right)\left(-\dfrac{1}{2} + i\dfrac{\sqrt{3}}{2}\right) = \dfrac{1}{4} - i^2\left(\dfrac{3}{4}\right) = 1$

b. $\left(-\dfrac{1}{2} - i\dfrac{\sqrt{3}}{2}\right)^2 = \dfrac{1}{4} + i\dfrac{\sqrt{3}}{2} + i^2\left(\dfrac{3}{4}\right) = -\dfrac{1}{2} + i\dfrac{\sqrt{3}}{2};$

$\left(-\dfrac{1}{2} - i\dfrac{\sqrt{3}}{2}\right)^3 = \left(-\dfrac{1}{2} - i\dfrac{\sqrt{3}}{2}\right)^2\left(-\dfrac{1}{2} - i\dfrac{\sqrt{3}}{2}\right)$

$= \left(-\dfrac{1}{2} + i\dfrac{\sqrt{3}}{2}\right)\left(-\dfrac{1}{2} - i\dfrac{\sqrt{3}}{2}\right) = \dfrac{1}{4} - i^2\left(\dfrac{3}{4}\right) = 1$

57. $a = 0, b = 1$ or $a = 0, b = -1$ **59.** $e^{\pi i} = \cos \pi + i \sin \pi$
$= -1 + i(0) = -1$

61. $\dfrac{e^{ix} - e^{-ix}}{2i} = \dfrac{\cos x + i \sin x - [\cos(-x) + i \sin(-x)]}{2i}$

$= \dfrac{\cos x + i \sin x - \cos x + i \sin x}{2i} = \dfrac{2i \sin x}{2i} = \sin x$

63. Using the results of Exercises 61 and 62,

$$\sin^2 x + \cos^2 x = \left(\dfrac{e^{ix} - e^{-ix}}{2i}\right)^2 + \left(\dfrac{e^{ix} + e^{-ix}}{2}\right)^2$$

$$= \dfrac{e^{2ix} - 2e^0 + e^{-2ix}}{4i^2} + \dfrac{e^{2ix} + 2e^0 + e^{-2ix}}{4}$$

$$= \dfrac{-e^{2ix} + 2 - e^{-2ix}}{4} + \dfrac{e^{2ix} + 2 + e^{-2ix}}{4} = \dfrac{4}{4} = 1.$$

Exercise Set 6.2

1. $13 - 4i$ **3.** $2 + 2i$ **5.** $5 - i$ **7.** $6 - 3i$ **9.** $-3 + 3i$

11. $-3i$ **13.** $\dfrac{3}{2} - \dfrac{9}{4}i$ **15.** $\dfrac{8}{85} + \dfrac{36}{85}i$ **17.** $\dfrac{9}{17} + \dfrac{2}{17}i$ **19.** $-\dfrac{6}{5} + \dfrac{11}{10}i$

21. $\dfrac{-i}{\sqrt{5}}$ **23.** $-\dfrac{2i}{\sqrt{2}}$ **25.** $\dfrac{4}{7} - \dfrac{\sqrt{5}}{7}i$ **27.** $\dfrac{18}{11} - \dfrac{6\sqrt{2}}{11}i$ **29.** $\dfrac{6\sqrt{2}}{7} - \dfrac{6\sqrt{5}}{7}i$

31. $\dfrac{2\sqrt{5}}{13} - \dfrac{4\sqrt{2}}{13}i$

Exercise Set 6.3

1. $(-1, 6)$ **3.** $(8, 1)$ **5.** $(0, 6)$ **7.** $-8 + 7i$ **9.** $-4 + 0i$ **11.** $7 - \sqrt{2}i$

13.   **15.**   **17.**

19.

21. $-1 + 6i$

23. $-4 - 5i$

25. $-1 + 4i$

27. $\sqrt{29}; 21.8°$

29. $2\sqrt{2}; 225°$

31. $7; 0°$

33. $5; 270°$

35. $3\sqrt{10}; 288.4°$

37. $4; 0°$

39. a. $a \neq 0, b = 0$ **b.** $a = 0, b \neq 0$ **c.** $b > 0$ **d.** $b < 0$

41.

43.

45.

Exercise Set 6.4

1. $4(\cos 120° + i \sin 120°)$

3. $10(\cos 210° + i \sin 210°)$

5. $2\sqrt{2}(\cos 315° + i \sin 315°)$

7. $8(\cos 180° + i \sin 180°)$

9. $-1 - \sqrt{3}i$

11. $-\sqrt{2} + \sqrt{2}i$

6

13. $\dfrac{1}{\sqrt{2}} - \dfrac{1}{\sqrt{2}}i$

15. $-\dfrac{\sqrt{3}}{4} + \dfrac{1}{4}i$

17. $-5\sqrt{2} + 5\sqrt{2}i$ **19.** $\dfrac{3}{2}\sqrt{3} - \dfrac{3}{2}i$ **21.** $-\sqrt{2} + \sqrt{2}i$ **23.** $-4 + 0i$

25. $-\dfrac{1}{5} - \dfrac{\sqrt{3}}{5}i$ **27.** $\dfrac{3}{2} - \dfrac{3\sqrt{3}}{2}i$ **29.** $4 + 4\sqrt{3}i$

31. $(a + bi)^2 = \rho(\cos\theta + i\sin\theta) \cdot \rho(\cos\theta + i\sin\theta)$; using Equation (1) in this section gives

$$(a + bi)^2 = \rho \cdot \rho[\cos(\theta + \theta) + i\sin(\theta + \theta)]$$
$$= \rho^2[\cos 2\theta + i\sin 2\theta].$$

33. $(a + bi)^4 = (a + bi)^3(a + bi)$; using the results of Exercise 32 and Equation (1) in this section, gives

$$\rho^3(\cos 3\theta + i\sin 3\theta) \cdot \rho(\cos\theta + i\sin\theta) = \rho^3 \cdot \rho[\cos(3\theta + \theta) + i\sin(3\theta + \theta)]$$
$$= \rho^4(\cos 4\theta + i\sin 4\theta).$$

35. From Section 4.1, $\sin(-\alpha) = -\sin\alpha$ and $\cos(-\alpha) = \cos\alpha$. Thus,

$$\cos(-\theta) + i\sin(-\theta) = \cos\theta - i\sin\theta,$$

the conjugate of $\cos\theta + i\sin\theta$.

Exercise Set 6.5

1. $\dfrac{27}{2} + \dfrac{27\sqrt{3}}{2}i$ **3.** $0 + 8i$ **5.** $-\dfrac{1}{8} + 0i$

7. $0 + \dfrac{1}{81}i$ **9.** $-\dfrac{1}{32} + \dfrac{1}{32}i$ **11.** $-\dfrac{1}{64} - \dfrac{\sqrt{3}}{64}i$

13. a. $\cos 40° + i\sin 40°$, $\cos 220° + i\sin 220°$
b. $\cos 30° + i\sin 30°$, $\cos 210° + i\sin 210°$

15. a. $2(\cos 40° + i\sin 40°)$, $2(\cos 160° + i\sin 160°)$, $2(\cos 280° + i\sin 280°)$
b. $\cos 20° + i\sin 20°$, $\cos 140° + i\sin 140°$, $\cos 260° + i\sin 260°$

17. a. $\cos 60° + i\sin 60°$, $\cos 150° + i\sin 150°$, $\cos 240° + i\sin 240$, $\cos 330° + i\sin 330°$
b. $\cos 75° + i\sin 75°$, $\cos 165° + i\sin 165°$, $\cos 255° + i\sin 255°$, $\cos 345° + i\sin 345°$

19. $\left\{\dfrac{1}{2} + \dfrac{\sqrt{3}}{2}i, -1, \dfrac{1}{2} - \dfrac{\sqrt{3}}{2}i\right\}$ **21.** $\left\{1, -\dfrac{1}{2} + \dfrac{\sqrt{3}}{2}i, -\dfrac{1}{2} - \dfrac{\sqrt{3}}{2}i\right\}$

23. $\left\{ \dfrac{1}{2^{1/3}} (\cos 110° + i \sin 110°), \dfrac{1}{2^{1/3}} (\cos 230° + i \sin 230°), \dfrac{1}{2^{1/3}} (\cos 350° + i \sin 350°) \right\}$.

25. In all cases, the graphs are spaced equally on the unit circle.

a.   **b.**   **c.**

d.   **e.**   **f.**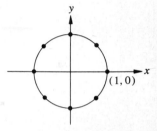

27. $\cos 3\theta = 4\cos^3 \theta - 3\cos \theta$

Exercise Set 6.6

1. $(3, 20°), (-3, 200°), (-3, -160°), (3, -340°)$ **3.** $(5, -90°), (-5, -270°), (-5, 90°), (5, 270°)$
5. $(-4, 60°), (4, 240°), (4, -120°), (-4, -300°)$
7. $(2\sqrt{3}, 2)$ **9.** $(-3/\sqrt{2}, -3/\sqrt{2})$ **11.** $(3\sqrt{3}/2, -3/2)$ **13.** $(4, 0^R)$
15. $\left(2\sqrt{2}, \dfrac{\pi}{4}^R\right)$ **17.** $\left(1, \dfrac{5\pi}{6}^R\right)$ **19.** $\rho \cos \theta = 4$ **21.** $\rho = \dfrac{8 \sin \theta}{\cos^2 \theta}$
23. $\rho = 4$ **25.** $\rho = 3 \sin \theta$ **27.** $x^2 + y^2 = 36$ **29.** $x^2 + y^2 = 4y$
31. $y = 2$ **33.** $x^2 - 8y - 16 = 0$

Exercise Set 6.7

1.   **3.**

5.

7.

9.

11.

13.

15.

17.

19.

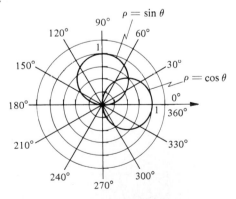

(0.7, 45°) and (0, θ) where θ has any measure

21.

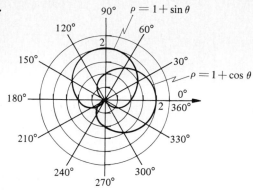

$\rho = 1 + \sin \theta$

$\rho = 1 + \cos \theta$

(1.7, 45°), (0.3, 225°) and (0, θ) where θ has any measure

23. $\{(1/\sqrt{2}, 45°)\}$; since ρ is a distance measured from the origin, it may be zero for more than one value of θ. In this case, $\rho = 0$ when $\theta = 0°$ in $\rho = \sin \theta$, and $\rho = 0$ when $\theta = 90°$ in $\rho = \cos \theta$.

25. $\left\{ \left(\dfrac{\sqrt{2}+1}{\sqrt{2}}, 45° \right), \left(\dfrac{\sqrt{2}-1}{\sqrt{2}}, 225° \right) \right\}$

27. ρ_1 and ρ_2 are the lengths of sides with included angle $\theta_1 - \theta_2$.
Using the Law of Cosines, $d^2 = \rho_1{}^2 + \rho_2{}^2 - 2\rho_1\rho_2 \cos(\theta_1 - \theta_2)$,

$$d = \sqrt{\rho_1{}^2 + \rho_2{}^2 - 2\rho_1\rho_2 \cos(\theta_1 - \theta_2)}.$$

6

Review Exercises

1. $x = -\frac{1}{3}, \quad y = -\frac{2}{3}$

2. $x = 1, y = 4; \quad x = -1, \quad y = -4$

3. $-8 + 8i$

4. $0 - i$

5. $-46 + 28i$

6. $22 - 14i$

7. i

8. $-i$

9. $\sqrt{10} + 3\sqrt{5}i$

10. 14

11. If $x = 3i$, then $(3i)^2 + 9 = 9i^2 + 9 = -9 + 9 = 0$;
if $x = -3i$, then $(-3i)^2 + 9 = 9i^2 + 9 = -9 + 9 = 0$.

12. If $x = 1 + 3i$, then $(1 + 3i)^2 - 2(1 + 3i) + 10 = 1 + 6i + 9i^2 - 2 - 6i + 10 = 0$;
if $x = 1 - 3i$, then $(1 - 3i)^2 - 2(1 - 3i) + 10 = 1 - 6i + 9i^2 - 2 + 6i + 10 = 0$.

13. $-15 - i$

14. $3 - 5i$

15. $\dfrac{7}{5} - \dfrac{13}{10} i$

16. $-\dfrac{13}{34} - \dfrac{8}{17} i$

17. $\dfrac{2i}{\sqrt{6}}$

18. $\dfrac{8}{9} - \dfrac{4\sqrt{5}}{9} i$

19. $(-4, 1)$

20. $(2\sqrt{3}, -2)$

21. $9 - 6i$

22. $3 + 4i$

23.

24.

25.

26.

27. $2\sqrt{5}, 26°\ 30'$

28. $\sqrt{29}, 248°\ 12'$

29. $\sqrt{2}\ (\cos 45° + i \sin 45°)$

30. $6(\cos 120° + i \sin 120°)$

31. $-\dfrac{7}{2} + \dfrac{7\sqrt{3}}{2}\ i$

32. $\sqrt{2} + \sqrt{2}i$

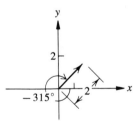

33. $-3\sqrt{2} - 3\sqrt{2}i$ **34.** $3\sqrt{3} + 3i$ **35.** $-\dfrac{8}{9} + 0i$ **36.** $-2\sqrt{2} + 2\sqrt{2}i$

37. $\dfrac{81}{2} + \dfrac{81\sqrt{3}}{2}\ i$ **38.** $0 - 8i$ **39.** $-\dfrac{1}{32} - \dfrac{\sqrt{3}}{32}\ i$ **40.** $-\dfrac{1}{256} + \dfrac{1}{256}\ i$

41. $\cos 15° + i \sin 15°,\ \cos 195° + i \sin 195°$

42. $5(\cos 110° + i \sin 110°),\ 5(\cos 230° + i \sin 230°),\ 5(\cos 350° + i \sin 350°)$

43. $\{\cos 45° + i \sin 45°,\ \cos 135° + i \sin 135°,\ \cos 225° + i \sin 225°,\ \cos 315° + i \sin 315°\}$

44. $\{10^{1/5}(\cos 48° + i \sin 48°), 10^{1/5}(\cos 120° + i \sin 120°), 10^{1/5}(\cos 192° + i \sin 192°),$
$10^{1/5}(\cos 264° + i \sin 264°), 10^{1/5}(\cos 336° + i \sin 336°)\}$

45. $(2, 250°), (-2, 70°), (2, -110°), (-2, -290°)$ **46.** $(-3, -330°), (-3, 30°), (3, 210°), (3, -150°)$

47. $\left(\dfrac{3}{2}, \dfrac{3\sqrt{3}}{2}\right)$ **48.** $(\sqrt{2}, -\sqrt{2})$ **49.** $\left(2\sqrt{2}, \dfrac{7\pi}{4}^R\right)$ **50.** $\left(2, \dfrac{7\pi}{6}^R\right)$

51. $\rho \cos \theta = 3$ **52.** $\rho^2(\cos^2 \theta + 3 \sin^2 \theta) = 5$

53. $x^2 + y^2 = 5y$ **54.** $y^2 - 4x - 4 = 0$

55.

56.

57.

58.

6

INDEX

W

Z

SUMMARY

1.1 Conversion Formulas

$$\frac{\alpha^\circ}{360} = \frac{\alpha^R}{2\pi}; \quad \alpha^\circ = \frac{180}{\pi}\,a^R; \quad \alpha^R = \frac{\pi}{180}\,\alpha^\circ$$

1.2 Trigonometric Ratios $(r = \sqrt{x^2 + y^2}, r \neq 0)$

$$\sin \alpha = \frac{y}{r} \qquad\qquad \csc \alpha = \frac{r}{y} \quad (y \neq 0)$$

$$\cos \alpha = \frac{x}{r} \qquad\qquad \sec \alpha = \frac{r}{x} \quad (x \neq 0)$$

$$\tan \alpha = \frac{y}{x} \quad (x \neq 0) \qquad \cot \alpha = \frac{x}{y} \quad (y \neq 0)$$

Basic Identities

$$\csc \alpha = \frac{1}{\sin \alpha}; \quad \sec \alpha = \frac{1}{\cos \alpha}; \quad \cot \alpha = \frac{1}{\tan \alpha}$$

$$\cos^2 \alpha + \sin^2 \alpha = 1$$

1.3 Functions for Special Angles

α°	α^R	$\sin \alpha$	$\cos \alpha$	$\tan \alpha$	$\csc \alpha$	$\sec \alpha$	$\cot \alpha$
0°	0^R	0	1	0	undef.	1	undef.
30°	$\frac{\pi^R}{6}$	$\frac{1}{2}$	$\frac{\sqrt{3}}{2}$	$\frac{1}{\sqrt{3}}$	2	$\frac{2}{\sqrt{3}}$	$\sqrt{3}$
45°	$\frac{\pi^R}{4}$	$\frac{1}{\sqrt{2}}$	$\frac{1}{\sqrt{2}}$	1	$\sqrt{2}$	$\sqrt{2}$	1
60°	$\frac{\pi^R}{3}$	$\frac{\sqrt{3}}{2}$	$\frac{1}{2}$	$\sqrt{3}$	$\frac{2}{\sqrt{3}}$	2	$\frac{1}{\sqrt{3}}$
90°	$\frac{\pi^R}{2}$	1	0	undef.	1	undef.	0
180°	π^R	0	-1	0	undef.	-1	undef.
270°	$\frac{3\pi^R}{2}$	-1	0	undef.	-1	undef.	0

$$\frac{1}{2} = 0.500$$
$$\sqrt{2} \approx 1.414$$
$$\frac{1}{\sqrt{2}} \approx 0.707$$
$$\sqrt{3} \approx 1.732$$
$$\frac{\sqrt{3}}{2} \approx 0.866$$
$$\frac{1}{\sqrt{3}} \approx 0.577$$
$$\frac{2}{\sqrt{3}} \approx 1.155$$

1.4 Signs of Trigonometric Ratios

Quadrant:	I	II	III	IV
	$x > 0,$	$x < 0,$	$x < 0,$	$x > 0,$
	$y > 0$	$y > 0$	$y < 0$	$y < 0$
$\sin \alpha$ or $\csc \alpha$	$+$	$+$	$-$	$-$
$\cos \alpha$ or $\sec \alpha$	$+$	$-$	$-$	$+$
$\tan \alpha$ or $\cot \alpha$	$+$	$-$	$+$	$-$

2.1 Trigonometric Ratios in a Right Triangle

$$\sin \alpha = \frac{\text{length of side opposite } \alpha}{\text{length of hypotenuse}}$$

$$\cos \alpha = \frac{\text{length of side adjacent to } \alpha}{\text{length of hypotenuse}}$$

$$\tan \alpha = \frac{\text{length of side opposite } \alpha}{\text{length of side adjacent to } \alpha}$$

$$\csc \alpha = \frac{\text{length of hypotenuse}}{\text{length of side opposite } \alpha}$$

$$\sec \alpha = \frac{\text{length of hypotenuse}}{\text{length of side adjacent to } \alpha}$$

$$\cot \alpha = \frac{\text{length of side adjacent to } \alpha}{\text{length of side opposite } \alpha}$$

2.2 Solution of Oblique Triangles

2∠'s and 1 side ⎞
              ⎟ **Law of sines:**
2 sides and an ∠ ⎟
opposite one of   ⎬ $\dfrac{\sin \alpha}{a} = \dfrac{\sin \beta}{b} = \dfrac{\sin \gamma}{c}$
the given sides   ⎠

2.3 Solution of Oblique Triangles

2 sides and the ⎞   **Law of cosines:**
 included ∠ ⎟ $c^2 = a^2 + b^2 - 2ab \cos \gamma$
            ⎬ $b^2 = a^2 + c^2 - 2ac \cos \beta$
   3 sides ⎠ $a^2 = b^2 + c^2 - 2bc \cos \alpha$

2.4 Circular Motion

Length of arc: $\qquad s = r \cdot \alpha^R$

Linear velocity: $\qquad v = \dfrac{s}{t} = \dfrac{r \cdot \alpha^R}{t} = r\omega$

Angular velocity: $\qquad \omega = \dfrac{\alpha^R}{t}$

3.1–3.3 Graphs of Trigonometric Functions

3.6 Inverse Functions

$$y = \mathrm{Sin}^{-1}\, x \quad \leftrightarrow \quad x = \sin y, \qquad -\frac{\pi}{2} \le y \le \frac{\pi}{2}$$

$$y = \mathrm{Cos}^{-1}\, x \quad \leftrightarrow \quad x = \cos y, \qquad 0 \le y \le \pi$$

$$y = \mathrm{Tan}^{-1}\, x \quad \leftrightarrow \quad x = \tan y, \qquad -\frac{\pi}{2} < y < \frac{\pi}{2}$$

$$y = \mathrm{Csc}^{-1}\, x \quad \leftrightarrow \quad x = \csc y, \qquad -\frac{\pi}{2} \le y \le \frac{\pi}{2} \quad (y \neq 0)$$

$$y = \mathrm{Sec}^{-1}\, x \quad \leftrightarrow \quad x = \sec y, \qquad 0 \le y \le \pi \quad \left(y \neq \frac{\pi}{2}\right)$$

$$y = \mathrm{Cot}^{-1}\, x \quad \leftrightarrow \quad x = \cot y, \qquad -\frac{\pi}{2} \le y \le \frac{\pi}{2} \quad (y \neq 0)$$